GNSR 2001

GNSR 2001

State of Art and Future Development in Raman Spectroscopy and Related Techniques

Edited by

Giacomo Messina

Department of Mechanics and Materials, Faculty of Engineering,
University "Mediterranea" of Reggio Calabria, Italy

and

Saveria Santangelo

Department of Mechanics and Materials, Faculty of Engineering,
University "Mediterranea" of Reggio Calabria, Italy

IOS
Press

Ohmsha

Amsterdam • Berlin • Oxford • Tokyo • Washington, DC

© 2002, The authors mentioned in the Table of Contents

All rights reserved. No part of this book may be reproduced, stored in a retrieval system, or transmitted, in any form or by any means, without the prior written permission from the publisher.

ISBN 1 58603 262 3 (IOS Press)
ISBN 4 274 90517 9 C3043 (Ohmsha)
Library of Congress Control Number: 2002106940

Publisher
IOS Press
Nieuwe Hemweg 6B
1013 BG Amsterdam
The Netherlands
fax: +31 20 620 3419
e-mail: order@iospress.nl

Distributor in the UK and Ireland
IOS Press/Lavis Marketing
73 Lime Walk
Headington
Oxford OX3 7AD
England
fax: +44 1865 75 0079

Distributor in the USA and Canada
IOS Press, Inc.
5795-G Burke Centre Parkway
Burke, VA 22015
USA
fax: +1 703 323 3668
e-mail: iosbooks@iospress.com

Distributor in Germany, Austria and Switzerland
IOS Press/LSL.de
Gerichtsweg 28
D-04103 Leipzig
Germany
fax: +49 341 995 4255

Distributor in Japan
Ohmsha, Ltd.
3-1 Kanda Nishiki-cho
Chiyoda-ku, Tokyo 101-8460
Japan
fax: +81 3 3233 2426

LEGAL NOTICE
The publisher is not responsible for the use which might be made of the following information.

PRINTED IN THE NETHERLANDS

Preface

Optical spectroscopy represents one of the most powerful and useful investigation tools. Due to the broad range of applications in scientific and technological research, its potential is enormous. Among the large variety of its branches, a leading role is played by Raman spectroscopy that, allowing the non-destructive material characterisation, is the most widely utilised diagnostic tool in research laboratories.

In Italy, there are several groups, working in universities, research laboratories and industries, active in this field. The cultural exchanges among spectroscopists may allow a profitable basis for a further deepening of the studied topics, aimed at a faster development of the spectroscopic investigation techniques.

An encounter opportunity for researchers working in the spectroscopy field is offered by the Conference organised by the National Group of Raman Spectroscopy and non-linear effects (GNSR).

The GNSR Meeting represents an appointment, usually recurring every two years. Its main purpose is to act as a common forum for spectroscopists, where the most recent and relevant Italian results and applications are presented. The GNSR Conference, hence, constitutes an opportunity for a stimulating exchange of ideas and experiences among the members of the lively scientific community involved, including a variety of scientists, such as physicists, chemists, engineers, architects and art historians, active in the field of Raman spectroscopy and non-linear effects.

Offering the possibility of both divulging assessed results and exploring the feasibility of new projects, the GNSR Meeting promotes the advancement of Raman spectroscopy and related techniques not only in research, but also in industry and education.

In 2001, the Meeting was organised and coordinated by the Physics Research Group of the Faculty of Engineering of the University "Mediterranea" of Reggio Calabria.

The guidelines of the Conference were the innovation and, thus, the new opportunities and challenges in the spectroscopic investigation techniques and the novel applications in the field of the scientific and technological research. The discussion was articulated in invited relations, oral and poster presentations.

The success of the Meeting, encountering a large and enthusiastic participation, and the level of the presented contributions induced us to conceive the idea of getting start to the series of the Conference Proceedings books. We believe that this book will be largely appreciated by the scientific community.

This volume contains the essence of what was presented at the seventeenth Meeting. The contributions submitted for the publication were rigorously selected, among those whose extended abstracts were previously accepted, by the Scientific Committee, for the presentation at the Conference.

The book, proposing a balanced mixture of new analysis methodologies, tools and applications, outlines synthetically the situation of Italian research in the field of Raman spectroscopy and non-linear effects, drawing its possible prospects and future developments.

Due to the wide variety and great diversity of the topics dealt with in the book contributions, including both theoretical and experimental aspects related to the application of Raman spectroscopy, and ranging from prospects for advancement in instrumentation to science and characterisation of highly-performing materials, covering organic and inorganic materials, in liquid or (crystalline, disordered or amorphous) solid phase, we preferred the contributions not to be classified in any way. Invited and accepted papers are all presented by ordering titles simply in alphabetic order.

The Scientific Committee and the Organising Committee decided and announced in the Closing Session that the "GNSR2003" Meeting will take place in Perugia, Italy.

We wish to thank all lectures and participants for their interesting and stimulating contributions to the success of the Meeting and of the present book. We are grateful for the generous support given by the sponsoring institutions, the Faculty of Engineering of the University of Reggio Calabria, the Department of Mechanics and Materials (MecMat) of the Faculty of Engineering, the Presidency of the Government of Region Calabria and by the National Inter-University Association for Material Science and Technology (INSTM), in particular for student grants.

Giacomo MESSINA
Saveria SANTANGELO

2001

Scientific Committee

Chiara Castiglioni, *Politecnico di Milano*
Giuseppe Compagnini, *Università di Catania*
Giacomo Messina, *Università di Reggio Calabria*

Organising Committee

Maria Grazia Donato, *Università di Reggio Calabria*
Giuliana Faggio, *Università di Reggio Calabria*
Giacomo Messina, *Università di Reggio Calabria*
Rosario Pietropaolo, *Università di Reggio Calabria*
Saveria Santangelo, *Università di Reggio Calabria*
Aldo Tucciarone, *Università di Roma Tor Vergata*

Conference Secretary

Maria Grazia Donato and Giuliana Faggio
Dipartimento di Meccanica e Materiali
Università "Mediterranea" di Reggio Calabria
Reggio Calabria, Italy

e-mail: gnsr2001@ing.unirc.it
web site: http://www.ing.unirc.it/attivita/conferenze/gnsr2001/

Contents

Art and spectroscopy: looking to paints and parchments

Erica Mannucci*, Giuseppe Zerbi
Dipartimento di Chimica Industriale e Ingegneria Chimica "G.Natta"
Politecnico di Milano, Piazza Leonardo da Vinci 32, 20133 – Milano
**Tel: + 39 – 02/23993230, Fax: + 39 – 02/23993231,*
E-mail: erica.mannucci@polimi.it

Abstract. The studies in the field of the Cultural Heritage conservation require today the help of various kinds of science and in particular the diagnosis at the chemical and molecular level of the components of the materials, thus allowing to characterize their condition of conservation.

It is obvious that in facing the problems of conservation of works of art any techniques of investigation must be non-destructive.

For this purpose, our research group is using optical diagnostics techniques (Fourier Transform Infrared Spectroscopy, FT-Raman and Raman spectroscopy) in order to understand and solve some of the scientific problems necessarily involved in any restoration work.

1. Introduction

The object of this work is to present some worked out cases of application of vibrational spectroscopy to the study of the materials in the field of the Cultural Heritage; our purpose is to show and discuss the potentiality of such analytical technique in this field.

In particular we will present the results obtained from the analysis of samples removed from:

1. "Last Supper" by Leonardo da Vinci in Milan,
2. the fresco from another "Last Supper" in the church of San Rocco of Inzago (Milan),
3. the parchments of the 16th century restored by the Ancient Book Laboratory of the Benedectine Monastery in Viboldone (Milan).

In order to help our spectroscopic investigation and to allow for a rapid identification of the substances in the samples, we have also constructed a database containing the spectra of several selected chemical species.

Our primary analytical tool has been the correlative study of the IR and Raman spectra by use of the typical "group vibrations" and the use of "frequency patterns" (i.e. the overall patterns in the spectra of selected frequency ranges).

The vibrational infrared absorption spectra were recorded with Fourier interferometers Nicolet 7000 (MCT detector) and Magna 560 (DTGS detector) .For the recording of the visible Raman spectra a XY Dilor Raman spectrometer has been used equipped with CCD detector "Spectrum One" (1024 x 128 pixel) liquid Nitrogen cooled. The Raman spectrometer has an Argon laser with $\lambda_{exc.}$ = 514,5 nm and an He–Ne laser with $\lambda_{exc.}$ = 632,8 nm. Fluorescence – free Raman spectra were obtained with excitation in the near – infrared region (1064 nm with an Nd: YAG laser) using a Nicolet 910 Fourier Transform Raman interferometer.

2. The "Last Supper" by Leonardo da Vinci

During the recent works of restoration of the "Last Supper" of Leonardo da Vinci, our research group was asked to support the restoration work by documenting with physico-chemical measurements the kind of chemical transformation and treatments the painting has been subjected during the work of conservation. This meant to carry out the chemical analysis of organic and inorganic compounds removed from the painting as very small samples. Thirteen samples were taken from different sites of the painting.

The description of the samples and the sites they were removed from, are shown in Figure 1. Their approximate average size is 0,5 mm.
The morphology of some of these samples consists of a few layers one on top of the other (Figure 2) which should correspond to the different phases of the making of the painting. In general one can distinguish the plaster in contact with the underlying wall, a layer of "preparation", an intermediate layer and finally the pigment.

Number	Sample	Sites of removal
1	3 layers	S.Simone's mantle
2	white + "preparation"	Table cloth (central zone)
3	3 layers before cleaning	S.Philip
4	3 layers after cleaning	S.Philip
5	wax	Table cloth
6	plaster + dark material	S.Jacob
7	dark material (after cleaning)	S.Jacob
8	"snake, background"	Lunette
9	"snake, blue"	Lunette
10	green leaves (monolayer)	Lunette
11	green leaves (monolayer)	Lunette
12	green leaves (2 layers)	Lunette
13	"red background"	Lunette

Fig.1 Samples removed from the "Last Supper" by Leonardo da Vinci

Fig.2. Section of sample 1 examined by SEM

Is necessary to specify that since the pictorial materials used in the "Last Supper" consist of an heterogeneous mixture of different organic and inorganic components, the spectra discussed in the present work give necessarily an average information about the whole composition of the sample. When further chemical details were necessary microsamples were studied by infrared or Raman with suitable microscopes added as accessories to our basic instruments; in such a way spatial resolution has been improved substantially.

Moreover, regarding the infrared analysis, in order to improve the signal to noise – ratio, we averaged over 512 spectra. Each spectrum has been recorded in the 4000–400 cm^{-1} range. The spectral resolution was 1 cm^{-1}. The infrared spectra were recorded on the samples embedded in KBr pellets.

As an example of the spectroscopic analysis we take the case of the spectra of a sample taken from St. Simon's mantle (Figure 3).
Comparing this infrared spectrum with those located as similar ones in our data base, it becomes apparent that peaks A E G L belong to the inorganic fraction of the substrate. Bands E (2510 cm^{-1}), G (1410 cm^{-1}) and L (870 cm^{-1}) can be assigned to the calcium carbonate occurring in the deepest layer of the sample. Band A (3695 cm^{-1}) indicates the presence of a fraction of Mg (OH_2). The dominating presence of calcium carbonate is confirmed also by the lines at 1085 cm^{-1} and 710 cm^{-1} of the FT-Raman spectrum (Fig.4)

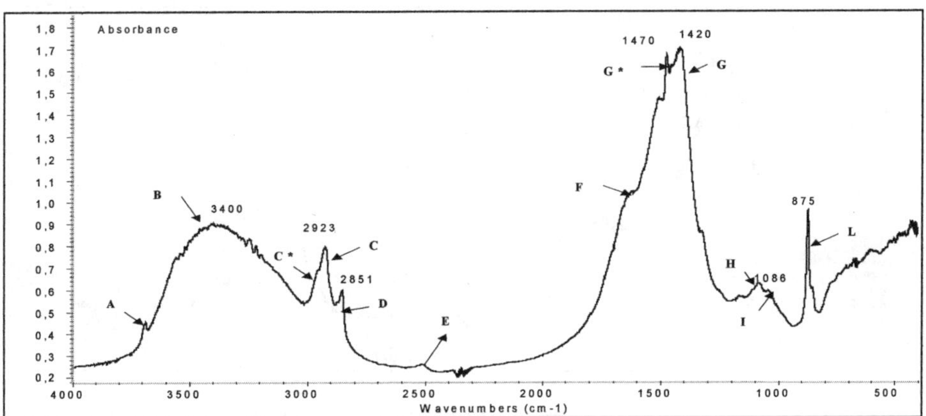

Fig.3 Sample 1 (St. Simon's mantle): FT-IR spectrum

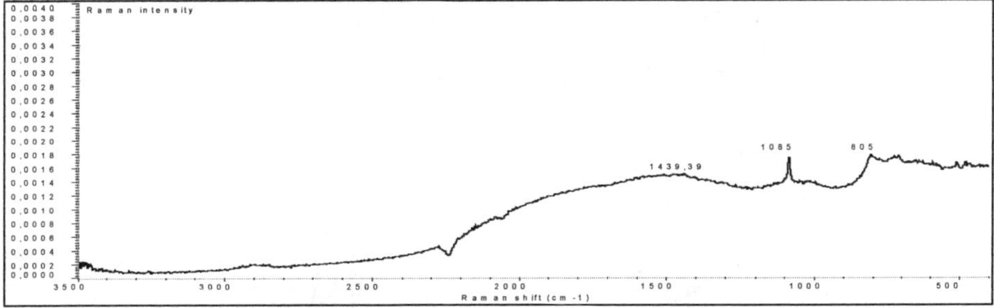

Fig.4 Sample 1 (St. Simon's mantle): FT-Raman spectrum

For the analysis of the organic fraction, characteristic bands (Figure 3) of the functional groups can be identified and used to recognize classes of chemical compounds. The exact nature of these compounds has been understood with the help of reference spectra. In the high frequency region of the IR spectrum the strong and broad peak B (~ 3400 cm $^{-1}$) indicates the presence of –OH groups; the triplet C*, C and D is associated to C-H stretching of methyl CH_3 and methylenic CH_2 groups which occur both in natural oils and in waxes necessarily used by the painter (or by the many restorers who took care of the Last Supper in the various centuries). Bending of CH_2 groups originate a band at 1470 cm $^{-1}$ (G*) and rocking of CH_2 groups a band at 720 cm $^{-1}$. The latter band is too weak to be observed in the samples taken from the Last Supper, but it is clearly visible in the infrared spectra of linseed oil and beeswax. A broad structure is observed between 1050 and 1100 cm $^{-1}$; a spectral region that usually indicates absorption due to alcohols, ethers and esters.

Let us focus on the bands H (1086 cm $^{-1}$) and I (1050 cm $^{-1}$) observed in the spectra of samples taken from St. Simon's mantle (Figure 3) and from St. Philip (Figure 5). In both spectra we notice the occurrence of bands B, C*, C and D and in addition of bands H and I.

The bands listed above coincide with the IR spectrum of Shellac (Figure 6) , organic resin used in the restoration works by M.Pelliccioli starting from 1946.

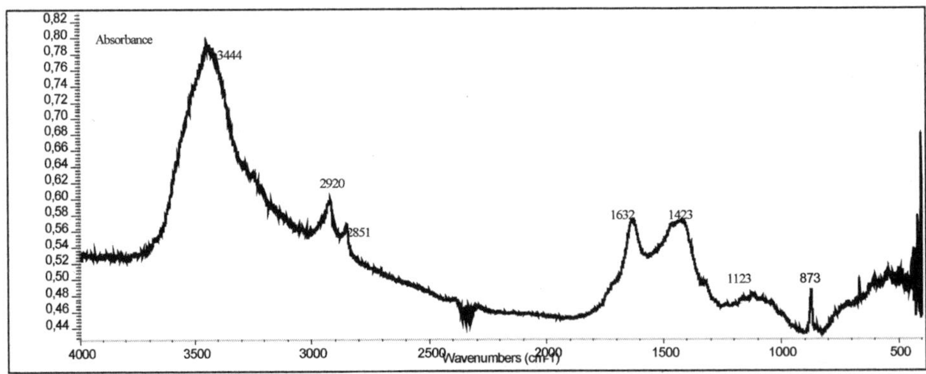

Fig.5 Sample 3 (St. Philip): FT-IR spectrum

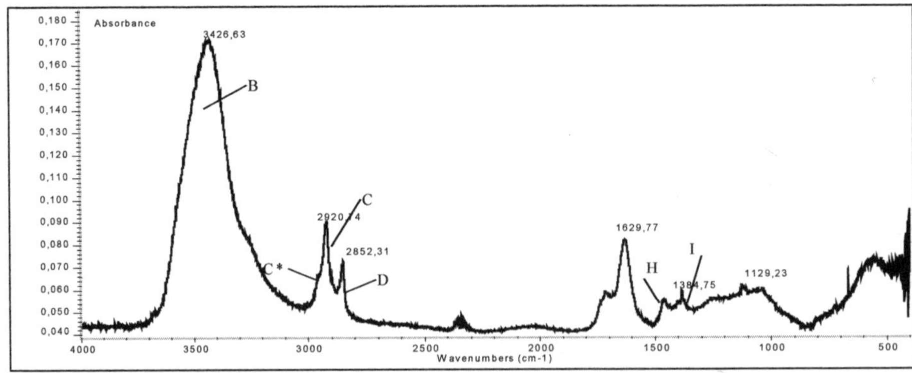

Fig.6 Shellac: FT-IR spectrum

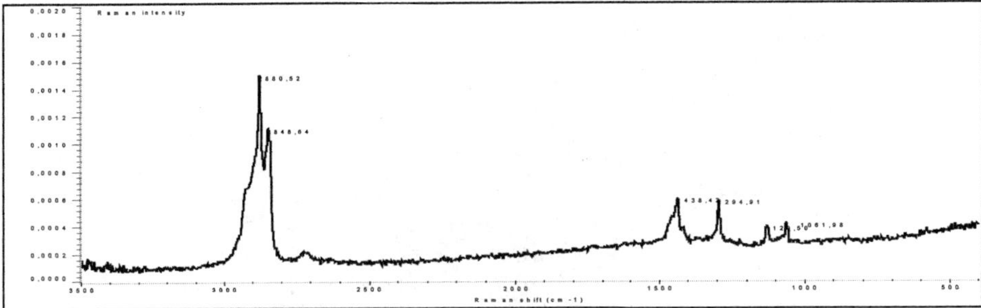

Fig.7 Sample 5 (Cloth): FT-Raman spectrum

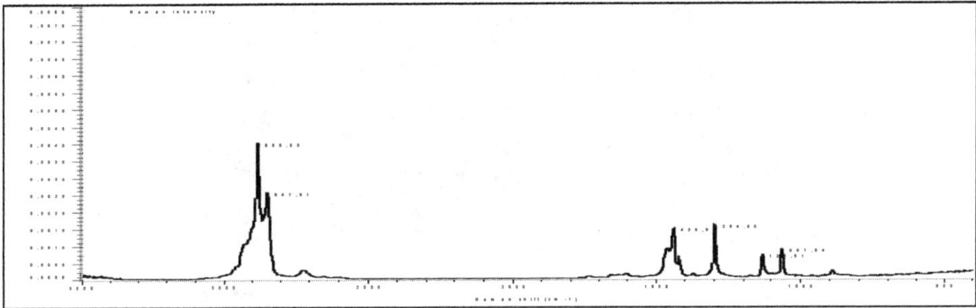

Fig.8 Beeswax: FT-Raman spectrum

We have already mentioned that the triplet of bands C*, C e D has to be associated to C-H stretchings of methyl CH_3 and methylenic CH_2 groups occurring both in the natural oils and waxes. The occurrence of beeswax on the painting is confirmed by the FT- Raman spectrum of the sample 5 (cloth: Figure 7) that shows the same bands of beeswax (Figure 8).

These few examples above show how easily vibrational spectroscopy can give scientific information about the chemical nature of pictorial material. Besides it allows to recognize some substances deposited or removed during the works of restoration. For further information about the analysis made on the samples removed from the "Last Supper" see [1]

3. The "Last Supper" in the San Rocco Church of Inzago (Milan)

Also in this case, during the restoration works (1999) on the "Last Supper" in the Church of San Rocco of Inzago (Milan) our study tried to identify with infrared and Raman measurements some of i) the inorganic components of the "support", ii) the organic components associate to binding materials and protective substances and iii) the pigments used in the fresco.

Altogether, 28 samples were taken from different sites of the painting. The sites from which the sample described in this paper were taken, are shown in Figure 9.

The infrared spectra were recorded on the samples embedded in KBr pellets; in some cases, in order to identify the organic substance, the samples were subjected to repeated extractions with different suitable solvent.

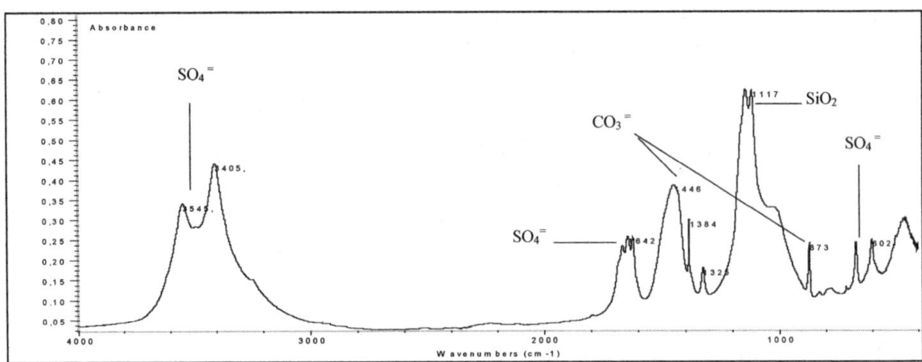

Fig.10 Sample 1: infrared spectrum

In order to identify the inorganic components, usually occurring in all the samples analyzed, we take the case of the infrared spectra recorded on sample 1 (sky blue pigment: Figure 10) and on sample 2 (plaster: Figure 11) .Figure10 (sample 1: infrared spectrum) shows the occurrence of the sulphate anions (gypsum: bands to 3400/3500 cm^{-1}; band to 1642 cm^{-1}; bands to 670 and 602 cm^{-1}), calcium carbonate (bands to 1446 and 873 cm^{-1}) and of silica (broad band near 1143/1117 cm^{-1}).

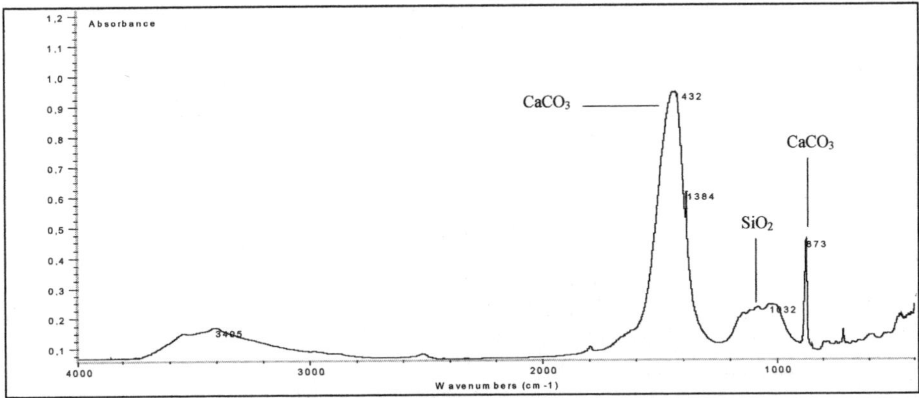

Fig.11 Sample 2: infrared spectrum

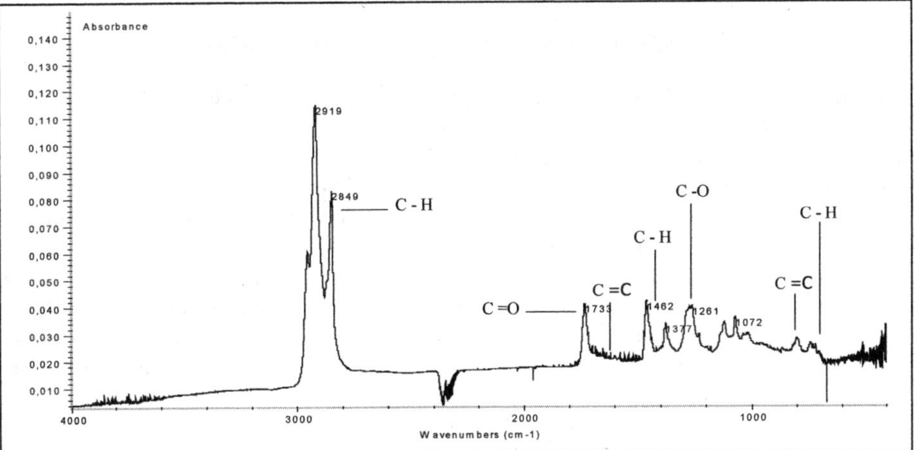

Fig.12 Sample 1: infrared spectrum of the organic residue from an extraction with CH_2Cl_2

The spectrum recorded on sample 2 (plaster: Figure 11) shows the occurrence of calcium carbonate and silica; the bands of gypsum are absent. Since these data have been observed also in other samples an overall analysis allows to conclude that the sulphate ion is located only on the surface of the painting.

For the analysis of the organic fraction we take the case of the samples obtained from the extraction in CH_2Cl_2.

In particular, we take the infrared spectra of the residues extracted from sample 1 (Figure 12) and sample 3 (Figure 13: blue pigment removed from the vault).

The first analysis of the spectrum recorded on sample 1 showed the dominant occurrence of one (or more) compounds characterized by a long polymethylenic chain like paraffins. The observation of the bands at 1733 cm^{-1} (> C=O) and 1261 cm^{-1} (C-O) restricts the field of polymethylenic to be considered and allows to identify a saturated (wax) or unsaturated (oil) ester.

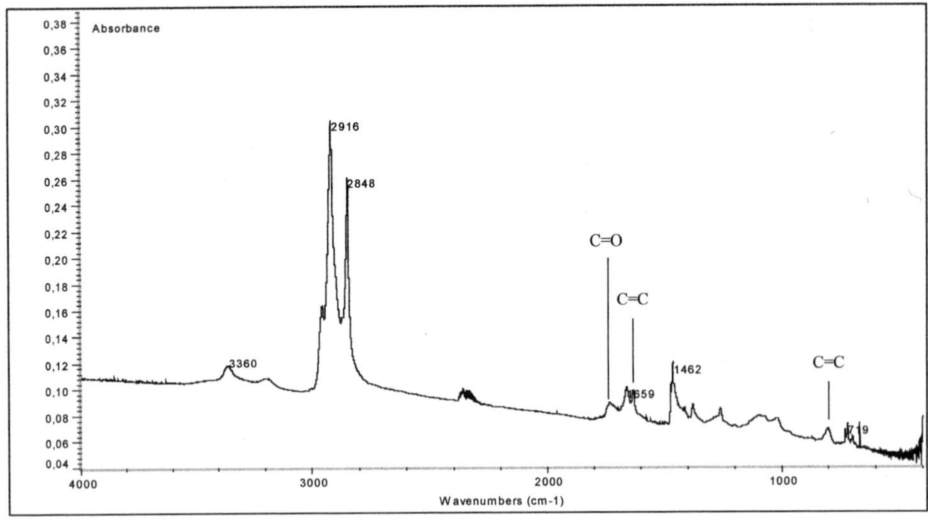

Fig.13 Sample 3: infrared spectrum of the organic residue from an extraction with CH_2Cl_2

The infrared spectrum recorded on the sample 3 (Figure 13) shows the very low intensity of the > C=O stretching and two bands at $1630 - 1650$ cm^{-1}.

These bands, including the one at 798 cm^{-1}, are associated with the presence of C=C bonds, typical of mono or poly – unsaturated polymethylenic compounds.

From chemistry we know that the C=C bond can, by reacting with the atmospheric oxygen, oxidize to form the >C=O bond.

An overall analysis of the spectra suggests that the sample taken from the vault of the church should be more recent if compared with sample 1. If this hypothesis is accepted for the time being, the problem is to determine what is the chemical nature of the original compound which must contain a polymethylene chain and one or more C=C bonds. The starting hypothesis that the material could be a wax or an oil cannot be any more accepted since both classes of materials are characterized by the occurring in the infrared of the >C=O bond, which indeed is almost totally absent in the spectrum recorded on sample 3.

A second hypothesis on the interpretation of the spectrum of sample 3 is that of the presence of a mixture consisting of a paraffin and probably of an additional compound of the class of the amines which can show absorption bands in similar regions of the spectrum.

The occurrence of the amines would justify the existence of natural proteinic compounds. In this case the compounds extracted from samples 1 and 3, are different. The analysis of this case is under further study.

The study and the identification of the pigments in frescoes turns out to be, generally speaking, particularly difficult, because the samples consist usually of an extremely heterogeneous mixture of materials in which the quantity of a pure pigment (sometimes particularly meaningful for the study) is relatively too small for its easy identification from the spectrum. In order to obtain accurate data, usually micro – spectroscopy analysis is used, which are, however, beyond the purpose of this paper .

For a general presentation of the various spectroscopic data we show an example of the identification of a pigment by macro-FT-Raman spectroscopy.

The spectrum was recorded on the gray pigment (sample 4) removed from the vault (Figure 14).

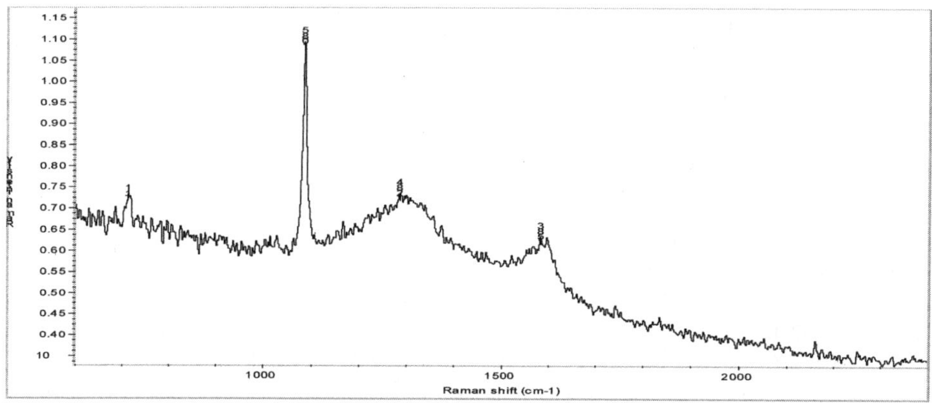

Fig.14 Sample 4: FT-Raman spectrum

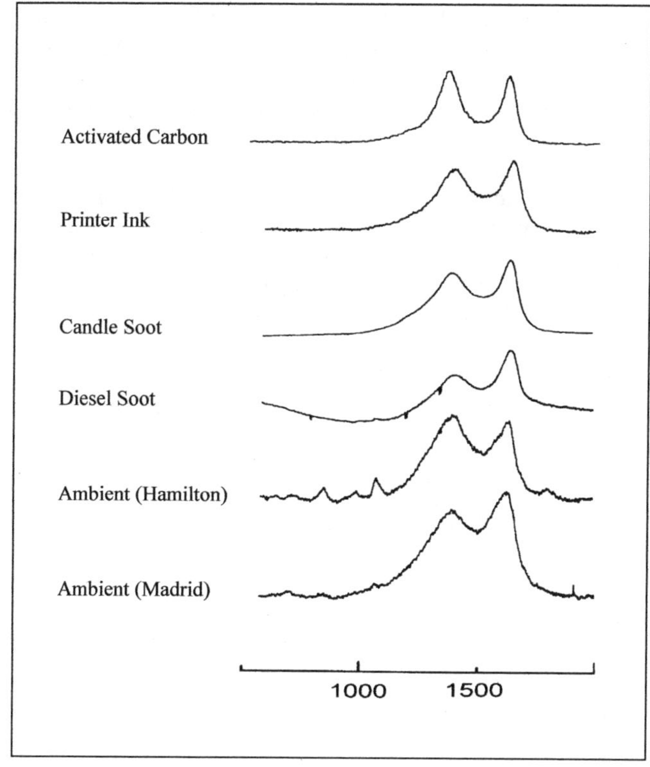

Fig.15 Raman spectra of carbon compounds [2]

Besides the peak at 1084 cm^{-1}, typical of calcium carbonate, two broad lines are observed at 1583 and 1284 cm^{-1}. These bands can be assigned to the mixture of carbonaceous materials (Figure 15), used in this case to obtain the gray color.

4. Parchments of the 16th century restored by the Ancient Book Laboratory

In this case, vibrational spectroscopy is used for the study [3] of chemical conservation of parchments from the Biblioteca Nazionale Universitaria di Torino, which owns many ancient books severely damaged by fire.

In our work Infrared and Raman spectroscopy were used for the characterization of intact, damaged and restored parchments.

The restoration process of burnt parchments is carried out in various steps. The original burnt specimens are left for a certain time in a glove-box at a certain humidity and softened in an atmosphere of water-ethanol-*n*-butanol. They are subsequently stretched in air at room temperature or dipped before stretching in a solution of water (48%), urea (2%), NaCl (2%) and ethanol (48%) and then washed with 1:1 water-ethanol mixture.

The request by the restorers is to understand the kind of interaction between the parchment and the solutions used for restoration.

Interesting and useful results are obtained with infrared spectroscopy as described below while. The recording of the Raman spectra with visible excitation at 514 nm is almost impossible by the strong fluorescence background. On the contrary fluorescence – free Raman spectra were obtained with excitation in the near- infrared region (1064 nm with Nd:YAG laser).

Let us first identify at the molecular level and using group frequencies, the chemical nature of the material examined. Figure 16 shows the infrared absorption spectrum of an intact parchment.

Moving from high to lower wavenumber, we assign the strong and broad band with ill-defined shoulders centered near 3424 cm $^{-1}$ to the stretching of OH and NH groups variously hydrogen bonded. The existence of NH groups is immediately confirmed by the occurrence of the medium-weak broad and ill-defined absorption between 500 and 600 cm^{-1} associated with the out-plane motions of the NH group.

The existence of the amide group, -CONH-, is indicated by the observation of a characteristic doublet (amide I and amide II) near 1643 and 1539 cm $^{-1}$. The weak band near 1239 cm –1 is the so called amide III band. The existence of the polypeptide chain is fully confirmed.

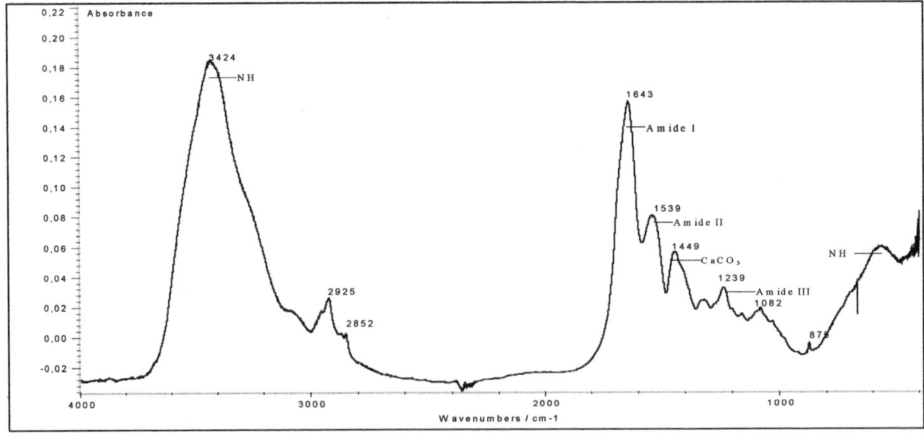

Fig.16 Infrared absorption spectrum of an intact parchment

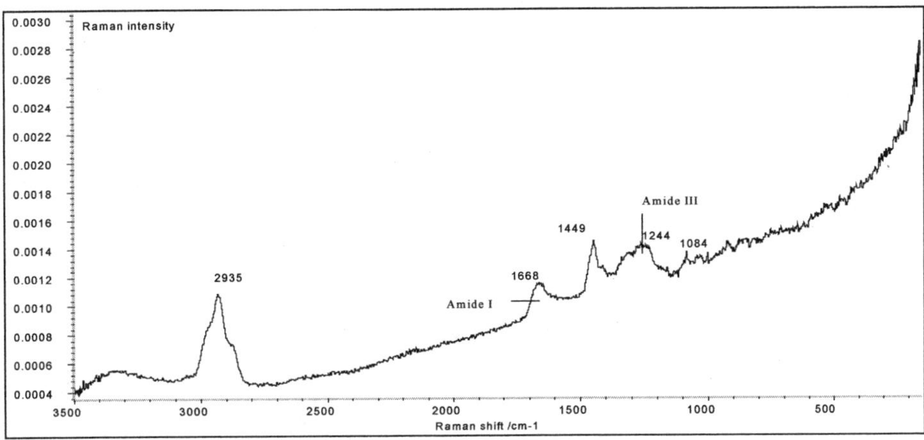

Fig.17 FT-Raman spectrum of an intact parchment

We then focused at a group of medium - weak lines at 2952, 2925 and 2852 cm^{-1} which certainly originate from the C-H stretching modes of –CH$_2$ and –CH$_3$ groups necessarily occurring in the amino acid residues which form the polypeptide chains.

In addition to the existence of the organic fraction of the material, the occurrence of an inorganic fraction is identified from the observation of the band at 1449 cm^{-1} together with the weak, but sharp, band near 875 cm^{-1}, both characteristic of the carbonate ion.

Figure 17 shows the FT-Raman spectrum of the same sample.

The line near 1668 cm^{-1} is the typical Raman active amide I line and the strong and ill-defined scattering near 1300 – 1250 cm^{-1} (amide III) can be associated with the stretching of the CN group. The breadth of the scattering in this wavenumber range is the result of the convolution of many vibrational transitions of the C-N oscillators in various geometric conformations and chemical environments.

The lines near 2935 and 2880 cm^{-1} are again assigned to the stretching of the -CH$_2$- and –CH$_3$ groups and the deformations modes of these groups probably produce the line observed near 1449 cm^{-1}. Contrary to the case of the absorption spectra in the infrared region, the existence of the inorganic fraction is not supported by many lines with the exception of the very weak scattering near 1084 cm^{-1} observed in the Raman spectrum of CaCO$_3$.

In order to understand the possible changes at the molecular level produced by the restoration process of burnt parchments, we thought it necessary to carry out a comparative analysis of the spectra on the same sample which had been subjected to the various treatments, namely original, burnt and restored.

Since differential scanning calorimetric (DSC) studies [4] have shown the occurrence of a few phase transitions at well defined temperatures, we recorded and analyzed temperature dependent infrared spectra in the temperature range 25 – 110 °C on the same sample in a KBr pellet with the infrared beam hitting the same spot.

Comparison of the spectrum of restored and original parchments shows an increase in the absorption in the OH stretching range (near 3550 cm^{-1}) and in the librational motions of the OH group near 600 cm^{-1}. The first conclusions is that the amount of water is larger in restored parchment than in the original one.

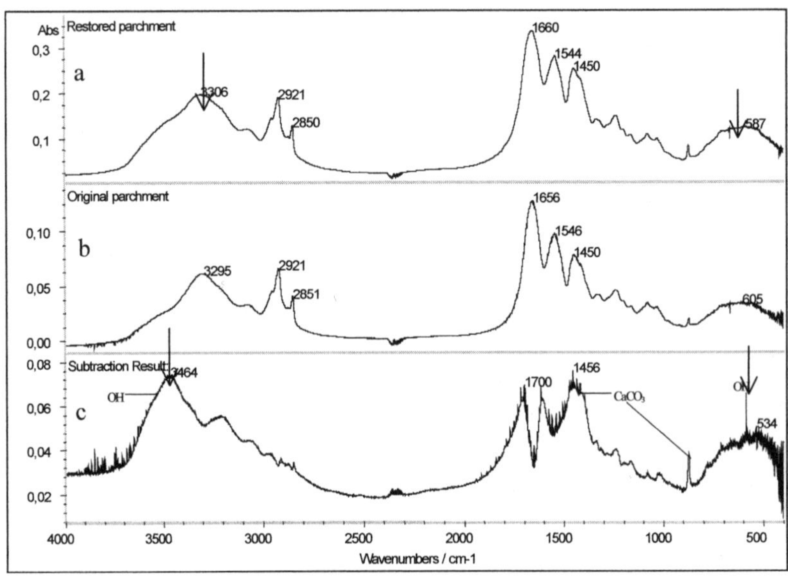

Fig.18 FT-IR spectra of parchment : a) original, b) restored and c) difference spectrum

The difference spectrum (Figure 18) clearly shows the absorption by the excess of water left in restored parchments. Moreover, this seems to contain a larger fraction of the inorganic material (carbonate ion)

We notice that in the difference spectrum the triplet of bands in the C-H stretching range near 2990 cm $^{-1}$ has been nicely compensated, thus showing, in the first approximation, the chemical similarity of the CH groups in the restored and original parchment.

Similar conclusions can be reached from the difference spectrum restored parchment – burned parchment (not shown), where again a larger amount of water in the restored parchment is unquestionably identified.

From these spectra it is not apparent what role is played, at the molecular and structural level, by the chemicals used in the restoration process. Various factors such as hydrogen bonds (from water, ethanol and urea), ionic concentration (NaCl) etc., can act locally or collectively, especially on the proteins inducing various structural changes.

In the spectra of the restored parchment no traces either of ethanol nor of urea are detected, indicating that either they have not been bound to the parchment material or, if present, their concentration is negligible.

DSC studies indicated that phase transitions occur as follows:
• Original sample: broad phase transition at 65°C and a broad, ill-defined additional transition near 85°C
• Burned sample: both endothermic and exothermic transitions are recorded
• Restored sample: transitions occur at approximately 65°C, 80°C, 90°C and 100°C.

We have recorded temperature dependent infrared spectra in the temperature range 25°C-110°C. The temperature dependent spectra of original parchment (Figure 19) show unquestionably that the wavenumbers of the C-H stretching change from 2921 and 2851 cm $^{-1}$ to 2927 and 2955 cm $^{-1}$, respectively, whereas the intensity ratio between the 1546 and 1450 cm $^{-1}$ bands changes slightly. At higher temperature one observes a progressive

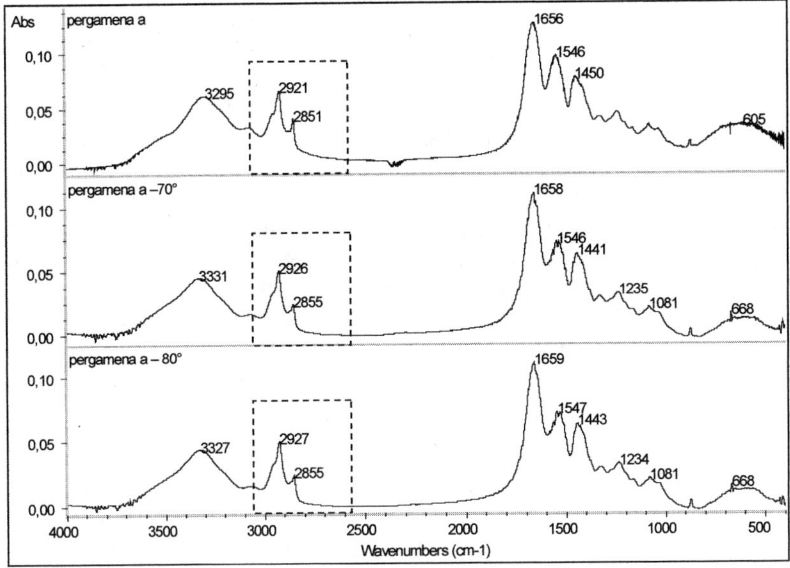

Fig.19 Infrared spectra (room temperature, 70°C, 80°C) of original parchment

decrease of the water content.

The changes in the C-H stretching with temperature imply a variation of the chemical nature of the CH oscillators. To a first approximation these changes can be ascribed either to the changes of the torsional angles of the molecular backbone in the protein molecule or to the changes of the "environment" inside or outside the protein molecules induced by the thermal treatment which , on its turn, also induces a decrease of the amount of water at higher temperatures.

The calorimetric transition near 60°C turns out from spectroscopy to be the most meaningful are and can be described with the help of the spectral changes observed at the same temperature in the infrared spectra. The spectrum indicates a loss of water and a concomitant structural change which probably takes place within the protein chain. The DSC peak must then be associated with a loss of water which produces a structural change.

It is then necessary to envisage that water plays a different role below and above 60°C. In particular, the fraction of absorbed at T > 60°C is likely to be involved in a simple process of swelling

5. Conclusions

In this work we presented a few cases in which vibrational spectroscopy (infrared and Raman) has been applied to the study of the materials relevant in the Cultural Heritage.

Both techniques allow to get interesting results and we think they can give a great contribution to Material Science in the restoration field.

The first part of this study allowed also to construct a database of organic and inorganic materials and its use, together with correlative analysis, allowed to recognize and characterize the samples examined.

In the Raman analysis a large number of samples showed a strong fluorescence emission. The problem has been partly solved using an exciting radiation with wavelength in the near – infrared region.

As a conclusion we do think that Raman and Infrared can be strongly suggested as analytical techniques in the field of Cultural Heritage. A big step forward in the field has been made by the availability of microscopes as accessories which reduce the size of the samples and increase the spatial resolution of such analytical studies.

The future seems even brighter with the very recent developments of near field techniques which reduce the spatial resolution to nanometers, thus allowing the identification and/or chemical mapping at nanometric scale.

Last, but not the least, portable Raman   spectrometers are at present being developed for the *in situ* analysis of materials with applications ranging from biology to the works of art.

References

[1]   E.Galbiati, E.Mannucci, G.Zerbi, Nuove Tecniche Diagnostiche per l'Analisi di Materiali Pittorici: l'Ultima Cena, *Tema* **4** (1998) 44-51

[2]   R.Escribano, J.J. Sloan, N.Siddique, N.Sze and T. Dudev, *Journal of Vibrational Spectroscopy* (in press)

[3]   E.Mannucci, R.Pastorelli, G.Zerbi, C.E.Bottani and A.Facchini, *J.Raman Spectrosc.* **31** (2000) 1089- 1097

[4]   D.Fessas, A.Schiraldi, R.Tenni, L.Vitellaro Zuccarello, A.Bairati, A. Facchini, Termochim. Acta **348** (2000) 129-7

GNSR 2001
G. Messina and S. Santangelo (Eds.)
IOS Press, 2002

15

Effect of the confinement on the structure of graphitic clusters: a study based on Raman spectroscopy of large polycyclic aromatic hydrocarbons

Chiara Castiglioni[*][(a)], Fabrizia Negri[(b)], Matteo Tommasini[(a)],
Eugenio Di Donato[(a,b)], Giuseppe Zerbi[(a)]

[(a)] *Dipartimento di Chimica Industriale e Ingegneria Chimica,*
Politecnico di Milano, Piazza Leonardo Da Vinci, 32 – 20133 Milano, Italy
[(b)] *Dipartimento di Chimica "G. Ciamician", Università di Bologna,*
via Selmi, 2 - 40126 Bologna, Italy
[*]Tel:+39-02-23993230, Fax:+39-02-23993231, E-mail:chiara.castiglioni@polimi.it

Abstract. A quantum chemical investigation of the Raman spectra of large polycyclic aromatic hydrocarbons (PAHs) as model for nanosized graphitic domains in carbon materials, is presented. The gradient corrected BLYP functional, in the framework of density functional theory was employed to compute equilibrium structures, vibrational force fields and Raman intensities of PAHs of different size (from 24 to 138 carbon atoms) and symmetry (D_{6h}, D_{3h}, D_{2h}). In addition, the equilibrium structures of larger PAHs were investigated by means of semiempirical calculations. The analysis of Raman activities in terms of local contributions from internal coordinates shows that the PAHs can be grouped in essentially two classes related to their "benzenoid" character. These results support the mechanism, recently proposed by us to explain the appearance of the D band in disordered graphitic materials.

1. Introduction

Very recently a new synthetic route has been developed [1] and PAHs of very large size have become available. These large polycyclic aromatic hydrocarbons can be considered as molecularly defined graphitic clusters, such as those present in defected graphite samples and other carbon materials. The availability of very large PAHs has suggested to us a new "molecular" approach for the structural characterization of a wide class of carbonaceous materials. Molecularly defined object can be indeed investigated in order to provide explanations for some not yet well understood experimental behaviors observed for disordered carbon materials containing nanosized sp^2 domains.

The Raman spectra of carbon materials containing a variable amount of sp^2 structures have been extensively studied [2-4]. Their first-order spectra show two main signatures: first, a band located at about 1580 cm^{-1}, which is known as graphite (G) band, since it is the only feature observed in the first-order Raman spectrum of highly ordered, crystalline graphite. The second band (D band) appears at about 1350 cm^{-1} for disordered samples of graphite (e.g. for nano-crystalline and micro-crystalline graphites) and in amorphous carbon materials containing sp^2 graphitic islands. Despite to the fact that these two main signatures have been known for a long time, the mechanism lying behind the activation of the D band is not yet completely understood.

Interestingly, the experimental spectra of PAHs [5,6] are also characterized by a few bands in approximately the same G and D regions. This is expected, given the similarity of their structure with graphitic islands, and offers the advantage of using molecules with a well known dimension and structure, to probe defects of similar size and structure in carbonaceous materials.

To assess quantitatively the correlation experimentally found between the Raman bands of graphite and the Raman active modes of PAHs, we carried out a quantum-chemical study of vibrational force field and Raman intensities of PAHs of different size and symmetry [7,8,9]. These calculations, performed at the BLYP/6-31G level of theory, confirmed the idea (earlier developed on the basis of classical vibrational dynamics calculations [5, 10]) that this correlation is due to the strong polarizability changes associated with two peculiar collective vibrational displacements, characteristic of a graphitic cluster, namely Z and $\mathcal{Q}$ vibrations [10].

The Raman active modes of PAHs in the 1600 cm^{-1} region correspond to vibrational eigenvectors with a large projection on the vibrational coordinate Z and give rise to the G band. In a similar way the normal modes (usually two or three) giving rise to strong Raman transitions in the D band region, show a large "content" of the vibrational coordinate $\mathcal{Q}$. In the case of a perfect 2-dimensional crystal of graphite the vibrational

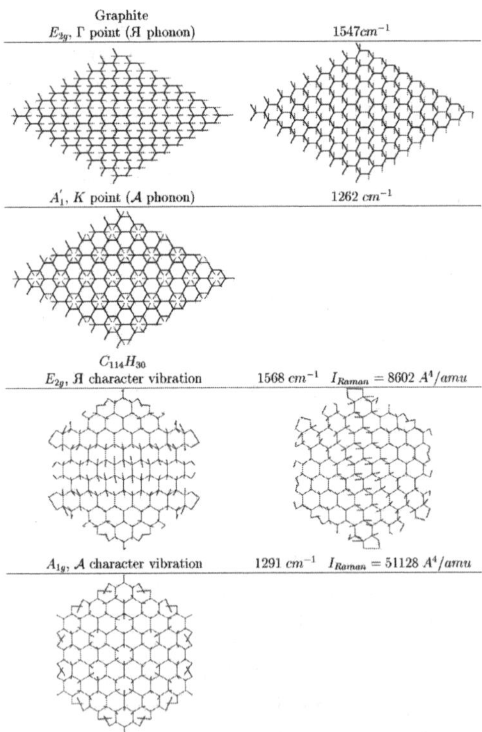

Fig.1 Upper panels: nuclear displacements associated to Z and $\mathcal{Q}$ phonons of a perfect 2-dimensional lattice of graphite. Lower panels: comparison with selected normal modes of C_{114} H_{30}, as obtained from BLYP/6-31G calculations.

coordinates z and $\mathcal{Q}$ correspond to two phonons of the crystal: the only one Raman active $\mathbf{k} = \mathbf{0}$ E_{2g} phonon and the totally symmetric A_1' phonon at $\mathbf{k} = \mathbf{K}$ [10].

In Figure 1 the nuclear displacements associated to these phonons are shown and compared with the computed eigenvectors of a large PAH ($C_{114}H_{30}$). The eigenvectors displayed for $C_{114}H_{30}$ have been selected on the basis of their very large Raman cross sections.

As it will be shown in the following, the understanding of the origin of the large Raman cross section of $\mathcal{Q}$ modes (i.e. those modes with a large projection on the vibrational coordinate $\mathcal{Q}$) allows to unravel the nature of the D band in graphitic materials.

2. Methods

The quantum chemical calculations were carried out at the B-LYP level of theory, a pure Density Functional Theory (DFT) method which uses the Becke exchange functional [11] and the Lee, Yang and Parr correlation functional [12]. The standard 6-31G basis set, as contained in the Gaussian 98 suite of programs [13] was employed to obtain equilibrium structures, vibrational force fields and Raman intensities. The PAHs investigated in this work, characterized by different symmetry, shape and size are shown in Figure 2.

In addition, the equilibrium structures of much larger PAHs were calculated with the help of the quantum consistent force field for π electrons (QCFF/PI) method [14], a well known and very efficient Hamiltonian to describe conjugated carbon systems [15].

The DFT computed Raman spectra of these molecules have been analyzed and compared with the experimental data (where available) obtained from the large

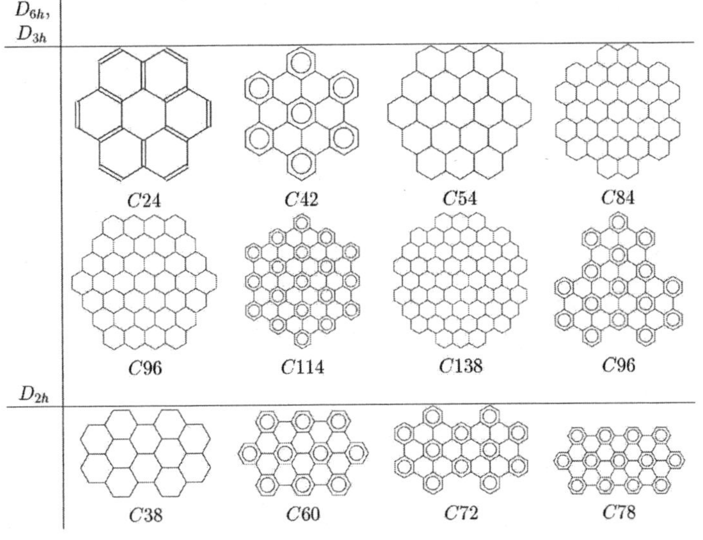

Fig.2 Molecular structure of the PAHs studied in this work.

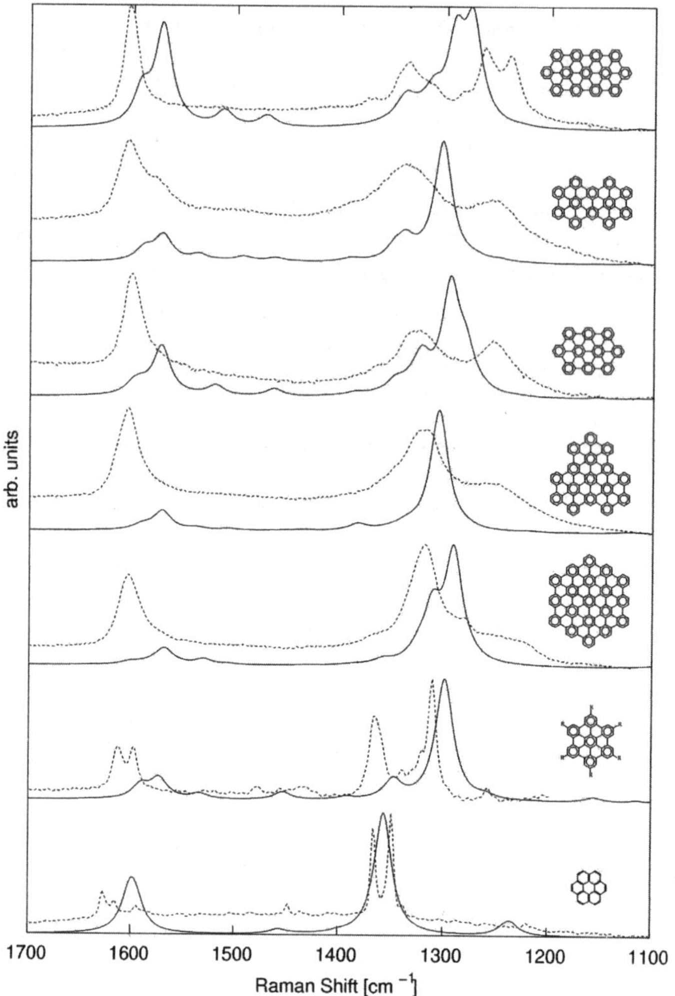

Fig.3 Comparison between calculated (solid line) and observed (dashed line) spectra for some PAH. A Lorentzian linewidth of 20 cm^{-1} was employed to obtain simulated spectra more easily comparable with the experimental ones.

PAHs recently synthesized [1]. Experimental and computed spectra compare very satisfactorily [6,7,9], as shown in Figure 3.

The analysis of the Raman intensities was carried out in terms of the local parameters $\partial\alpha/\partial R_t$ (R_t = internal vibrational coordinates), as obtained by the suitable transformation of the polarizability derivatives with respect to the Cartesian nuclear displacements, $\partial\alpha/\partial x_i$, directly obtained from DFT calculations.

The relationship:

$$\frac{\partial\alpha}{\partial Q_j} = \sum_t \frac{\partial\alpha}{\partial R_t} L_{tj} \tag{1}$$

where L_{tj} is an element of the eigenvector describing the normal mode Q_j in terms of internal valence coordinates, allows to evaluate the contribution to the Raman intensities due to selected localized internal degrees of freedom [10]. In this way we have shown that the strong Raman intensity of transitions in the D band region is almost completely determined by CC stretching contributions [8,9].

This behavior suggests to analyze the CC stretching Raman parameters ($\partial\alpha/\partial R_{CC}$) in the belief to clarify the physical origin of this band.

3. Molecular structure of PAHs.

From Figure 2 it can be seen that the PAHs investigated in this work can be grouped not just according to their symmetry, but also according to their topology.

In Figure 2 $C_{42}H_{12}$, $C_{114}H_{30}$, $C_{60}H_{22}$, $C_{72}H_{26}$, $C_{78}H_{26}$, $C_{96}H_{30}$ are represented as "all-benzenoid" molecules with Robinson rings [16] drawn inside carbon rings containing aromatic sextets. For all these molecules, the "all-benzenoid" character, theoretically predicted according to the Clar rule [17] is clearly reflected by the geometry of the ground state. The optimized structures show an alternation pattern where the bonds belonging to aromatic sextets are systematically shorter than the bonds linking those rings. The alternation is less pronounced in the region closer to the center of the molecule.

The analysis of the optimized structures of the PAHs which do not belong to the above mentioned class of "all-benzenoid" molecules, allows to identify two additional classes that we arbitrarily label "markedly non-benzenoid" and "partially-benzenoid" PAHs. $C_{96}H_{24}$ can be described as a markedly "non-benzenoid" structure, with a bond length pattern which follows the opposite trend with respect to the "all-benzenoid" PAHs, as if they were made by six membered "holes" (quasi-single CC bonds) linked by shorter (quasi-double) CC bonds. $C_{84}H_{24}$ has been found to have a partially benzenoid structure, with the presence of both aromatic sextets and isolated quasi-double bonds. Compare for instance Figure 4.

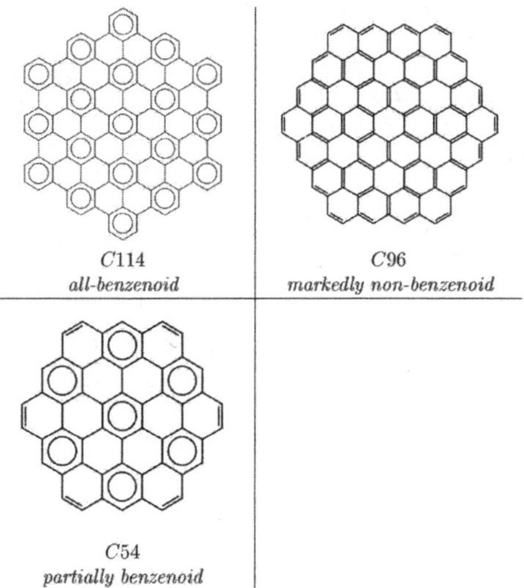

Fig.4 Schematic representation of the three classes of PAHs identified according to the ground state optimized geometry.

On the other hand, $C_{54}H_{18}$ and $C_{138}H_{30}$ show a bond length pattern quite similar to that of "all-benzenoid" PAHs. According to the computed equilibrium structures, Robinson rings can be drawn also in this case, starting from the ring at the center of the molecule: $C_{54}H_{18}$, and $C_{138}H_{30}$ can be properly described as if they were obtained starting from the structures of $C_{42}H_{12}$ and $C_{114}H_{30}$ respectively (see Figures 2 and 4). In other words, the calculations suggest that the dominant Clar formula for these PAHs is that of the preceding (in terms of number of carbon atoms) all-benzenoid PAH. A marked equalization of the bonds (found in the case of $C_{138}H_{30}$) can be connected with its only partially benzenoid character and with its large dimensions.

A quantitative information relative to CC bond lengths obtained from quantum-chemical calculations is reported in Figure 6. In the latter Figure, the BLYP/6-31G computed CC bond lengths of D_{6h} PAHs are plotted against the distance of the center of each CC bond from the molecular center of mass. In Figure 5 we depicted the largest PAH considered in our semiempirical computations (see below) in order to facilitate the reading of Figure 6 along with Figures 7, 8 which follow. The solid circles in Figure 5 collect points whose distance from the center of mass is 5, 10, 15 and 20 Å, respectively. In addition, four dashed circles separated by 1 Å are enclosed between two solid circles. Thus, for the large PAH shown in Figure 5, as well as for all the smaller PAHs considered in this work, it is easy to extract the approximate distance from the molecular center of a selected CC bond, by simply looking at the closest circle (solid or dashed) in Figure 5. The corresponding CC bond length can be extracted from Figures 6-8 by looking at the point in the graph, whose abscissa corresponds to the distance extracted from Figure 5.

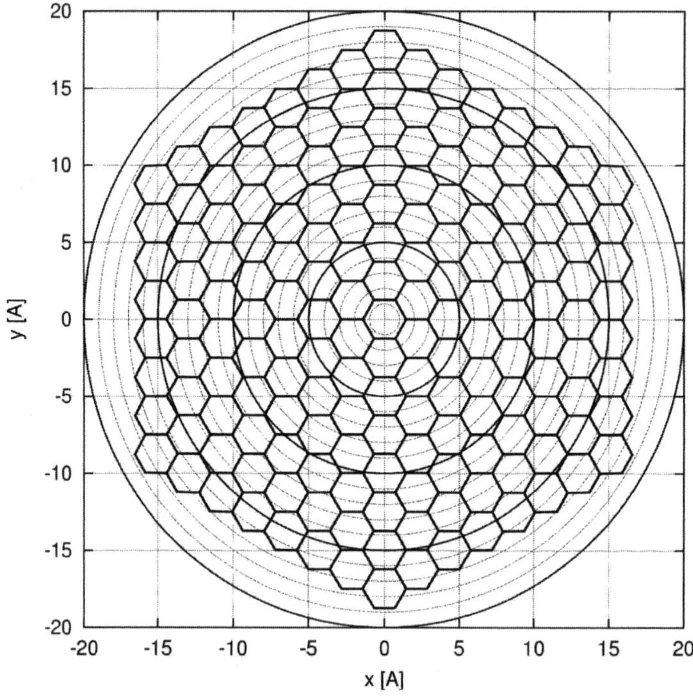

Fig.5 Identification of the CC bond distance from the molecular center for D_{6h} PAHs. Each bond can be identified through the distance of its midpoint from the center of the molecule. The dashed circles are equally spaced by 1 Å and the solid ones by 5 Å.

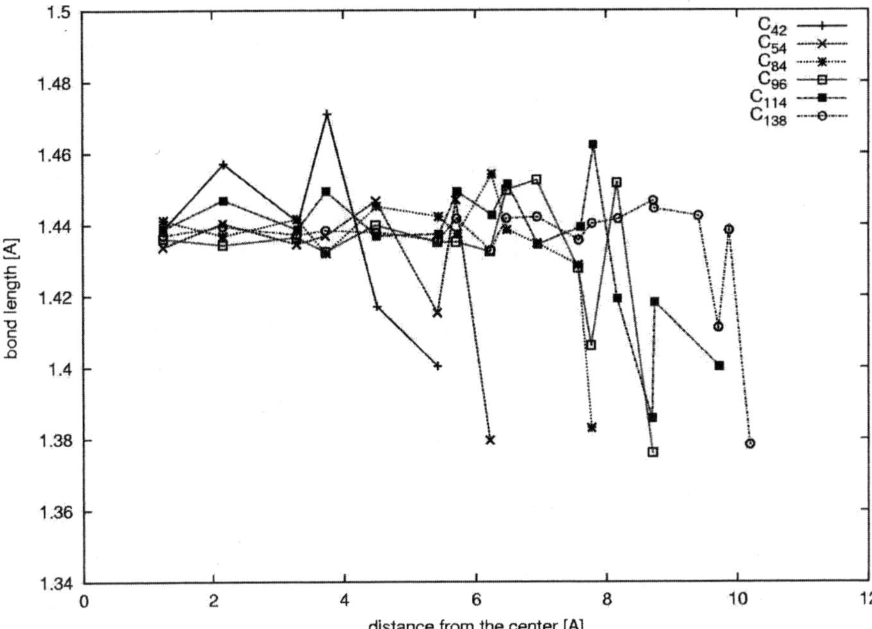

Fig.6 Comparison of CC bond lengths (from BLYP/6-31G optimized geometries) in D_{6h} PAHs. CC bond distance from the center defined according to Fig. 5.

As anticipated, the equilibrium structures of very large PAHs were computed by employing the QCFF/PI Hamiltonian. Fourteen D_{6h} PAHs were investigated. Three of them belong to the class of all-benzenoid PAHs, and their peripheral topology is of armchair type (two examples are C_{42} and C_{114} in Figure 2). The peripheral topology of seven PAHs is of zigzag type (see for instance C54, and C96 in Figure 2) and finally, the remaining four PAH have peripheral topologies characterized by a mixture of armchair and zigzag patterns. The computed bond lengths are collected in Figures 7a, 7b.
The four graphs reported in Figures 7a, 7b correspond to the three different peripheral topologies described above. Moreover, the PAHs with a zigzag edge give rise to two separated graphs showing in Figure 7a, since three of them show the CC bond alternation typical for all-benzenoid PAHs, while the remaining four show the opposite trend, typical for markedly non-benzenoid PAHs. Interestingly, similarly to the more sophisticated and computationally expensive DFT calculations, QCFF/PI results follow the same trend as far as bond length alternation is concerned. Indeed, the alternation is more marked in this case, especially for large PAHs. This result is not surprising, since DFT calculations are known to overestimate the equalization of bond lengths, as for instance in polyenes. In this sense, the results presented in Figures 7a, 7b indicate that the QCFF/PI method can be considered as a suitable alternative to DFT calculations to study the structural and electronic properties of very large PAHs.

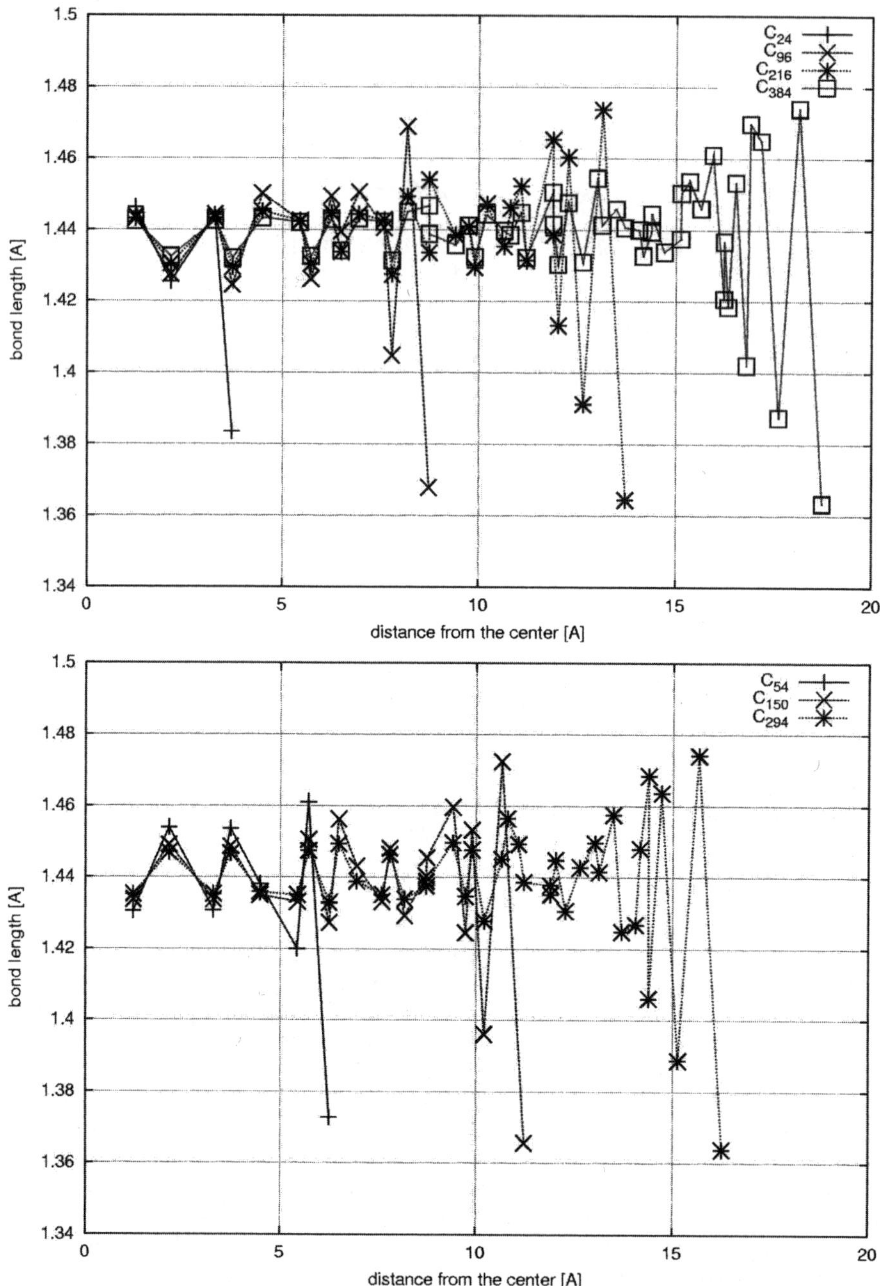

Figure 7a Comparison of CC bond lengths (from QCFF/PI optimized geometries) in D_{6h} PAHs. The data for markedly non-benzenoid PAHs, with a zigzag edge, are shown in the upper panel; the data corresponding to partially-benzenoid PAHs with a zigzag edge are shown in the lower panel. The CC bond distance from the center is defined according to Fig. 5.

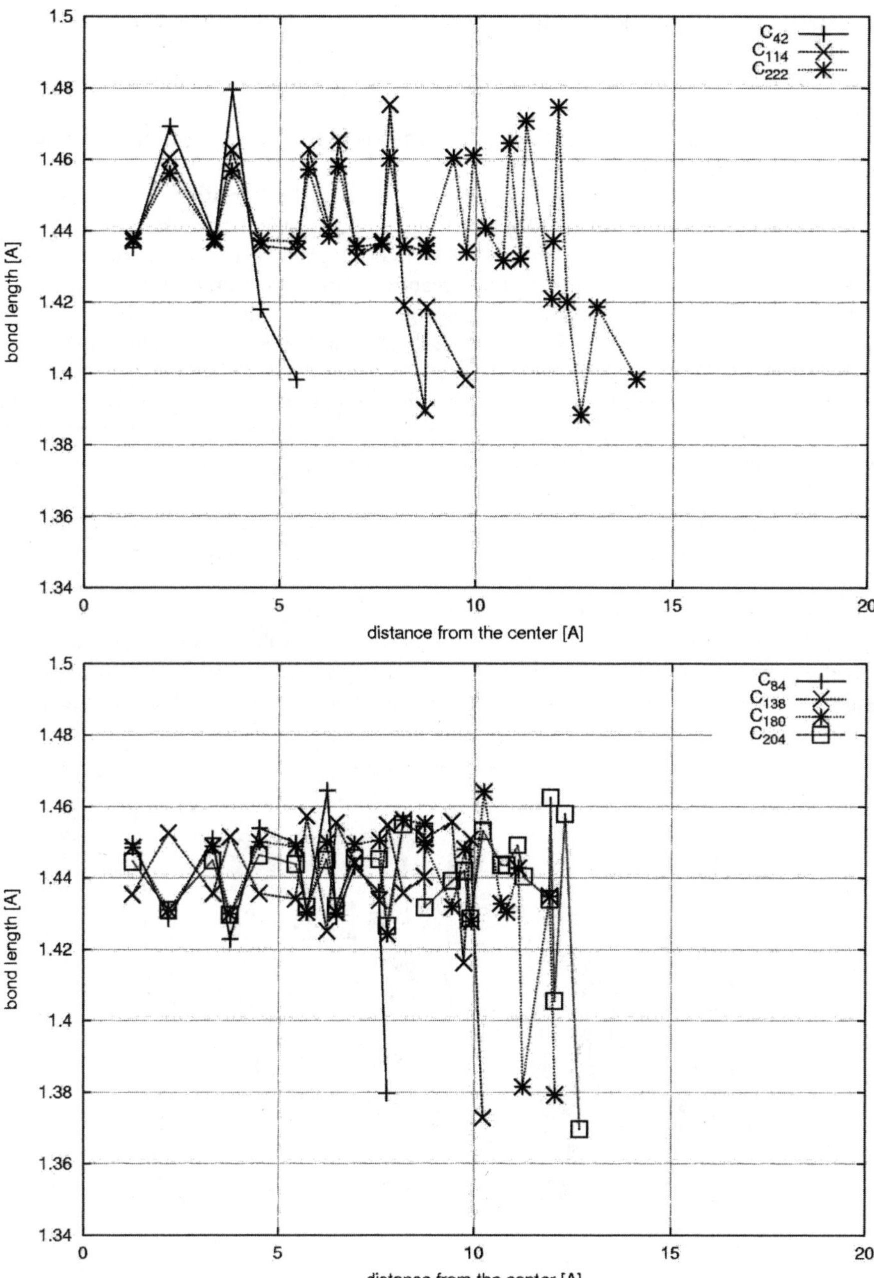

Fig.7b Comparison of CC bond lengths (from QCFF/PI optimized geometries) in D_{6h} PAHs. The data for all-benzenoid PAHs, with armchair edge, are shown in the upper panel; the data corresponding to PAHs with a partially armchair and partially zigzag edge are shown in the lower panel. The CC bond distance from the center is defined according to Fig.5.

4. Analysis of computed polarizability derivatives CC stretching contributions

The analysis of the computed Raman intensity in PAHs [8,9] brings to the following conclusions:

i) Within a given class of PAHs, the total Raman activity increases (more than linearly) with increasing the size of the molecule.

ii) For D_{6h}, D_{3h} PAHs the Raman spectra are dominated by transitions in a region near 1300 cm^{-1}, associated to normal modes belonging to totally symmetric species.

iii) For D_{2h} molecules the features near 1600 cm^{-1} becomes relatively more important with respect to those at about 1300 cm^{-1}, as the "anisotropy" of the system increases.

iv) The inspection of the computed CC stretching contributions to Raman intensity [6,8,9], always show that the strongest Raman bands belonging to totally symmetric species are mainly due to polarizability changes associated to stretchings of CC bonds.

According to observation iv) we expect that the origin of the intensity behavior in the D band region can be rationalized by means of the direct analysis and comparison of the local CC stretching parameters ($\partial\alpha/\partial R_{CC}$) which enter the expression for Raman intensity (see eq. 1).

This analysis can be performed by inspecting the values of the invariant $\alpha' = 1/3$ Tr ($\partial\alpha/\partial R_{CC}$), which is plotted against the CC bond number (see Figure 5), for all D_{6h} PAHs, in Figure 8.

The analysis of these data brings to the following observations:

1) The CC stretching polarizability tensors show larger values than the polarizability derivatives with respect to the other internal coordinates (CCC and CCH bendings). For instance the α' values of CC stretching tensors are from one to two orders of magnitude

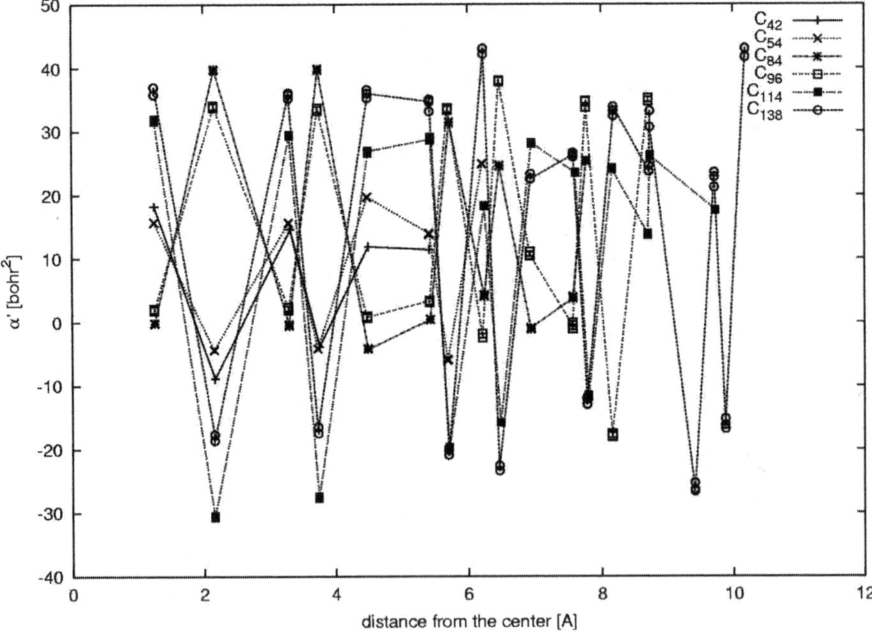

Fig.8 Comparison of α' values ($\alpha' = 1/3$ Tr ($\partial\alpha/\partial R_{CC}$)) for D_{6h} PAHs. Bond numbering as described in Fig.5.

larger than the α' values of CCC and CCH bending tensors. A direct consequence of this fact is point iv) above.

2) The absolute values of α', reported in Figure 8 increase while passing from smaller to larger molecules. This fact suggests that the polarizability change associated with a given CC stretching is not just a "local" phenomenon, but involves the delocalized π system as a whole. Indeed, the increase of the available number of conjugated π electrons results in a relevant increase of the value of the stretching tensors. This fact allows to justify point i).

3) A close inspection of Figure 8 shows a peculiar trend of the α' values while passing from bond to bond. Following the bond identification depicted in Figure 5 and by comparison with Figure 6 one can realize that the magnitude and sign of α' change according to the different character of the CC bond which stretches (in general, shorter bonds have positive values of α'). In the case of "all-benzenoid" molecules (see for instance $C_{114}H_{30}$) we can group the CC bonds into two different sets characterized by a positive and negative value of α', respectively. In the two sets we find bonds involved in the aromatic sextets (positive α') and bonds which link the "aromatic" rings (negative α'). $C_{54}H_{18}$ follows the same rule, due to its "partially-benzenoid" character.

In the case of markedly non-benzenoid PAHs (such as $C_{84}H_{24}$ and $C_{96}H_{24}$) we find the opposite trend for α', while following the same bond sequence toward larger distances from the center of the molecule. A very small, positive value for the inner ring (first bond), a very large positive value for the second bond, a vanishing α' value for the third bond, and so on. This pattern can be can be related to the existence of two different sets of bonds with a different electrical behavior.

The results commented on in 3) are noteworthy since they indicate that the electronic structure of all the PAH considered is very far from the one expected in the case of a perfect, infinite graphene sheet. In this last case, because of symmetry, the values of α', relative to any CC bond must be identical. One can conclude that the confinement of the conjugated electron system (owing to the finite size of the molecules) dramatically changes the bond property with respect to polarizability derivatives. This peculiarity is certainly reflected by the BLYP/6-31G optimized molecular geometries, which indeed correlate well with the trend of the α' parameters.

A second consequence directly originating from the nature of the α' parameters is the fact that vibrations with a large $\mathcal{R}$ character are very strongly Raman active for all the molecules considered. Let us consider a collective vibrational coordinate $\mathcal{R}$, confined in a molecular domain (ideal coordinate $\mathcal{R}$). This coordinate can be described analytically starting from the totally symmetric phonon with wave-vector $\mathbf{K}$ of graphite [8,10] (Figure 1). When it is applied to benzenoid PAHs the ideal $\mathcal{R}$ coordinate describes a collective breathing motion of the aromatic rings (all the rings breathe in phase).

To understand why the breathing vibration $\mathcal{R}$ determines a very large change of the molecular polarizability we can consider the expression of the tensor $\partial\alpha/\partial A$ which, for PAHs belonging to D_{6h} symmetry group, reads [9]:

$$\frac{\partial\alpha}{\partial A} \approx \sum_i \frac{N_i}{K_i}\begin{pmatrix} \alpha' & 0 & 0 \\ 0 & \alpha' & 0 \\ 0 & 0 & 0 \end{pmatrix} \tag{2}$$

Here the sum is extended over all the sets of symmetry equivalent CC bonds, N_i is the number of CC bonds in the given i-th set and $K_i = 4$ if the i-th set contains bonds belonging to the rings which breathe during $\mathcal{R}$ and $K_i = -2$ otherwise.

If we now turn back to the conclusions drawn from the analysis of Figure 8 with regard to the sign and the magnitude of the α' invariant, we can state that:

- For all-benzenoid systems and for partially-benzenoid PAHs positive α' values enter Eq. 2 with a positive K_i coefficient and negative α' values enter Eq. 2 with a negative K_i coefficient, thus building up, in a cooperative way, a large polarizability change during the $\mathcal{R}$ vibration.

- For markedly "non-benzenoid" systems all the positive and large α' enter with a negative K_i coefficient giving rise to a cooperative negative change of the molecular polarizability. Small α' (positive or negative), all belonging to the rings which breathe during $\mathcal{R}$ (positive K_i) give a negligible contribution to $\partial\alpha/\partial A$.

A similar reasoning applied to D_{2h} PAHs allows to explain in a parallel way the large Raman activity in the D spectral region (1300 cm^{-1}) also observed for these molecules. More important, the conclusion above also allows to understand the activation of the D line in disordered graphitic materials. Indeed, if we transfer the information so far collected from the quantum-chemical calculations on PAHs, to the graphitic clusters (e.g. those embedded in amorphous carbon materials), we expect that also in confined (finite size) graphitic domains some kind of structural relaxation will lead to the formation of aromatic sextets (or eventually to structures like those of non-benzenoid PAHs). If this happens we expect to find, also in these clusters, two families of CC bonds with markedly different α' values. Making use of this concept and following a very simple symmetry based reasoning, it can be shown that the immediate consequence of this "dimerization" of the graphite lattice, is the Raman activation, in the graphitic domains, of a totally symmetric **K** phonon of graphite [8]. This phonon has exactly the shape of the $\mathcal{R}$ vibration.

5. Conclusions

The analysis in terms of local contributions to Raman intensities clarifies the origin of the large Raman cross section for the few bands at about 1300 cm^{-1}, observed (and calculated) for all the PAHs examined. It is shown that the values of the local stretching polarizability tensors (α' parameters) can be used as a very sensitive probe of the character of CC bonds in PAH molecules. More specifically, they allow to easily identify the presence of aromatic sextets in "all-benzenoid" PAHs. Moreover, the α' correlate with the equilibrium bond lengths that results from quantum-chemical calculations of equilibrium structure.

On the other hand, the molecular approach here presented has a wider relevance since it indicates that the effect of the confinement in graphitic materials (i.e. the presence of nanosized clusters of condensed aromatic rings) can induce a structural relaxation, with respect to the structure of a perfect graphite crystal with all the CC bonds perfectly equivalent. The spectroscopic signature of this relaxation, which can be described as a "dimerization" of the lattice, is just the appearance of the D line in the Raman spectrum.

Acknowledgements

We are deeply indebted to Prof. K. Müllen (Max Planck Institut für Polymerforschung) who has provided us with the samples of structurally defined PAHs studied in this work. The availability of these materials has opened new horizons to our work. This research has been supported by the University of Bologna (Funds for selected research topics: Project "Materiali innovativi"), by CNR (Project "Applicazioni di spettroscopie ottiche", Project: "Nanotechologies" and "Progetto finalizzato materiali speciali per tecnologie avanzate: Thin layers molecular materials for electronics and nonlinear optics") and from MURST (Project "Analisi della struttura vibrazionale di spettri elettronici", ex 60%; Project "Supramolecular devices", ex 40%)

References

[1] M.D. Watson, A. Fechtenkötter, K. Müllen, *Chem. Rev.* **101** (2001) 1267; A. Stable, P. Herwig, K. Müllen, J. P. Rabe, *Angew. Chem. Int. Ed. Engl.* **34** (1995) 1609; A. Fechtenkötter, K. Saalwächter, M. a. Harbison, K. Müllen, H. W. Spiess, *Angew. Chem. Int. Ed. Engl.* **38** (1999) 3039

[2] F. Tuistra, J.L. Koenig, *J. Chem. Phys.* **53** (1970) 1126 ; R. J. Nemanich and S. A. Solin, *Phys. Rev. B* **20** (1979) 392; Y. Kawashima, G. Katagiri, *Phys. Rev. B* **52** (1995) 10053 ; M. S. Dresselhaus, G. Dresselhaus, P. C. Eklund, Science of Fullerenes and Carbon Nanotubes, Academic Press, San Diego 1996

[3] Póksic, M. Hunfhausen, M. Koós, and L. Ley, *J. Non-Cryst. Solids* **227-230** (1998) 1083 ; M. J. Matthews, M. A. Pimenta, G. Dresselhaus, M. S. Dresselhaus, and M. Endo, *Phys. Rev. B* **59** (1999)6585

[4] A.C. Ferrari, J. Robertson, *Phys. Rev. B* **61** (2000) 1; E. Riedo, E. Magnano, S. Rubini, M. Sancrotti, E. Barborini, P. Piseri, P. Milani, *Solid State Communications* **116** (2000) 287

[5] C. Mapelli, C. Castiglioni, E. Meroni, and G.Zerbi, J. Mol. Struct. **480-481** (1999) 615

[6] C. Castiglioni, C. Mapelli, F. Negri, G. Zerbi, J. Chem. Phys., **114** (2001) 563

[7] M. Rigolio, C. Castiglioni, G. Zerbi, F. Negri, J. Mol. Structure, **79** (2001) 563

[8] C. Castiglioni, F. Negri, M. Rigolio, G. Zerbi, J. Chem. Phys., in press 2001

[9] F. Negri, C. Castiglioni, M. Tommasini, G. Zerbi, J. Phys. Chem., submitted

[10] C. Mapelli, C. Castiglioni, G. Zerbi, and K. Müllen, Phys. Rev. B **60** (1999)12710

[11] A. D. Becke, Phys. Rev. A **38** (1988) 3098

[12] C. Lee, W. Yang, R.G. Parr, Phys. Rev. B **37** (1988) 785

[13] Gaussian 98, Revision A.3, M.J. Frisch, G.W. Trucks, H.B. Schlegel, G.E. Scuseria, M.A. Robb, J.R. Cheeseman, V.G. Zakrzewski, J.A. Montgomery Jr., R.E. Stratmann, J.C. Burant, S. Dapprich, J.M. Milla, A.D. Daniels, K.N. Kudin, M.C. Strain, O. Farkas, J. Tomasi, V. Barone, M. Cossi, R. Cammi, B. Mennucci, C. Pomelli, C. Adamo, S. Clifford, J. Ochterski, G.A. Petersson, P.Y. Ayala, Q. Cui, K. Morkuma, D.K. Malick, A.D. Rabuck, K. Raghavachari, J.B. Foresman, J. Cioslowski, J.V. Ortiz, B.B. Stefanov, G. Liu, A. Liashenko, P. Piskorz, I. Komaromi, R. Gomperts, R.L. Martin, D.J. Fox, T. Keith, M.A. Al-Laham, C.Y. Peng, A. Nanayakkara, C. Gonzalez, M. Challacombe, P.M.W. Gill, B.G. Johnson, W. Chen, M.W. Wong, J.L. Andres, C. Gonzalez, M. Head-Gordon, E.S. Replogle, J.A. Pople, Gaussian Inc., Pittsburgh, PA, 1998.

[14] A. Warshel, M. Karplus, J. Am. Chem. Soc. **94** (1972) 5612.

[15] F. Negri, G. Orlandi, J. Phys. B, **29**, (1996) 5049-5063.

[16] J.W. Armit, R. Robinson, J. Chem. Soc. **103** (1925) 1604

[17] E. Clar, "The aromatic sextet", Wiley, London 1972

GNSR 2001
G. Messina and S. Santangelo (Eds.)
IOS Press, 2002

Fast elementary photophysical processes in organic molecules

Paolo Foggi[(a,b)]*, Giacomo Forti[(a)], Frederik V.R.Neuwahl[(a)]

a)LENS, Largo E.Fermi2, 50125 Firenze,Italy
b)Dipartimento di Chimica, Università di Perugia
** E-mail: foggi@unipg.it; Phone: +39-75-5855580 ; Fax: +39-75-5855598,*

Abstract. Ultrashort laser pulses with duration below 100 fs make it possible to investigate the primary processes occurring in the excited electronic states of any molecular system. Among the laser techniques nowadays utilized to investigate these processes, transient absorption (TA) spectroscopy with UV-visible pump, probe pulses has a good flexibility and allows to attain the maximum time resolution. The aim of this contribution is to show few examples of fast intramolecular elementary processes occurring on time scales around 1 ps after excitation of small organic molecules. In all the examples the dynamical processes are fully resolved by utilizing the TA technique.

1. Introduction

Photophysical processes are those processes resulting from an electronic excitation of a molecule and, strictly speaking, not involving any chemical change. They occur on time scales which typically span several orders of magnitude from 10^{-14} s to more than 10^{-7} s.[1] Since the introduction of ultrafast laser sources it has been possible to fully investigate the various time regimes and characterize not only the processes occurring on the lowest, generally long living, excited electronic state but also on higher energy states.

The photoexcitation generates a coherence in the system. The primary process is therefore dephasing which leads to an out of equilibrium population of electronic and vibrational states. Coherent control during the excitation process is one of the most promising developments of ultrafast photophysics and photochemistry.[2] However such techniques require generally very short pulses (10 fs) with well defined shapes and phases. With 100 fs pulses most of the coherent processes are averaged out. On the contrary, with such pulses, population (energy) relaxation processes can be characterized with a high level of accuracy.

Following the dephasing, various alternative relaxation pathways can be available. In Fig.1 we schematize some of them. Rigorously speaking this processes are purely intramolecular only in isolated molecules. In solution, most of them are modulated by the interaction with the solvent. However, the energy is generally transferred intermolecularly at low frequencies: the solvent can be regarded as a heat sink.

The excitation process brings to vibrationally hot molecules. Even when the energy of the exciting radiation is equal to the 0-0 electronic transition the molecule has to relax some extra energy due to the difference between the minima of the ground and the excited states. Fast rearrangements and vibrational relaxation (VR) are the primary energy relaxation processes.

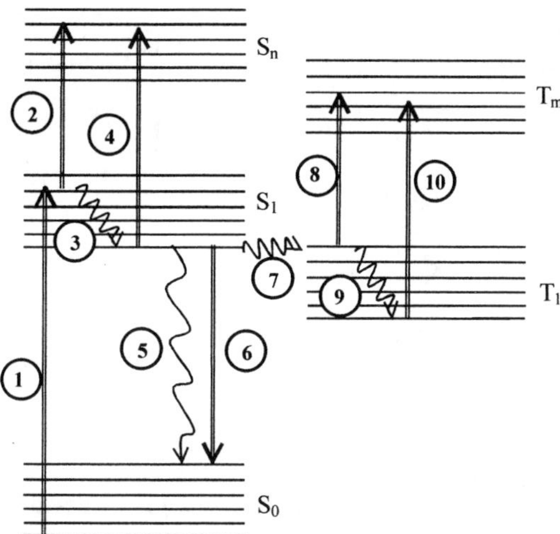

Fig.1 Intramolecular photophysical processes and their time ordering. Up arrows indicate absorption processes: **1** excitation process, **2** TA from unrelaxed state, **3** VR, **4** TA from relaxed S_1 state, **5** $S_1 \rightarrow S_0$ IC, **6** spontaneous emission, **7** ISC, **8** TA from unrelaxed triplet state, **9** VR, **10** TA from relaxed T_1 state

However strong coupling with other states of the system can compete. The internal conversion (IC) generally follows VR processes. It is described as a horizontal process involving the interaction between a vibrational level of the excited electronic state and high energy vibrational levels belonging to lower energy electronic states. The combined effect of VR and IC brings rapidly most of the photo-excited molecules
into the lowest excited singlet state (S_1). This state is generally the most long living excited state of the molecule from which energy relaxation toward the ground level occurs both radiatively and non-radiatively.

As depicted in Fig.1, one of the possible decay pathways for S_1 is the intersystem crossing (ISC). This process generally occurs with very long time constants. It is forbidden by rules of conservation of angular momentum. However in the presence of singlet states with $n\pi^*$ nature ISC can be very fast (ElSayed's rule)[1].

This paper deals with the study of photophysical processes occurring in some aromatic hydrocarbons. We review some recent results obtained in our laboratory obtained with transient absorption (TA) spectroscopy. We choose three molecules to illustrate both the potentiality of the TA technique and what are the temporal regimes attainable with it.

2. The experimental apparatus

Fig.2 schematizes the experimental set up developed at LENS[3-5]. A mode-locked Ti:sapphire laser provides 70 fs (14 nm broad) pulses in the spectral range 750 – 850 nm with an average power of 0.5 W at 80 MHz. The output is amplified in a regenerative amplifier. The amplified pulses are 90 fs long, tunable in the range 780 – 820 nm with an average power of 0.7 W at 1 kHz. Tunable light from 240 nm in the UV down to 10 μm in the middle infrared can be attained by means of different non-linear optical mixing

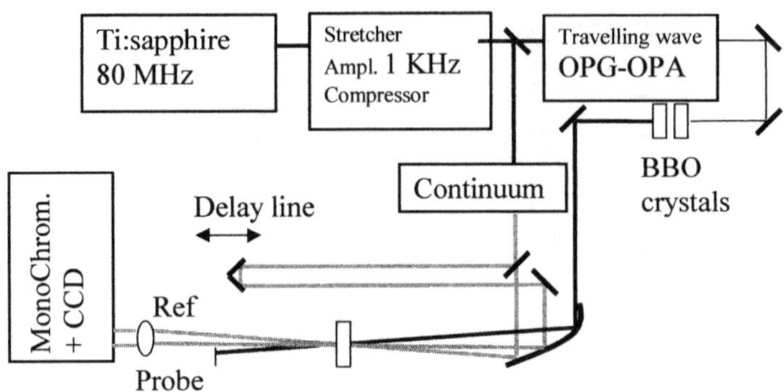

Fig.2 The femtosecond laser set up

processes. A small portion of the fundamental (2 mW) is focussed into a 1 mm thin CaF_2 plate to generate a continuum of white light.

As indicated in Fig.2 the probe pulse is split into two identical pulses. One, passing through the volume of the sample excited by the pump, probes the molecular changes occurring after the excitation process, the other, passing through a different volume of the sample, acts as the reference beam. The probe is linearly polarized at 54.7° with respect to the direction of the excitation polarization. This prevents to measure any orientational dynamics which can interfere with that occurring in the excited states[6].

The set up is completed with a delay line with 1 μm resolution and a flat field spectrograph equipped with a thermoelectrically cooled, CCD camera which collects simultaneously the reference and the probe spectra.

3. Results and discussion

This section is divided into three parts. The first one concerns the study of the ultrafast $S_2 \rightarrow S_1$ IC in pyrene. The second contains the summary of the study by TA spectroscopy of the first two excited states of azulene. Some aspects of the dynamics occurring in S_2 and S_1 states are discussed. The third example demonstrates the capability of TA spectroscopy in measuring the ultrafast ISC process in molecules containing carbonyl groups.

3.1 Pyrene.

Pyrene is an aromatic molecule strongly fluorescing from S_1 state. The S_1 lifetime is very long and do not need high temporal resolution in order to be measured.

On the contrary higher energy levels, being strongly coupled to lower energy states, have very short life times. For example, no emission spectrum from S_2 or higher states can be measured even in the vapour. The fluorescence quantum yield of those states is very low. Therfore time resolved fluorescence technique cannot be utilized. The TA spectroscopy, being sensitive to the population can detect rapid processes such IC among high energy levels [7]. In the present case we are able to detect the formation of S_2 state and to follow its decay towards S_1. In the spectrum is firstly observed a band at 580 nm which forms within the pulse duration and decays with a single exponential law with time constant of 110 fs. After few tens of fs it is possible to observe the transient spectrum due to S_1.[8]. The rise time of the TA signal due to

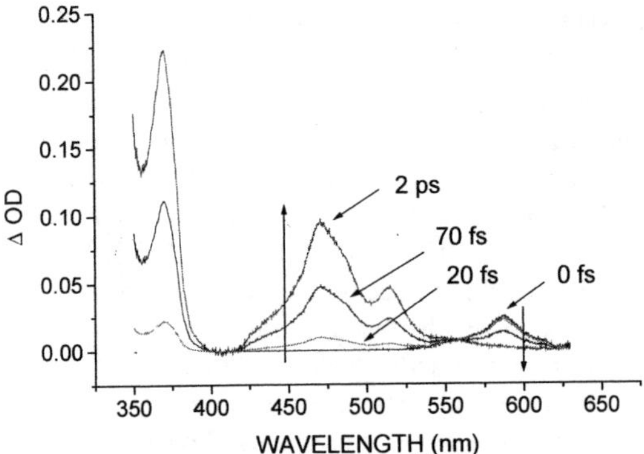

Fig.3 The TA spectra of pyrene after excitation at 340 nm, corrected by the white continuum dispersion. The band due to the transient from S_2 forms instantaneously and decays with a single exponential law with τ =110 fs. The bands due to S_1 grow with the same time constant. A well defined isosbestic point is observed at 565 nm. The arrows indicate the direction of the spectral evolution.

the growing of population in S_1 coincides exactly with the decay of S_2.

3.2 Azulene.

Azulene is the most investigated among the known examples of closed-shell polyatomic molecules exhibiting $S_0 \leftarrow S_2$ fluorescence. S_1 is not fluorescent and is short living (~1 ps). In recent years the dynamics of S_1 and S_2 states have been characterized by time-resolved techniques. [9-12]

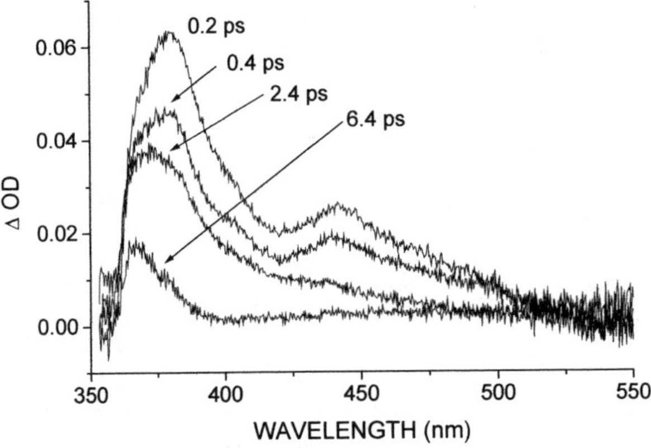

Fig.4 The transient absorption of S_1 state of azulene. After the decay of the S_1 state a band corresponding to an hot vibrational state of the ground electronic level is still visible

3.2.1 *The S_1 state.*

Several pump-probe experiments with sub-picosecond resolution have been performed to measure the lifetime of the short lived S_1 state in solution.[9] In solution and   at   room temperature the S_1 state is almost non-fluorescent and its lifetime has been
determined to be of the order of 1 ps depending on the solvent polarity. Qualitatively it has been suggested that the radiationless decay involves low frequency C-C stretching and C-C bending modes[13]. However, up to now there are few experimental evidences that such a relaxation mechanism rules the decay of the state. [11,14]

In our work it has been possible to record the transient spectrum from the S_1 state (see Fig.4) and its temporal evolution. In the spectral region investigated (350-650 nm) the spectrum shows a broad structure with peaks at 380 nm and 440 nm and, additionally, shoulders at 366 nm, 405 nm and 480 nm. The intensity of the bands, except for the shoulder at 366 nm, decreases with an exponential law with time constant of 1.2 ps in cyclohexane and 1.4 ps in acetonitrile. These values are in good agreement with those previously reported for the lifetime of the S_1 state.

The disappearance of the band at 380 nm allows the clear observation of a peak at 366 nm which lasts for longer times and whose intensity decays with an exponential law with time constant of 6 ps. The longer lifetime and the position in the spectrum suggest that this band is due to a hot vibrational state producing a longer wavelength absorption in the spectrum of the $S_0 \rightarrow S_2$ transition. The band is about 1500 cm^{-1} lower than the S_0-S_2 0-0 origin. In the fluorescence spectrum the most intense band occurs exactly 1500 cm^{-1} from the origin and is attributed to one of the C-C stretching modes of a_1 symmetry. This result suggests that the decay mechanism involves specific vibrational modes.

3.2.2 *The S_2 state*

As previously mentioned the S_2 state of the molecule is fluorescent. Its decay has a time constant of more than 1 ns[10]. At very short times and in polar solvents (acetonitrile, methanol) an additional dynamics is observed (see Fig.5). The effect is ascribed to solvent dynamics triggered by the change in the permanent dipole moment of azulene passing from the ground to the excited state (0.8 D and 0.4 D respectively according to ab-intio calculations[15]).The dynamics in methanol is exponential with a time constant of about 2 ps,

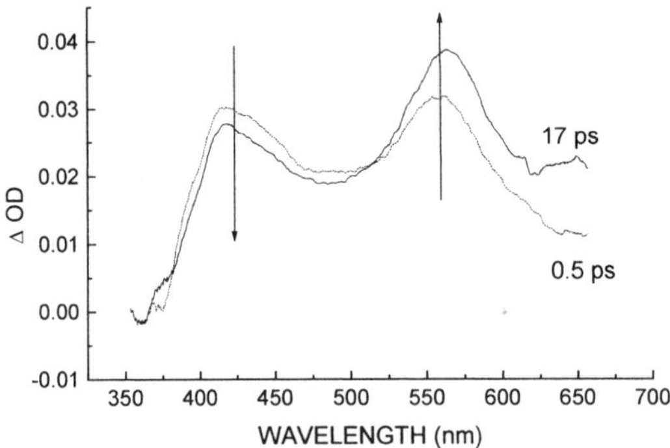

Fig. 5 The transient spectra of S_2 state of azulene in methanol. Two strong bands can be clearly detected. The arrows indicate the direction of the spectral evolution at positive delays.

in perfect agreement with the dynamics measured on the pure solvent with light scattering techniques. The variation of intensity of the two bands can be interpreted as due to the different symmetry (direction of the transition moment) of the $S_2 \rightarrow S_m$ transitions. The lowest frequency band is ascribed, according to ab-initio calculations, to a $A_1 \rightarrow B_2$ transition while the higher frequency to a $A_1 \rightarrow A_1$ transition [15]. When the azulene molecules are in the ground state the solvent molecule are closer to them. The potential experienced by the dipole moments oscillating parallely and perpenducularly to the average direction of the dipoles of the solvent changes during the solvent relaxation process. This can affect the intensity of the absorption.

3.3 *Isophtaldialdehyde*

Isophtaldialdehyde is a dicarbonilic aromatic compound and can be formally considered a benzaldehyde derivative. Interest on this particular compound has raised because of its presence as a "chromophore" in *benzylamidic[2]catenane* systems.[16] The present study aimed the possibility of utilizing some electronic property of this molecule in order to investigate the dynamical properties of the catenane. *[2]catenanes* are, at present, the smallest among the interconnected ring systems. Recent studies concern the existence of a relationship between the topological properties of the *[2]catenanes* (due to the mechanical interconnection of the two rings) and light absorption. The steady state UV-visible absorption spectrum of the *[2]catenane* results chiefly from the chromophore.

The existence of two carbonilic groups in isophtaldialdehyde should, in principle, help the opening of non radiative electronic relaxation pathways by spin orbit coupling. By the study of the photophysical dynamics of isophtaldialdehyde excited states we expect to characterize the ISC process and to measure the time constants.

The excited states are generated by a ultrashort (100 fs) pulse whose central wavelength is 260nm. The steady state absorption spectrum of isophtaldihaldehyde presents an

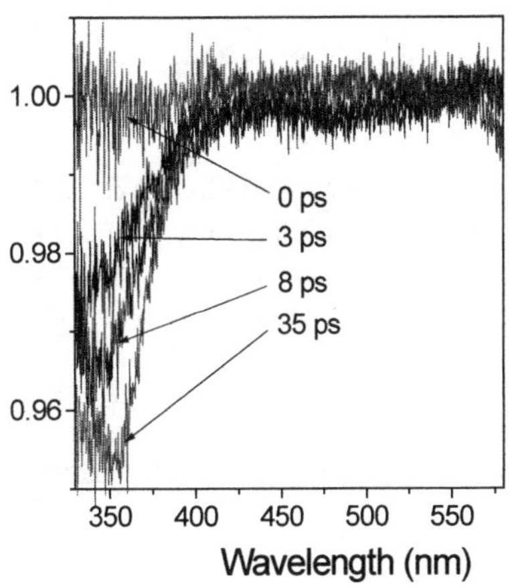

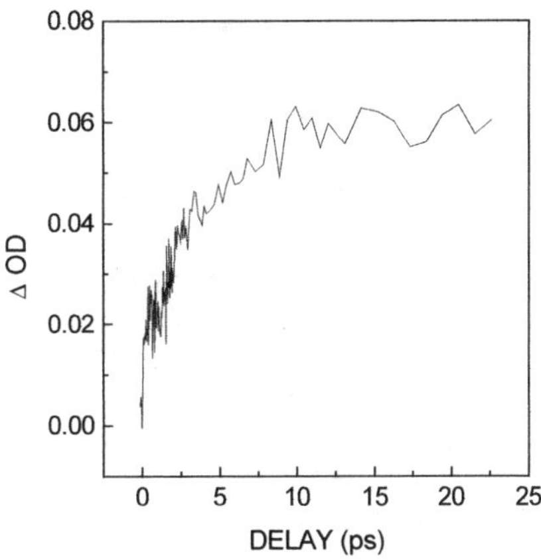

Fig.6 The TA spectrum of isophaldialdehyde (left) and the dynamics of formation of the triplet measured at 350 nm (right).

absorption maximum at 225 nm with a shoulder at 250 nm. The two bands are attributed to (π–π*) transitions localised on the benzenic ring. Weak absorption in the 270-320 nm region is due to *(n-π*)* transitions. The presence in the molecule of two carbonilic groups suggests the existence of two almost degenerate *n* levels. Selective preparation of one of the two lowest energy (n-π^*) singlet levels is possible but, because of the low extinction coefficient of the transitions, generation of an excited states concentration capable to generate a detectable transient absorption signal would require too high a concentration of the sample, which may cause aggregate formation. On the contrary a vibrationally cold singlet (π–π^*) excited state (S_3) is easily populated by the 260 nm excitation pulse.

The transient absorption spectrum of isophtaldialdehyde in acetonitrile solution presents an intense band at 345 nm and a broader and weaker one at 480 nm. The temporal evolution of the spectra shows that the two bands intensify without changing their shape and that they have the same temporal behavior. Both the bands reach the maximum of intensity in less 30 ps and keep their structure up to the nanosecond time scale. The kinetics of the transient absorption signals detected at 345 nm and 480 nm differ only for the intensity of the absorption maximum. The spectrum is attributed to a long half-life excited state. The rise of the transient absorption signal is at least one order of magnitude slower than the time duration of the instrumental function of the experiment. This means that the absorption cannot be attributed to the state prepared by optical excitation. Kinetics extracted at 345 nm and 480 nm provide double exponential kinetics curves with time constants of 3 ps and 15 ps.

The electronic relaxation pathway of the S_3 excited state can be described as an internal conversion process that leads to the population of S_2 (n-π*) state. According to a model proposed by Hochstrasser et al.[17] for benzophenone the excess electronic energy of S_2 will flow through two different and concomitant pathways: a) internal conversion to S_1 b) intersystem crossing to T_1. The kinetic constants of this two concomitant processes will result from the sum of the constants of the two separate ones. Moreover the S_1 states undergoes a El Sayed's rule permitted relaxation process to T_1. This kinetic scheme provides for time behavior of the concentration of the triplet state a simple solution

$$[T_1] = C_1 \cdot \left[1 - e^{(k_{IC} - k'_{ISC}) \cdot t} \right] + C_2 \cdot \left(1 - e^{k_{ISC} \cdot t} \right)$$

According to this model the kinetics recorded at 345 nm and 480 nm are interpreted as due to the combined effect of two processes both leading to the triplet state. The recorded kinetics are well described by the convolution of a 230 fs gaussian instrumental function with a double exponential rise curve. The largest time constant extracted from this simple model gives an upper limit for the IC process between S_2 and S_1 but there is no direct information about the ISC between S_2 and T_1. More interesting is the second time constant which gives us an estimate of the time scale of the dominant ISC process between S_1 and T_1. The ISC relaxation pathway is very fast and comparable with that measured in benzophenone.[17]

References

[1]   M.Klessinger and J.Michl, *"Excited States and Photochemistry of Organic Molecules"* VCH, New York, 1995.

[2]   H.Kawashima, M. M. Wefers, and K. A Nelson Annu. Rev. Phys. Chem. 46, 627, (1995).
[3]   P.Foggi, L.Bussotti EPA Newsl. 66, 1, (1999).
[4]   F.V.R. Neuwhal, L. Bussotti, P. Foggi
        in *Research Advances in Photochemistry and Photobiology*, Vol. 1, 77,
        (Global Research Network, Trivandrum, Kerala, India 2000).
[5]   P.Foggi, L.Bussotti, F.V.R. Neuwahl Int.J.Photoenergy 3, 103, (2001).
[6]   G.Fleming *Chemical Application of Ultrafast Spectroscopy*
        (Oxford University Press, New York 1986).
 [7]   F.V.R. Neuwhal, P. Foggi Laser Chem. 19, 375, (1999).
[8]   P. Foggi, L. Pettini, I. Santa, R. Righini, S. Califano
        J. Phys. Chem. 99, 7439, (1995).
[9]   B.D.Wagner, M.Szymanski,R.P.Steer J.Chem.Phys. 98 , 301, (1993).
[10]   B.D.Wagner, D.Tittelbach-Helmrich, R.P.Steer J.Phys.Chem. 96, 7904, (1992).
[11]   U.Sukowski, A.Seilmeier, T.Elsaesser, S.F.Fischer J.Chem.Phys. 93, 4094, (1990).
[12]   L.Ciano, P.Foggi, P.R.Salvi J.Photochem.Photobio.A 105, 129, (1997).
[13]   A.L.Sobolewski Chem.Phys. 115, 469, (1987).
[14]   A.Seilmeier, U.Sukowski, W.Kaiser, S.F. Fischer
        in *Ultrafast Phenomena V*, 454 (Springer-Verlag Berlin 1986).
[15]   P.R.Salvi, L.Moroni, F.V.Neuwhal, P.Foggi to be published.
[16]   D.A.Leigh, A.Murphy, J.P.Smart, M.S.Deleuze, F.Zerbetto J.Am.Chem.Soc. 120, 6458, (1998).
[17]   R.N.Hochstrasser, H.Lutz, G.W.Scott Chem.Pys.Lett. 24, 162, (1974).

GNSR 2001
G. Messina and S. Santangelo (Eds.)
IOS Press, 2002

Forensic applications of Raman spectroscopy: investigation of different inks and toners

S. Savioli, D. Bersani, P. P. Lottici
INFM and Department of Physics of the University of Parma
Parco Area delle Scienze 7, 43100 Parma

M. Placidi
Jobin-Yvon s.r.l.
Via Cesare Pavese 35/AB, 20090, Opera (MI)

Ten. Col. Luciano Garofano
Scientific Investigation Rep.- Carabinieri
Parco Ducale 3, 43100, Parma

Abstract. MicroRaman spectroscopy has been used in order to characterise different ballpoint pen inks, ink-jet printers and toners by using different excitation wavelengths. The results are discussed in connection with SERRS effect.

1. Introduction

The differentiation and identification of inks is a problem often encountered by expert in the examination of documents being either hand-written or printed. In fact, the widespread use of ink-jet printers has made necessary for document examiners to find a method of linking a questioned document to individual inkjet printer. The examinations generally used in the sphere of inks, TLC (Thin Layer Chromatography), l'HPLC (High Performance Liquid Chromatography), GC/MS (Gas Chromatography/Mass Spectrometry) and PyGC/MS (Pyrolysis Gas Chromatography), even supplying a very clear and well-defined compositional description of the ink, have some disadvantages: slowness in the examinations, especial preparation of the sample and, above all, destruction of the sample. Furthermore, these techniques do not give satisfactory results with toners. A largely used spectroscopic technique is the FTIR that, however, requires a laborious sample preparation that usually implies a document modification. The micro-Raman Spectroscopy has already been shown to be an important tool in forensic context such as explosive and drug fields. The distinctive advantages of this technique are several: high discriminant power between substances even very similar one another, unnecessary preparation of the samples, quickness of the analysis. Most important, the undestructiveness of samples, which gives the opportunity to repeat the examination, plays a crucial role in a field where the sample is a proof and, as it, has not to be modified by examinations. Aim of this work is that to verify whether the Raman spectroscopy, even through the SERRS technique (Surface Enhanced

Resonant Raman Scattering), allows distinguishing among several ballpoint pen inks, ink-jet print inks and laser print toners, used in written documents.

2. Inks and Toners

An ink is a pigment (substance which colours other materials) scattering or a dye (a soluble or insoluble colouring matter) inside a vehicle. This vehicle allows the adhesion of the pigment or the dye to the final substratum. Generally, there are some additives in the inks, as antimycotics and ingredients to be used in particular environmental conditions. Some inks can also include some especial markers to allow the production time and brand identification. A toner is a dry ink formed of plastic resins (the main ingredient), dyes, waxes, making flowing agents, agents which provide it with a charge and, only in the magnetic toners, magnetic fibres.

3. Experimental

Normally, samples analyzed have been typed or handwritten characters on common paper sheets. The paper used is Fabriano Copy Performance A4, 210x297 mm- 80g/m².
Raman spectra have been obtained using a micro-Raman Labram Jobin-Yvon, equipped with an integrate confocal microscope, CCD detector, holographic notch filter and motorized XY stage for mapping facilities at micrometric lateral resolution. Excitation is provided at 632.8 nm by a He-Ne laser and by laser diode at 784.8 nm. In order to avoid sample damaging neutral filters has been used in order to reduce the laser power on the sample. Paper background signal has been deeply studied in order to distinguish the paper background signal from the inks and toner spectra. SERRS effect has been obtained by deposition of small drop of a silver colloidal solution, obtained with citrate reduction of silver nitrate, on the investigated area.

4. Results and Discussion

4.1 Ballpoint Pens Inks

Ten different black inks and seven coloured inks have been analysed. In Fig. 1, spectra obtained by two ballpoint pens of different marks, Brio and Duo marks, have been reported. Spectra allow clearly distinguishing the two inks: in case of the Brio pen, graphite peaks are very evident (a 1328 cm^{-1} e 1586 cm^{-1}), and they are present in many of the black inks analysed. The most part of black ballpoint pens have shown distinctive and univocal spectra. Even in the few examinations performed on other colour inks (see for example Fig. 2) good results have been obtained with characteristic spectral fingerprints.

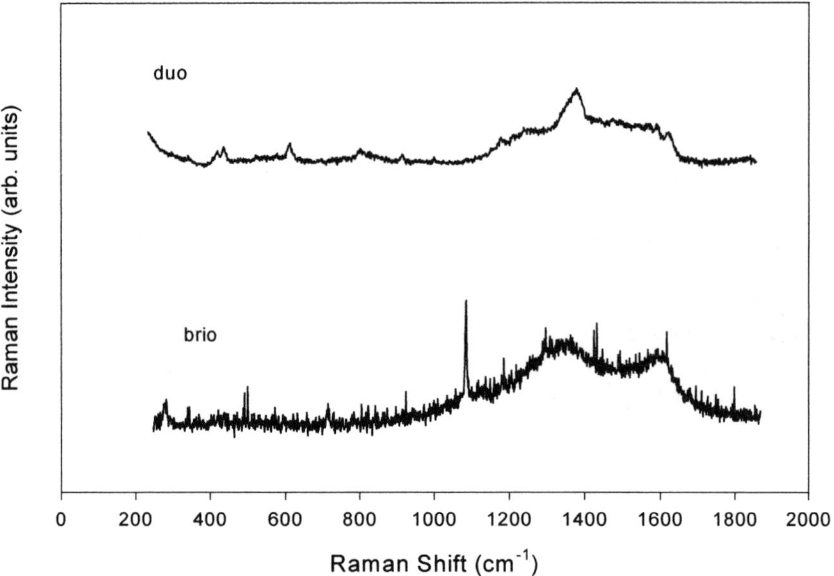

Fig.1 Raman spectra of ball-point pen black ink Brio and Duo obtained with λ= 632.8 nm.

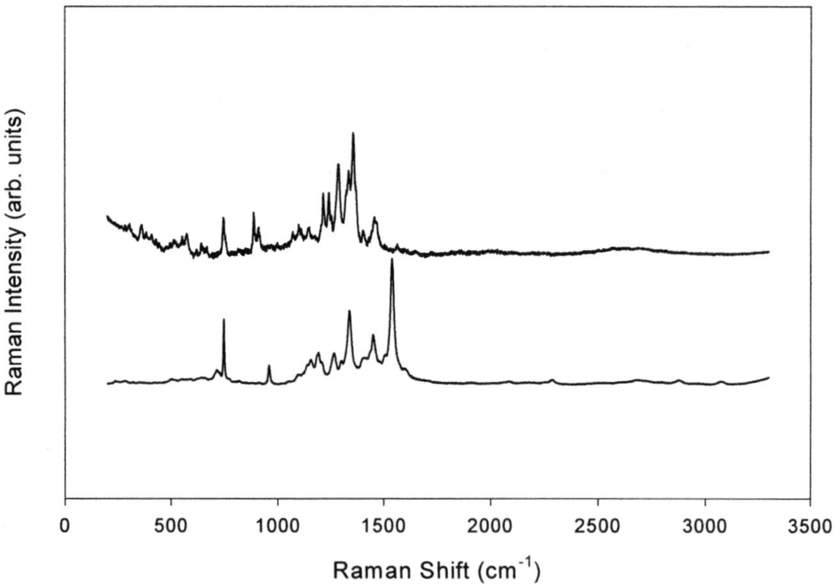

Fig.2 Raman spectra of (bottom) green ink Stabilo and (top) gold ink Trattopen (λ= 632.8 nm).

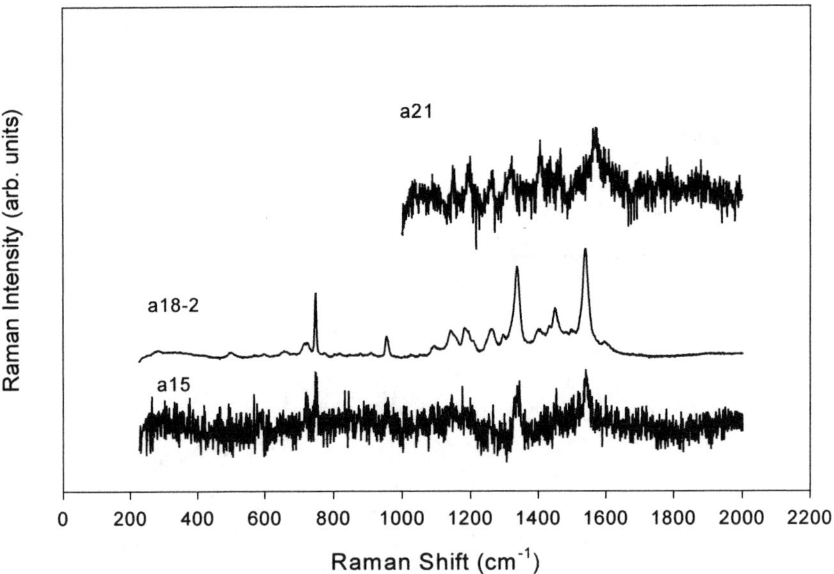

Fig.3 Raman spectra of Canon printers : A15 = ink BC-05 (02) from Canon BJC 250; A18 = Ink BJC-250 from Canon BJC 250; A21= Ink BC21 from Canon BJC 4550.

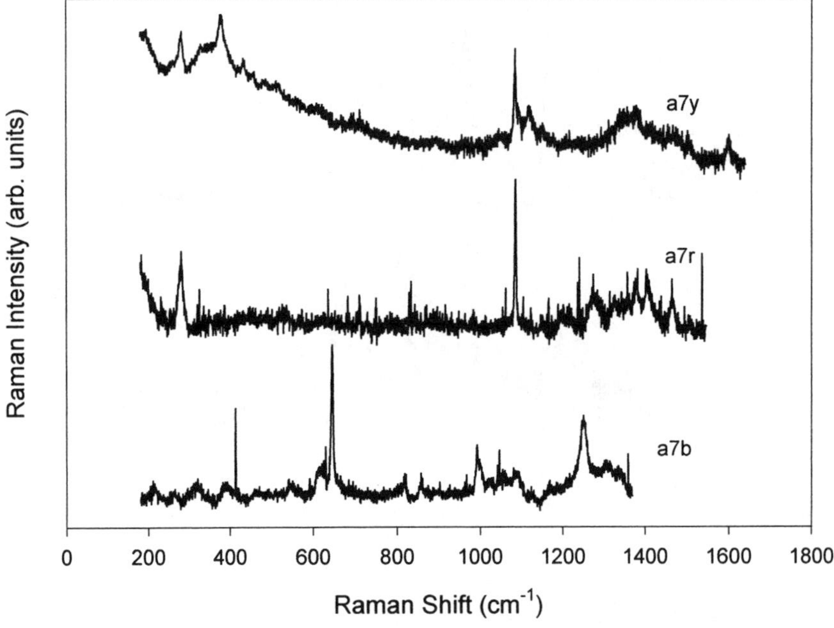

Fig.4 Coloured ink HP 51629A (y = yellow, r = mag, b = cyan) from HP Deskjet 670C (λ= 784.8 nm). Peak at 1085 cm^{-1} arise from the calcium carbonate present in the paper.

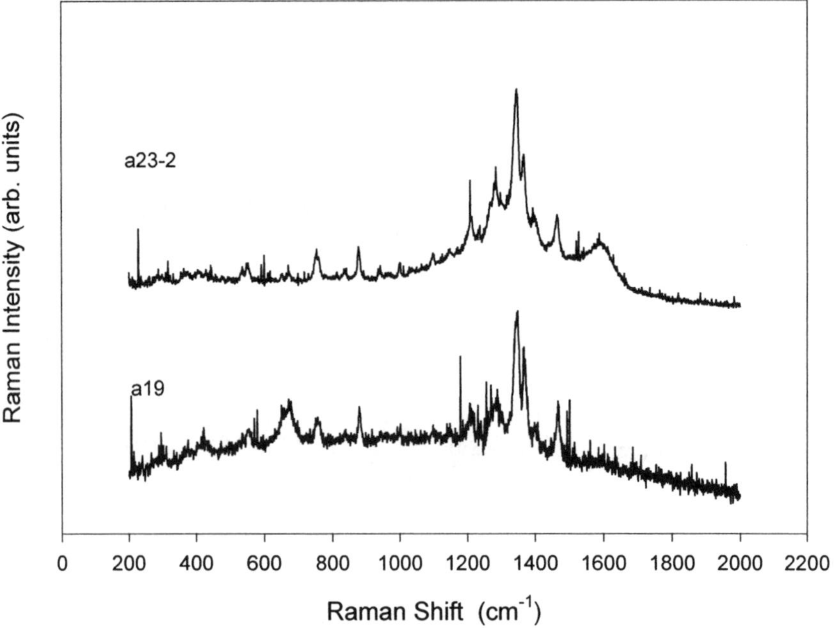

Fig.5 Raman spectra from not identified mark toners reported as a23 e a19. λ = 632.8 nm.

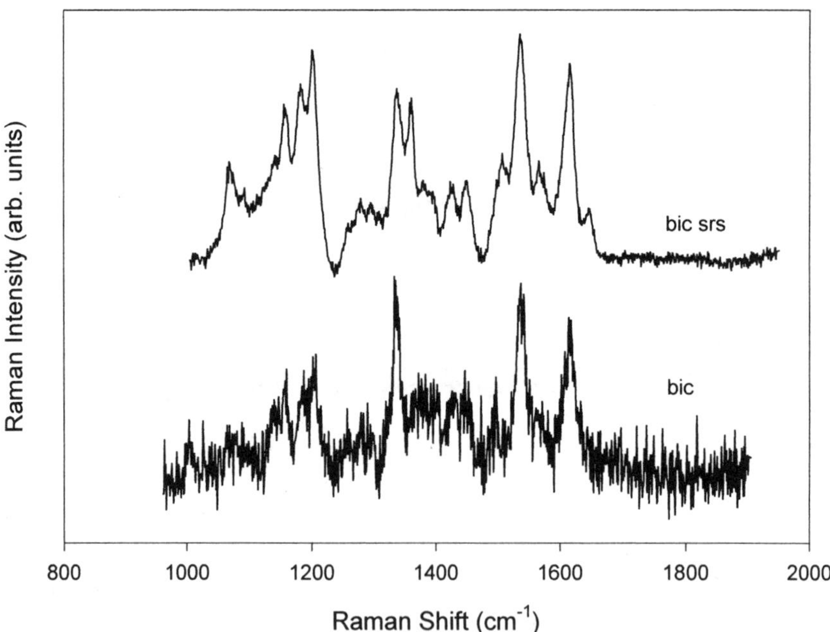

Fig.6 Raman spectra of black ball point ink Bic obtained at λ= 632.8 nm with (up) and without (without) silver colloids

4.2 Ink-Jet printers

Twelve different black inks and more than twenty coloured inks, coming from three-colour and four-colour printers, have been analysed.

Also spectra obtained by black inks for Ink-Jet have almost always provided us with elements that allow us distinguishing one another. Inks with a different mark but referring to the same printer have exhibited different spectral features. In Figure 3, spectra of three different inks marked Canon for Bubble-Jet printers have been reported. Strong similarities between spectrum A15 and A18 can be observed with regard to the peak positions, but really the signal to noise ratio is very different in the two cases, which could constitute a further distinctive element. Similar results have also been obtained by the coloured inks (Fig. 4). In this case the excitation at 784.8 nm is needed in order to remove the strong fluorescence background characteristic of these samples. The four-colour printer inks have in general shown greater spectra differences among different marks, with regard to the three-colour printer inks. In this last case, in fact, the most part of the cyan and the yellow have shown very similar spectra and only the magenta look different in a significant manner.

4.3 Toner

Spectra obtained by laser printer toner are different one another, more than for the peak frequencies, for their relative intensity, by evidencing the presence of common components, but in different proportion. Even though the spectral range comprised between 1200-1600 cm^{-1} shows the very similar structure in most of the analysed toners. Nevertheless, the presence of weak spectral features, in different ranges, allows characterising them as evidenced in Fig. 5 regarding to two toners of unknown mark laser printer. However, not all the toners have given a useful Raman signal; in some cases, only a broad florescence band, without any characteristic Raman feature, has been obtained. In these cases, neither the use of the laser line excitation at 784.8 nm gives appreciable results. On the contrary to what happens for the ink-jet printer, that give analogous spectral features using the same ink on different printers, slightly different spectra are obtained by using the same toner on different laser printers. Such a behaviour can be ascribed to different toner deposition temperature.

4.4 SERRS

One of the commonest problem encountered in the Raman analyse of inks is the fluorescence background arising from dyes and pigments. Moreover, the use of NIR laser line excitation can solve this problem. On the other hand, it is a well-known fact that SERS (Surface Enhanced Raman Scattering) allows a strong improvement of Raman signal (up to 10^6) together with a quenching of the fluorescence background. This effect is even more impressive if SERRS (Surface Enhanced Resonançe Raman Scattering) conditions are achieved by using the proper laser line excitation. Furthermore, SER(R)S condition allows reducing the laser power of the sample preventing sample damaging. In Fig. 6, it is evident the validity of SERRS use in the ballpoint ink analysis.

Obviously, there are some differences in the quality of two spectra: SERRS highly increased the signal to noise ratio, allowing the view even of weak peaks.

Strikingly enough, differences in spectral quality appear evidently. In fact, the signal to noise ratio is strongly improved in samples treated with silver colloid, where SERRS effect is present, and, moreover, weak spectral features become evident in these treated samples. In any case, the arising of SERRS enhanced anomalous peaks, as the one at 1857 cm -1 of Figure 6, could give problems to the comparison with spectra obtained on samples not treated with silver colloid.

Generally, Raman spectra improvement has been obtained in ballpoint ink Raman analysis by using silver colloid. On the contrary, no improvement has been observed in toner or ink-jet printers.

5. Conclusions

Raman micro-spectroscopy is confirmed as a powerful not invasive and not destructive tool analysis in order to discriminate ballpoint inks, either black or coloured, ink-jet inks and laser printer toners. Only in a few cases no reliable spectra has been obtained. Most important, differences in the same trademark ink have been observed.

In coloured ink-jet analysis, NIR excitations play a crucial role in order to eliminate fluorescence background. SERRS contribute given good result in ballpoint ink analysis. Minor results have been obtained in the toner and printer ink analysis.

References

[1] R.L. Brunelle, A sequential multiple approach to determining the relative age of writing inks, *International Journal of Forensic Document Examiners* 1 (1995) 94-98

[2] R.L. Brunelle, A.A. Cantu, A critical evaluation of current ink dating techniques, *Journal of forensic Sciences* 32 (1987) 1511-1521

[3] S.O. Vikman, Applicability of FTIR and Raman Spectroscopic methods to the Study of ink-jet and electrophotographic prints. In: International Conference on Digital Printing Technologies, Helsinki University of Technology, 2000

[4] K. Kneipp, S. Wang, Single Molecule Detection Using Surface-Enhanced Raman Scattering (SERS), *Physical Review Letters* 78 (1997) 1667-1672

[5] A. Jimenez, S. Sandoval Micro-Raman spectroscopy: a powerful technique for materials research, *Microelectronics Journal* 31 (2000) 419-427

[6] M. Claybourn, M. Ansell, Using Raman Spectroscopy to Solve Crime: Inks, Questioned Documents and Fraud

[7] S.R.Emory, W.E.Haskins, S.Nie, Direct observation of Size-Dependent Optical Enhancement in Single Metal Nanoparticles, *J. Am. Chem. Soc.* 120 (1998) 8009- 8010

[8] K.Arya, R.Zeyher Theory of Surface-Enhanced Raman Scattering. In Apllied Phisics- Light Scattering in Solids IV

[9] J. J.Laserna, Modern Techniques in Raman Spectroscopy, University of Malaga Editor, Spain

[10] P. Vandenabeele, L. Moens, H. G. M. Edwards, R. Dams, Raman spectroscopic database of azo pigments and application to modern art studies, *Journal of Raman Spectroscopy* 31 (2000) 509-517.

[11] R.M. Seifar, J.M. Verheul, F. Ariese, U.A. Brinkman, C. Gooijer Applicability of Surface- Enhanced Resonance Raman Scattering for the discrimination of ballpoint pen inks, *Analyst* 126 (2001) 100-115

[12] J. Levinson Questioned Documents: a Lawyer's Handbook , Academic Press, New York, 2001

GNSR 2001
G. Messina and S. Santangelo (Eds.)
IOS Press, 2002

High-frequency features in Raman spectra of reactively sputtered *a*-CN:H thin films

G.Messina, S.Santangelo*

INFM, Dipartimento di Meccanica e Materiali, Facoltà di Ingegneria, Università "Mediterranea", località Feo di Vito, 89060 Reggio Calabria, Italy
* E-mail: santange@ing.unirc.it tel.: +39.(0)965.875305; fax:+39.(0)965.875201

G.Fanchini, A.Tagliaferro

INFM, Dipartimento di Fisica, Politecnico di Torino, corso Duca degli Abruzzi 24, 10129 Torino, Italy

A.Tucciarone

INFM, Dipartimento di Scienze e Tecnologie Fisiche ed Energetiche, Università di Roma "Tor Vergata", via Tor Vergata 110, 00133 Roma, Italy

Abstract The high-frequency region of the Raman spectra of reactively-sputtered hydrogenated amorphous carbon-nitride (*a*-CN:H) thin films is analysed in order to gain further information about the film structural properties, studied, as usually, by monitoring the evolution of the D- and G- bands. By the aid of the complementary infrared (IR) film characterisation, excluding any appreciable hydrogen incorporation within the films, the broad asymmetrical Raman band centred at about 3000 cm^{-1} is ascribed to the second order of the D- and G- bands. The finite-crystal-size effects, due to the very small dimension of graphitic cluster islands, are suggested as responsible for the detection of both overtone- and combination-bands, as well as for the VDOS softening, resulting in the considerable broadening of the second-order components. The found dependence of the second- on the first-order integrated-intensity is explained semi-quantitatively in terms of thermal population of the involved phononic states.

1. Introduction

A considerable scientific and technological interest concerns the synthesis of the hypothetical β-C$_3$N$_4$ [1] and super-hard (sp^3) compounds. Actually, nitrogen incorporation results in strongly improved characteristics of both amorphous carbon (*a*-C) and hydrogenated amorphous carbon (*a*-C:H films) [2-5]. The non-destructive characterisation of the derived materials is still entrusted to Raman spectroscopy [2-4,6,7]. However, the analysis very often privileges the spectral range dominated by the D- and G- bands, meanwhile the features observable above 2400 cm^{-1}, probably due to their controversial assignment, are frequently ignored.

The high-frequency region of the Raman spectra of reactively-sputtered a-CN:H thin films is here accurately analysed in order to gain further information about the film structural properties. By the aid of the complementary IR film characterisation, indicating no appreciable hydrogen-incorporation within the films, the broad asymmetrical Raman band centred at ~3000 cm^{-1} is ascribed to the D- and G- second order. The finite-crystal-size effects, due to the very-small dimensions of the graphitic-cluster islands, are suggested as responsible for both overtone- and combination- band detection, as well as for the VDOS softening, resulting in the considerable second-order component broadening. The found dependence of the second- on the first- order integrated-intensity is explained semi-quantitatively in terms of thermal population of the involved phononic states.

2. Experimental details

2.1 Film deposition

Thin films of a-CN:H were deposited on (100) c-silicon substrates by a conventional 13.56 MHz r.f. diode sputtering system, operating with an Ar/He mixture, with the addition of N_2 and H_2 as reactive gases. The graphite target (99.999% purity, 20 cm in diameter) was placed at 25 mm from the grounded electrode holding the substrate, mantained at 100°C, during the deposition. He and Ar flows were 30 and 70 sccm (standard cubic centimetre per minute), respectively ; N_2 and H_2 flows, Φ_{N_2} and Φ_{H_2}, varied between 3.0 and 20. sccm and between 2.2 and 5.0 sccm, respectively. The r.f. power, W_{rf}, ranged between 180 and 300 W; the total pressure, p_{tot}, between 20 and 38 mtorr. For further details see Tab.1 and ref. *[8]*.

2.2 Film characterisation

The Raman spectra were recorded, at RT, in the 200-3600 cm^{-1} region by using a Jobin Yvon Ramanor U-1000 double monochromator equipped with an electrically cooled Hamamatsu R943-02 photo-multiplier as a detector, and photon counting electronics. Excitation wavelength was 514.5 nm. The S/N ratio was improved by recording multiple scans. A power density of ~20 W/mm^2 at the sample surface was utilised in order to prevent sample annealing. In order to monitor the structural modifications produced by the different growth conditions through the evolution of the main spectral features, Gaussian bands, superimposed to a Gaussian photoluminescence background, were considered and their frequency position, width (FWHM) and intensity chosen by a least-square best-fit method.

The IR absorption spectra were recorded in the 450-4000 cm^{-1} range on a Perkin-Elmer FTIR-2000 spectrometer (1 cm^{-1} resolution) by using a bare substrate as a reference. Up to 64 interferograms were recorded and averaged to improve the S/N ratio.

3. Results and discussion

The Raman spectra of the investigated a-CN:H films (Fig.1) are largely dominated by the well-known D- and G- bands. The relative fitting parameters are listed in Tab.2 together

Tab.1 Growth conditions of the investigated a-CN:H samples: W_{rf} and p_{tot} respectively indicate the r.f. power and the total pressure. Φ_{N2} and Φ_{H2} denote the flows of the reactive gases (N_2 and H_2).

Film #	W_{rf} (W)	p_{tot} (mTorr)	Φ_{N2} (sccm)	Φ_{H2} (sccm)
1	300	20	20.	0.0
2	300	21	7.0	5.0
3	300	38	4.2	3.0
4	200	30	4.2	2.2
5	180	38	3.0	2.2

Tab.2 Frequency position (ω_D and ω_G) bandwidth (σ_D and σ_G) of the D- and G- bands of the investigated a-CN:H samples. The average size of graphitic cluster islands (L_C) are also reported.

Film #	ω_{D-1} (cm)	σ_{D-1} (cm)	ω_{G-1} (cm)	σ_{G-1} (cm)	L_C (nm)
1	1363	302	1563	187	1.6[*]
2	1375	313	1572	147	1.9[*]
3	1394	371	1563	149	1.4
4	1387	365	1562	148	1.6
5	1396	394	1564	140	1.3

with the average size of the graphitic cluster islands, as obtained from the D/G intensity ratio according to ref.s *[9]* and *[10]*. Briefly, the results evidence both the film clusterisation-degree and structural-disorder level to be strongly influenced by the changes in the deposition conditions. A detailed analysis of the low-frequency region of the spectra, as well as the evolution of D- and G- bands with changing the growth variables for films grown under similar conditions is reported elsewhere *[11]*. The attention is here focused on the broad asymmetrical and nearly-featureless band, observed between 2400 and 3400 cm^{-1}.

It is worthwhile noticing that the analysis of Raman spectra is rarely extended to this spectral region and, eventually, the features there-observed are not mentioned at all *[12,13]* or the interest is mainly centred on the background *[14]*. In a few cases an assignment is proposed for the features above 2400 cm^{-1}: in amorphous carbon, they are, at times, briefly reported as the D- and G- second-order *[15,16]*; meanwhile, in hydrogenated amorphous carbon, the presence of the Raman band at ~3000 cm^{-1} is assumed as indicative of the hydrogen incorporation into the film *[17,18]*.

Aiming at clarifying the origin of the Raman structure at ~3000 cm^{-1}, the IR absorption spectra are examined and compared to Raman spectra (Fig.2). A satisfactory correspondence is found in the region below 2400 cm^{-1}; contrarily, no analogy is found above 2400 cm^{-1}. The absence, between 2855 and 3050 cm^{-1}, of the bands originating from the sp^2 and sp^3 C-H$_x$ stretching modes *[5]* suggests no appreciable hydrogen-incorporation to occur in the investigated samples. The broad asymmetrical Raman feature centred at ~3000 cm^{-1} is consequently attributed to the D- and G- band second-order.

The shape of the second-order spectrum of crystalline graphite has been demonstrated to be sensitive to the average size, L_C, of the graphitic cluster islands. As L_C decreases below ~2.5nm, the wave-vector selection-rule is released with consequent appearance of additional features and noticeable broadening *[19]*.

As for the investigated *a*-CN:H films, the L_C values, obtained from the D/G intensity ratio as $[(I_D/I_G)/0.55]^{1/2}$ (denoted by a *) *[9]* or as $4.4/(I_D/I_G)$ *[10]*, range between 1.6 and 1.9 nm (Tab.2) *[20]*, thus suggesting the possibility that finite-crystal-size effects can be observed in the second-order spectrum of the present samples. Actually, three broad Gaussian lines are necessary to reproduce the asymmetrical and nearly-featureless band at 3000 cm^{-1}: two of these centred at the frequency-positions approximately expected for the two overtones

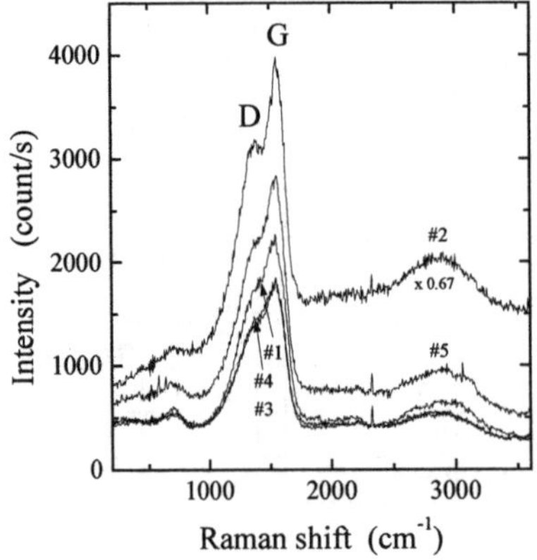

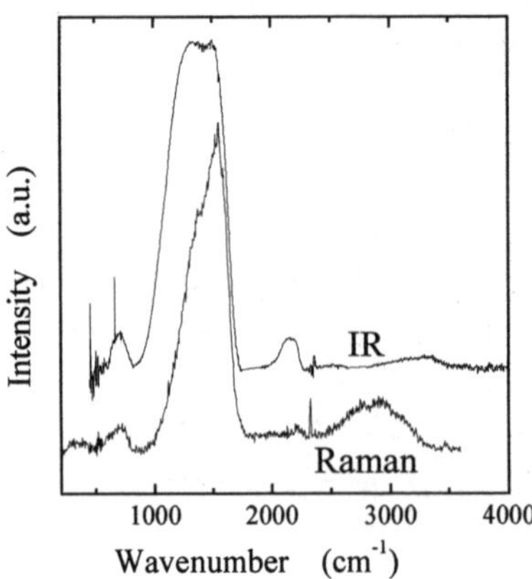

Fig.1 As-measured Raman spectra of the investigated *a*-CN:H films. The intensity of the D- and G- bands progressively decreases on going from samples #1 to #3; the spectra of samples #3 and #4 are nearly superimposed.

Fig.2 Comparison between IR- and Raman- spectra, after background subtraction. The spectra shown refer to sample #3.

(namely, $\sim2\omega_D$ and $\sim2\omega_G$) and the third at the frequency-position roughly coincident with that of a combination band (namely, $\sim\omega_D+\omega_G$). These findings would further support our attribution.

However, in order to further test the validity of our assignment, the integrated intensity of the structure, ascribed, on the basis of the non-detection of bonded hydrogen, to the second order of the D- and G- bands, $^{II}I_{D+G}$, is plotted (Fig.3) *vs* the sum of the integrated intensities of the D- and G- bands, $^{I}I_{D+G}$. A nearly-quadratic increase of $^{II}I_{D+G}$ with $^{I}I_{D+G}$ is found. The observed dependence would, hence, confirm that, even though the incorporation of a very little hydrogen-amount into the investigated films cannot be entirely excluded, the band at 3000 cm⁻¹ has to be basically regarded as the D- and G- second order structure.

In the following, a semi-quantitative explaination is proposed for the observed $^{II}I_{D+G}$ dependence on $^{I}I_{D+G}$.

As well known, the light-scattering spectrum of a single crystal shows the frequencies ω_i of the modes with wavevector $\boldsymbol{q}_i$ equal to the scattering vector $\boldsymbol{q}$ (crystal-momentum selection-rule). The probability $\Pi_{n_i \to n_i+1}$ for a transition in which one phonon of frequency ω_i is created (namely, $n_i \to n_i+1$, $n_i = (e^{\hbar\omega_i/K_BT} - 1)^{-1}$ being the mean occupation number of the involved mode before the transition)

$$\Pi_{n_i \to n_i+1} \propto (n_i+1) \tag{1}$$

enters in determining the intensity $I(\omega_i)$ in the first-order emission spectrum; meanwhile, the probability $\Pi_{n_i \to n_i+1,n_j \to n_j+1}$ for a transition in which two phonons of frequencies ω_i and ω_j are created, given by the product of the two corresponding one-phonon transition,

$$\Pi_{n_i \to n_i+1,n_j \to n_j+1} = \Pi_{n_i \to n_i+1} \cdot \Pi_{n_j \to n_j+1} \propto (n_i+1)(n_j+1) \tag{2}$$

eventually contributing to the intensity of the corresponding second-order feature of the spectrum.

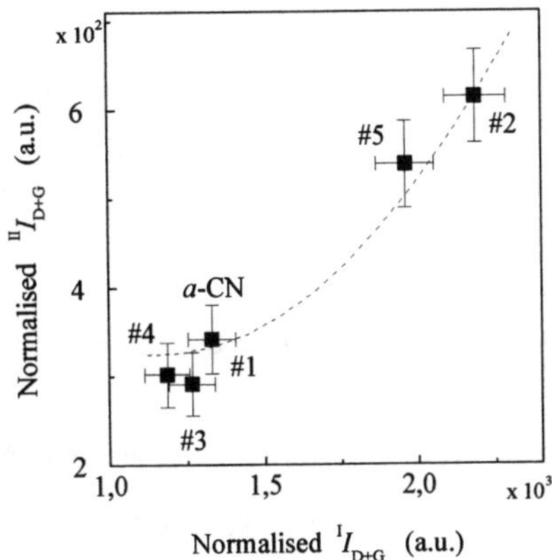

Fig.3 Integrated intensity of the second-order ($^{II}I_{D+G}$) vs the integrated intensity of the first-order ($^{I}I_{D+G}$) of the D- and G- bands (all the intensities are normalised to the film thickness).

In an amorphous material, the short correlation-length, deriving from the translational-symmetry loss, breaks the usual wavevector selection-rule for Raman modes, allowing all normal modes to participate in the light-scattering process. As a consequence, the first-order Stokes-component intensity of Raman spectrum at frequency ω becomes

$$I(\omega) = \Sigma_b \, C_b(\omega) \cdot G_b(\omega) \cdot [n(\omega)+1]/\omega,$$

where $n(\omega)$, $C_b(\omega)$ and $G_b(\omega)$ and respectively denote the thermal population of the involved mode, the band-dependent Raman coupling-coefficient and the vibrational density-of-states of the band b in the disordered network [21]. However, without further going into details, if, in order to explain semi-quantitatively the observed dependence of the second- on the

first- order integrated intensity of the D- and G- bands, according to the above interpretation of the second-order structure in terms of overtones and combination, the creation is basically considered of two phonons of i) equal-frequency ω_1, and ii) different-frequencies ω_1 and ω_1, the transition probabilities, according to eqs.(1) and (2), result proportional to $(n_1+1)(n_1+2)$ and $(n_1+1)(n_2+1)$, in cases i) and ii), respectively.

Hence, terms such as (n_D+1) and (n_G+1) enter in determining the first-order integrated-intensity $^{I}I_{D+G}$; meanwhile, terms such as $(n_D+1)(n_D+2)$ and $(n_G+1)(n_G+2)$ (overtones) and $(n_D+1)(n_G+1)$ (combination) contributing to the second-order integrated-intensity $^{II}I_{D+G}$, finally determining the nearly-quadratic dependence observed (Fig.3).

4. Conclusion

Reactively-sputtered hydrogenated amorphous carbon-nitride thin films are characterised by Raman spectroscopy and the crucial role of the deposition conditions in determining the film clusterisation-degree and structural-disorder level is briefly evidenced. Particular emphasis is given to the analysis of the high-frequency region of the spectra, about which few and often contradictory indications can be found in literature.

The broad asymmetrical and nearly-featureless band, detected in Raman spectra between 2400 and 3400 cm^{-1}, is here investigated by the aid of the complementary IR characterisation technique. On the basis of the indications coming from the analysis of the IR absorption spectra, evidencing no appreciable hydrogen-incorporation within the films, the Raman feature centred at $\sim$3000 cm^{-1} is ascribed to the second-order of the D- and G- bands.

The very-small dimensions of the graphitic-cluster islands are then shown to be responsible for the occurrence of finite-crystal-size effects, resulting in the detection of a combination band together with the D- and G- band overtones, as well as in the considerable broadening of all the observed second-order components. The found nearly-quadratic dependence of the second- on the first- order integrated-intensity is finally semi-quantitatively understood in terms of thermal population of the involved phononic states.

References

[1] A.Y.Liu, M.L.Cohen, *Phys. Rev.* B**41** (1990) 10727-10734

[2] K.G.Kreider, M.J.Tarlov, G.J.Gillen, G.E.Poirier, L.H.Robins, L.K.Ives, W.D.Bowers, R.B.Marinenko, D.T.Smith, *J. Mater. Res.* **10** (1995) 3079-3083

[3] J.Koskinen, J-P.Hirvonen, J.Levoska, P.Torri, *Diamond Relat. Mater.* **5** (1996) 669-673

[4] S.R.P.Silva, J.Robertson, G.A.J.Amaratunga, B.Rafferty, L.M.Brown, J.Schwan, D.F.Franceschini, G.Mariotto, *J. Appl. Phys.* **81** (1997) 26-2634

[5] Y.H.Cheng, Y.P. Wu, J.G.Chen, X.L.Qiao, C.S.Xie, *Diamond Relat. Mater.* **8** (1999) 1214-1219

[6] M.M.Lacerda, D.F.Franceschini, F.L.Freire Jr., G.Mariotto, *Diamond Relat. Mater.* **6** (1997) 631-634

[7] C.Lenardi, M.A.Baker, V.Briois, L.Nobili, P.Piseri, W.Gissler, *Diamond Relat. Mater.* **8** (1999) 595-600

[8] G.Fusco, F.Giorgis, C.F.Pirri, A.Tagliaferro, E.Tresso, C.De Martino, P.Rava, *Defects and Diffusion Forum Vols.* **134-135** (1996) 3-14

[9] A.C.Ferrari, J.Robertson, *Phys. Rev.* B**61** (2000) 14095-14107

[10] F.Tuinstra, J.L.Koenig, *J. Chem. Phys.* **53** (1970) 1126-1130

[11] G.Messina, A.Paoletti, S.Santangelo, A.Tagliaferro, A.Tucciarone, *J. Appl. Phys.* **89** (2001) 1053-1058

[12] M.Y.Chen, D.Li, X.Lin, V.P.Dravid, Y.W.Chung, M.S.Wong, W.D.Sproul, *J. Vac. Sci. Technol.* A**11** (1993) 521-524

[13] J.A.McLaughlin, B.Meenan, P.Maguire, N.Jamieson, *Diamond Relat. Mater.* **5** (1996) 486-491

[14] J.Vyskocil, P.Široký, V.Vorlícek, V.Perina, *Diamond Relat. Mater.* **5** (1996) 466-470

[15] A.V.Stanishevsky, L.Yu.Khriachtchev, *Diamond Relat. Mater.* **5** (1996) 1355-1358

[16] Q.Wang, D.D.Allred, J.González-Hernández, *Phys. Rev.* B**47** (1993) 6119-6121

[17] M.K.Fung, W.C.Chan, Z.Q.Gao, I.Bello, C.S.Lee, S.T.Lee, *Diamond Relat. Mater.* **8** (1999) 472-476

[18] A.K.M.S.Chowdhury, D.C.Cameron, M.S.J.Hashmi, J.M.Gregg, *J. Mater. Res.* **14** (1999) 2359-2363

[19] R.J.Nemanich, S.A.Solin, *Phys. Rev.* B**20** (1979) 392-401

[20] The Tuinstra-Koenig *[10]* or the Ferrari-Robertson *[9]* relationship has been used depending on the D/G intensity ratio exceeded or not the critical value of 2.2 *[9]*, respectively. The obtained average sizes are consistent with the preliminary results of small-angle x-ray scattering measurements performed on samples deposited under similar conditions.

[21] R.Shuker, R.W.Gammon, *Phys. Rev. Lett.* **25** (1970) 222-225

GNSR 2001
G. Messina and S. Santangelo (Eds.)
IOS Press, 2002

49

Hydration effect of Poly(Ethylene Oxide) by Raman Scattering, Viscosity and Acoustic Measurements

C. Branca, S. Magazù[*], G. Maisano, F. Migliardo, P. Migliardo,
G. Romeo

*Dipartimento di Fisica and INFM, Università di Messina, P.O. Box 55
Papardo,98166 S. Agata Messina*
[*]*Phone: +39-090-6765025, Fax: +39-090-395004, E-mail:magazu@dsme01.unime.it*

Abstract. We report on Raman scattering, viscosity and ultrasonic measurement on Poly(Ethylene Oxide) (PEO) in aqueous solution. The analysis of the D-LAM (acronym for Disordered Longitudinal Acoustic Mode) spectral contribution on the pure, reveals an olygomer-polymer transition for a polymerization degree of about 13. In aqueous solutions the frequency increase towards values corresponding to the crystal ones and the sharpening of the D-LAM spectral contribution, indicate that the addition of water destroys the intermolecular interactions and stiffens the coil structure. The temperature analysis of the Raman D-LAM band, reveal that the solvent power of water increases up to T= 45°C, decreasing at higher temperature .It will be shown that, interpreted in conjunction with ultrasonic data, these apparently far findings, provide a single coherent mechanism capable of encompassing the structural properties of our systems. In order to study the hydration effect of PEO and its dependence on the molecular weight, we report on viscosity and compressibility measurements on aqueous solutions of PEO at different concentration and temperature values. In particular, ultrasonic technique allows evaluating the hydration number for polymer samples at different polymerization degree.

1. Introduction

Recently Poly(Ethylene Oxide) (PEO) has received a growing attention not only from the applicative point of view. The simpleness of its structure and the unusual solubility in water [1,2], in fact, make it a precious model system [3] for studying the interaction mechanisms of water with hydrophilic surfaces [4] and macromolecules. The chemical structure, $H-(O-CH_2-CH_2)m-OH$, of this synthetic polymer includes two terminal groups, H and OH, which play an important role in short compounds. The hydrophobic ethylene units and the hydrophilic oxygens, which alternate along the chain, are responsible of its amphoteric character [5]. The similarity of the ether oxygen spacing (2.88 Å) with that of the oxygens in water (2.85 Å), could explain the polymer solubility in water [5,6], which persists in all proportions at temperatures lower than the boiling point of water [2]. Above this point it presents a miscibility gap that, by diminishing the polymerisation degree, m, shifts towards greater temperatures and vanishes [7] for m<48. PEO is commercially available in an extremely broad Molecular Weight (Mw) range. PEO with m<150 is generally called Poly(Ethylene Glycol), or briefly PEG.

A wide variety of experimental techniques have been employed to investigate the conformation of PEO in the crystalline and molten state. In the isolated Ethylene Glycol (EG) molecule, the basic entity of PEO, many equilibrium configurations can be obtained by rotation of the two CH_2OH groups around the C-C axis. IR, Raman and computer simulation studies [8] reveal, also in the liquid phase, the existence of an intra-molecular H-bond which makes the gauche conformation the most energetically favoured; in addition, the OH interacting groups give rise to links, via hydrogen bonding, among adjacent chains [8]. The findings so far agree with the conformational assignment for the crystalline state to internal rotation about the $O-CH_2$, CH_2-CH_2 and CH_2-O bonds of trans-gauche-trans (tgt) respectively. Crystalline PEO is retained to present a helical conformation that contains seven structural units CH_2-CH_2-O with two helical turns per fibre identity period (19.3 Å) [9,10]. The structure of this polymer in the melt or in solution, has continued to intrigue investigators over the years. Pursuit of understanding has led various theoretical studies to quite different results, while vibrational spectroscopy has revealed to be difficult to apply with confidence, when a broad distribution of a large variety of conformations is involved. It has become clear, however, that a strongly disordered conformation is favoured in the molten state, while in aqueous solution the tgt conformation is stabilized, due to hydrogen bonds between the ether-oxygens chain and water molecules [7,11].

The purpose of the present work is to show how the joint employment of Raman scattering, viscosity and compressibility measurements can furnish complementary information on the structural length scales of PEO, both in the melt and in solution. The analyses of the Raman D-LAM (acronym for Disordered Longitudinal Acoustic Mode) spectral contribution, of viscosity and adiabatic compressibility data, rejoining apparently far indicators, afford the opportunity to characterize the dependence of the polymer structural parameters on polymerisation degree, solvent content and temperature. Therefore, we report on viscosity and compressibility measurements of PEO aqueous solutions at different nominal molecular weights, concentration and temperature values.

2. Experimental set-up

We examined high purity samples, purchased from Aldrich-Chemie, of Ethylene Glycol (EG) and PEO with average Mw of 106, 200, 300, 400, 600, 900, 1000, 1540, 3400 Da, both in melt and in aqueous solution at different concentrations. The wide range of commercially available Mws, makes PEO an ideal system for such kind of studies. The manufacturer provided the Mw average values and polydispersities. To investigate the role played by the interactive OH end groups, we also examined the CH3 terminated species, corresponding to the same degree of polymerisation of PEO. The solutions were freshly prepared and slowly filtered with 0.22 μm PTFE filters.

Polarized (I_{VV}) and depolarised (I_{VH}) spectra were obtained by a SPEX Ramalog 5 triple monochromator in a 90° scattering geometry in the (-40°C ÷ 80°C) temperature range. To avoid fluorescence, which screens the Raman signal when short-wavelength laser excitations are used, the 5145 Å line of an Argon laser was chosen. The laser power was maintained at approximately 5 W. The detection apparatus consisted of a photon counting system whose outputs were processed on line by a computer. The scattered photons were automatically normalized for the incoming beam intensity in order to ensure good data reproducibility. The optical purity of the samples ensured to collect data with good signal-to-noise ratio and with high reproducibility. The samples were sealed in optical quartz cells and then mounted in an optical thermostat which stabilizes temperature within ±0.1 °C. The spectral range covered was -100 cm^{-1}÷1600 cm^{-1}, with an instrumental resolution, from 0.2 to 4 cm^{-1}, depending on the examined spectral range. The spectra at different resolutions were subsequently numerically matched and corrected for the density ρ, for the refractive index n and for local field effects. These corrections correspond to a normalization of the intensity by the factor $n\rho^{-1}$ $(n^2+2)^{-4}$, with n(T) and ρ(T) taken from the literature. The spectral information which matters most in this study are the isotropic scattering intensities, calculated from the parallel and perpendicular components of the scattered light. In an ordinary (non-resonant) Raman effect, for linearly polarised excitations, The depolarisation ratio $\rho = I_{VH}/I_{VV}$, which as well known varies in the range $0 \leq \rho \leq 3/4$, being nearly equal to 3/4 for fully depolarised bands, has been also evaluated.

As far as shear viscosity is concerned, we examined high purity samples, purchased from Aldrich-Chemie, of PEG with average Molecular Weight, M_w, of 200, 400, 600, 1000 and 2000 Da in aqueous solution at different concentrations. Measurements were performed on PEO/water solutions at a temperature of 25° C by means of a standard Ubbhelohde viscometer. The viscometer, mounted in a suitable bath, which stabilizes temperature within ±0.02°C, was chosen with long flow time in order to minimize the kinetic energy correction. Before measuring viscosity, the samples were stabilized for sufficient time at the given temperature. Experimental data turn out to be reproducible with an indetermination lower than one part per thousand.

Sound velocity measurements were performed on polymeric aqueous solutions at a temperature of 5° C, 20° C, 25° C, 40° C and 70° C by pulse echoes technique using a home-made thermoregulated (±0.01°C) acoustic interferometer working at a frequency of 3 MHz, purposely projected to assure accuracy of the velocity experimental measurements better than ± 0.1%. The electronic equipment consisted of a standard Matec Inc. apparatus and the measurements were performed using the echoes overlapping method. We have checked, by performing measurements at some values of concentration and for frequencies from 3 MHz to 20 MHz, that the sound velocity is almost frequency-independent in this range indicating that we are not in the presence of possible relaxation processes.

Auxiliary density measurements on polymer/water mixtures, necessary to evaluate of the shear viscosity of the solutions from the kinematic one and of compressibility coefficient, were performed using a standard picnometer technique.

3. Results and discussion

The polarized and depolarised spectra measured in a Raman experiment are essentially connected with the Fourier transform of the polarizability tensor autocorrelation function :

$$J_{VV}(\underline{Q},t) \approx < \sum_{i,j,V,V'} \left(\underline{\varepsilon}_s \tilde{\alpha}_I^V(0)\underline{\varepsilon}_I \right)\left(\underline{\varepsilon}_s \tilde{\alpha}_I^{V'}(t)\underline{\varepsilon}_I \right) q_i^V(0)q_j^{V'}(t)\exp i\underline{Q}\left[\underline{r}_i(t) - \underline{r}_j(0)\right] > \tag{1}$$

where $\underline{\varepsilon}_s$ and $\underline{\varepsilon}_I$ are the scattered and incoming polarization vectors, $\tilde{\alpha}^v = \left(\dfrac{\partial\tilde{\alpha}}{\partial q} \right)_{q=q_v}$, q_v

being the vibrational normal coordinate and $\underline{r}_i$ the position of the i scattering particle [12]. By polarization analysis the different polarization character of the spectral contribution can be evidenced, through the evaluation of the isotropic contribution:

$$I_{iso} = I_{VV} - 4/3\ I_{VH}$$

The strongly polarised LAM of polymers is commonly attributed to a complex of contributions representing the polymer skeletal bending and stretching vibrations. According to the elastic rod model, which is appropriate for crystalline systems, the LAM frequencies are related to the stem length by the formula [13]:

$$\omega_{LAM} = kv = \frac{n}{l_K}\sqrt{\frac{E}{\rho}} \qquad (2)$$

where n is an odd integer (we will be concerned with n=1 since Raman activity is associated with a change in polarizability, and even values of n give no such change); E, ρ and l_K are the polymer elastic modulus, density and stem-length.

Although an approximation, relation (2) gives important information, in that the LAM frequency is inversely proportional to the stem length undergoing this vibration and directly proportional to the propagating velocity $v = \sqrt{\dfrac{E}{\rho}}$ of the longitudinal perturbation.

In non-crystalline systems the presence of conformational disorder dramatically changes the low-frequency Raman spectrum and the narrow LAM band, characteristic of the crystalline state, is replaced by the much broader polarized band associated to disordered longitudinal acoustic mode (DLAM) [14]. In such cases l_K represents the length below which the coil can be treated as rigid; it is essentially connected with the energy difference, $\Delta\varepsilon$, corresponding to

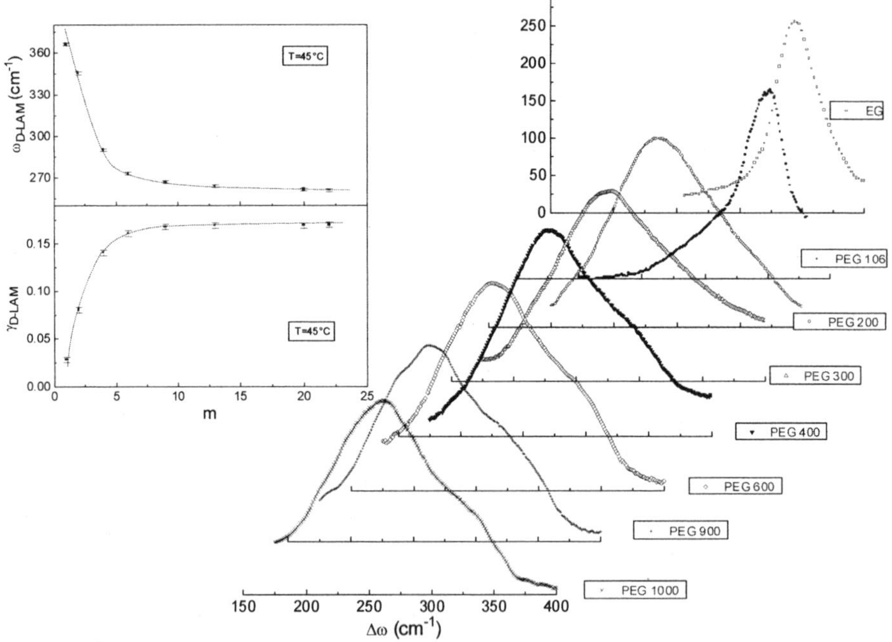

Fig.1 Isotropic spectra of the D-LAM contribution as a function of polymerisation degree at T=45°C. In the insert we report D-LAM centre frequency (ω_{D-LAM}) and width (γ_{D-LAM}) as a function of polymerisation degree at T=45°C; the dashed lines are guides for the eye.

the minima of the trans-gauche conformations. For the polyethylene chain, as well as for similar linear polymers, it results: $l_K = l_0 \exp\left(\dfrac{\Delta\varepsilon}{K_B T}\right)$, where l_0 is of order of a few Angstroms. Increasing the ratio $\dfrac{\Delta\varepsilon}{k_B T}$, the percentage of trans conformations rises, namely the chain becomes locally less flexible [15,16]. The D-LAM spectra still depends on the statistical distribution of all-trans chain segments, but in contrasts with 1/m dependence observed for LAM modes of ordered chains, a different experimental relation is found: $\omega_{D\text{-LAM}} = \omega_0 + B/m^2$, ω_0 and B being constants [17].

In Fig. 1 we present the isotropic Raman spectra relative to the D-LAM contribution for EG and PEO with average M_w of 106, 200, 300, 400, 600, 900, 1000 corresponding to values of m=1, 2, 4, 6, 9, 13, 20, 22. For m=1 and 2, the bands turn into narrow, quite symmetric and well defined spectral contributions, due to the low disorder along few link chains. For increasing values of m, however, the bands broaden and then develop into a wide band for longer chains, assuming practically the same peak frequency and width for m>13.

To extract relevant information, we firstly tried different symmetric fitting functions. However, since apart from the shortest chains, the spectra both in the pure and in solution present strongly non-symmetrical profiles, we adopted a different fitting criterion. As above pointed out, the D-LAM contribution results from a non homogeneous overlapping of sub-bands by the many different conformers existing in the liquid. So any substantial difference in the t-g distribution is reflected in the D-LAM band and can give rise to non-symmetric profiles [18,19].

The non symmetric function, which furnishes the best fitting results, is the log-normal function, firstly used for characterizing the chain length distribution of linear polymers, such as those we are dealing with:

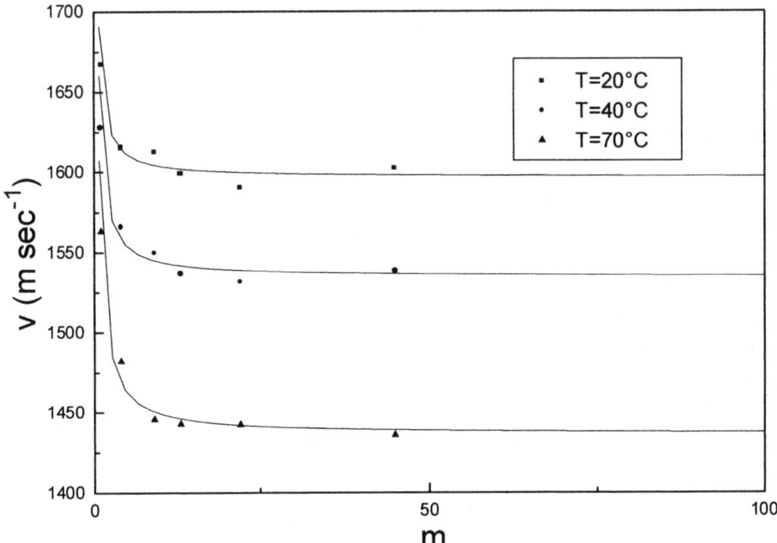

Fig. 2 Ultrasonic velocity as a function of polymerisation degree at three temperatures. The continuous lines are fit results.

$$I(\omega) = A \exp\left[-0.5\left(\frac{\ln\left(\dfrac{\omega}{\omega_{D\text{-}LAM}}\right)}{\gamma_{D\text{-}LAM}}\right)^2\right] \tag{3}$$

where A is the amplitude, $\omega_{D\text{-}LAM}$ the band centre frequency and $\gamma_{D\text{-}LAM}$ the width of the distribution. The characteristic frequency $\omega_{D\text{-}LAM}$ and width $\gamma_{D\text{-}LAM}$, obtained by the fitting procedure as a function of m, are reported in fig.1.

As above indicated, both are strongly affected by the chain length in the low M_w range, whereas, by increasing m, a crossover to an almost constant behaviour is observed for m>13. The decrease of $\omega_{D\text{-}LAM}$ can be explained by taking into account that the amenable to test propagating velocity v of the disordered longitudinal acoustic mode within the coil is, in a sense, connected with the sound velocity measured in the system: $v = \sqrt{\dfrac{1}{\beta\rho}}$, β being the system compressibility. The $1/m^2$ dependence of the ultrasonic velocity, see fig. 2, could justify the experimentally observed dependence of $\omega_{D\text{-}LAM}$ on the polymerisation degree. On the other hand the trend of $\omega_{D\text{-}LAM}$, $\gamma_{D\text{-}LAM}$ and v is common to other physical quantities of the system. Taken together, these observations evidenciate an oligomer-polymer transition for $M_w \geq 600$; this value, in a sense, defines the minimum length for when a polymer is polymeric in behaviour [20].

As fig.2 shows for PEG 600, by increasing the water content, the most striking features revealed in the D-LAM spectra are the remarkable frequency increase (over 10 cm^{-1}) towards values corresponding to the crystal ones, and the significative sharpening of the spectral contribution. Such evidences confirm that PEO in water tends to assume, in respect to the melt case, a more ordered conformation, closer to the crystalline one. The picture that emerges is that adding water molecules, a certain number of these bonds themselves, by H-bond, to the oxigens of the oxirane groups, so promoting the formation of more rigid hydrated polymeric coils. This gives rise to the $\omega_{D\text{-}LAM}$ increase, and, on a macroscopic scale, to the diminishing of the system compressibility, see fig. 3. It is plausible to hypothesise that the compressibility of the polymeric coil is closely connected with the compressibility of the entire system until a full hydration of the polymer chain is reached. A further addition of water molecules, in fact, introduces an amount of bulk water and, as a consequence, the measured system compressibility beyond the hydration values, corresponds to the average value between that of the hydrated polymer and that of bulk water. This hydration process destroys the intermolecular interactions among the polymeric chains and promotes the sharpening of the D-LAM contribution, until, essentially, only one intense band, corresponding to the isolate hydrated polymer coil is found. In this frame, as above stressed, the amount of water which signals the crossover to the $\omega_{D\text{-}LAM}$ and $\gamma_{D\text{-}LAM}$ plateau values should correspond to the full hydration of the polymer.

To get information on the hydration number, we report the viscosity data of the PEO/water solutions at different Molecular Weight, Mw, of 200, 400, 600, 1000 and 2000 Da are presented in Fig.4 as a function of concentration, c (grams of solute per cm^3 of solution), at T=25° C. It can be seen from this figure that the viscosity η of the investigated solutions depends on the average molecular weight of PEO and increases in a non-linear manner with concentration.

Following a relatively simple geometric model, based on Einstein's viscosity law [21, 22], considering the chain molecules as rigid spheres, the relative viscosity, η_r, can be expressed

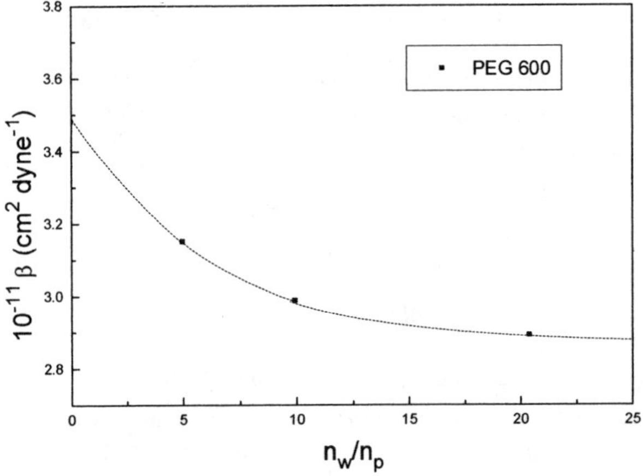

Fig.3 PEG 600 compressibility behaviour as a function of the water/polymer molar ratio; the dashed line is a guide for the eye.

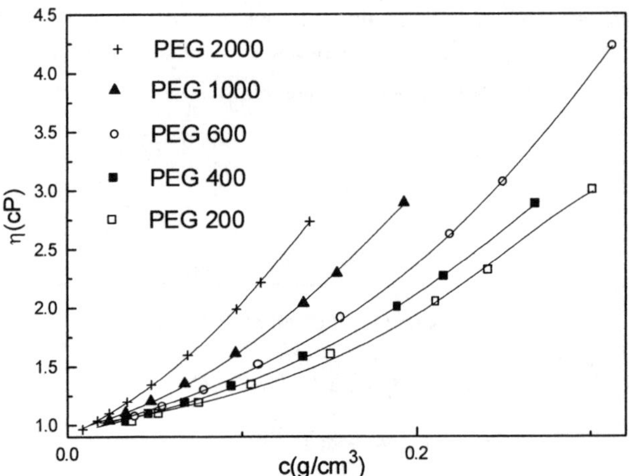

Fig.4 Concentration dependence of viscosity for aqueous solutions of PEG of nominal molecular weights: 200, 400, 600, 1000 and 2000 g/mol at T=25° C

as: $\eta_r=\eta/\eta_o=1+2.5\phi$, where η and η_o are the solution and solvent viscosity, respectively and ϕ is the volume fraction that these spheres occupy. In this way a calculation of solvation numbers from the change of solution viscosity with solute concentration can be achieved. The specific volume of hydrated solute can be obtained by the following equation:

$$\bar{v}_\eta = \frac{\eta_{sp}}{c}\left(\frac{1}{f+\eta_{sp}}\right) \qquad (3)$$

where $\eta_{sp} = \dfrac{\eta - \eta_0}{\eta_0}$ is the specific viscosity, c is the concentration, gram of solute per cm^3 of solution, and f=2.5 is the numerical factor from Einstein equation. As remarked by Linow [21] the value of specific volume does not depend on the f, which means that for the model used, the shape of the solute particles is not involved in the calculation of the hydration number. On the other hand, from the density of solution, d, and the density of solvent, d$^\circ$, the apparent specific volume of the solute, can be obtained:

$$v_\phi = \frac{1000(d^\circ - d)}{mdd^\circ M_W} + \frac{1}{d} \tag{4}$$

where m is the molality and M_W is the molecular weight of solute.

From eq. 3 and eq. 4 we can evaluate the volume of solvent bound per gram of solute, $v_\eta - v_\phi$, related to the viscosity solute-solvent interaction strength parameter, M, by the following equation:

$$M = \frac{\rho(\bar{v}_\eta - \bar{v}_\phi)M_W}{M_{wo}} \tag{5}$$

being M_{wo} the solvent molecular weight. Thus, all volume changes connected with the hydration process are included. From the extrapolation at infinite dilution of eq. 5, the hydration number, n_h^η, i.e., the number of moles of bonded water molecules in the inner hydration sphere per mole of solute, can be calculated:

$$n_h^\eta = \lim_{c \to 0} M \tag{6}$$

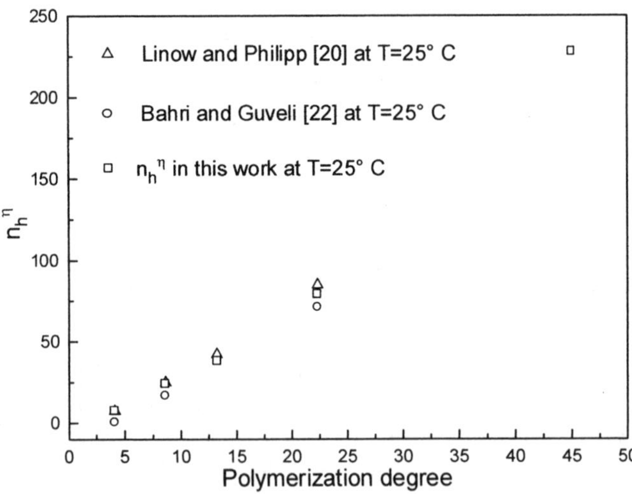

Fig.5 Comparison of hydration numbers for poly(ethylene glycol)s of nominal molecular weights: 200, 400, 600, 1000 and 2000 g/mol determined by viscometry found in this work and literature values.

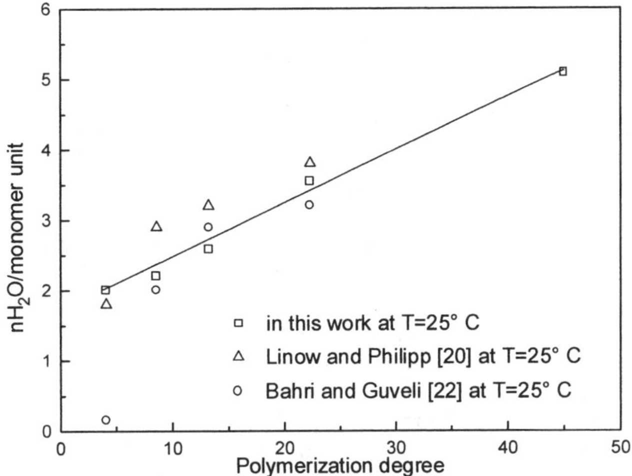

Fig.6 Comparison of average numbers of moles of water bound per mole of monomeric unit of PEG of M_w: 200, 400, 600, 1000 and 2000 g/mol determined by viscometry found in this work and literature values. The continuous line is fit result for data found in this work.

The behaviour of hydration number for different polymerisation degree is shown in Fig. 5. It can be seen that n_h^{η} increases with the average molecular weight of the solute; the same behaviour has been also observed by other authors [23, 24].

In Fig. 6 the number of moles of bonded water per monomer unit of polymer is shown together with the values reported from Linow [21] and Bahri [24]. As we can see, there is not a fully agreement with Bahri's data [24]. However, we consider his data not reliable since it is reasonable retain that the basic hydration of PEG is not satisfied until two water molecules have been added to each $-CH_2CH_2O-$ group [25].

In order to get information on the solute-solvent interaction strength, density and ultrasonic velocity measurements were performed on polymeric aqueous solutions as a function of concentration at 5° C, 20° C, 25° C, 40° C and 70° C. Through the evaluation of the adiabatic compressibility coefficient by the Lorentz relationship $\beta = 1/\rho v^2$, we are able to get information on the polymeric hydration number. We assume, following a molecular model [26], that the volume of the solution V can be partitioned into two contributions: the hydration volume, V_H, where significant interactions between the polymer and water occur, and the bulk water volume, V_w. Following this model the volume of the solution can be written as $V = n_p V_H + \left(n_w - n_p n_H^{\beta}\right)V_w$, where n_p and n_w are the mole numbers of polymer and of water, respectively. Taking the derivative with respect to pressure at constant entropy, we find:

$$V\beta = -n_p \frac{\partial V_H}{\partial P} + \left(n_w - n_p n_H^{\beta}\right)V_w \beta_w \qquad (7)$$

where: $V = \left(\left(n_w M_w + n_P M_p\right)/\rho\right)$ with ρ the solution density. Under the hypothesis of negligible compressibility for hydrated units, $(\partial V_H / \partial P) = 0$, we obtain for the hydration number the formula:

$$n_H^\beta = \lim_{n_p \to 0} \frac{n_w V_w \beta_w - V\beta}{n_p V_w \beta_w} \tag{8}$$

being β and β_w the adiabatic compressibility of the solution and of water, respectively. It is well known that the main drawback in calculating hydration numbers from ultrasonic measurements is the assumption that the compressibility of polymer together with its closely associated water molecules, is negligible [26].

Inherent in this model is the further assumption that the compressibility of water molecules near, but not inside the primary hydration shells, is the same as that of pure water.

In Fig. 7, we report the values of the number of bonded water molecules as a function of the polymerisation degree at the investigated temperatures. . The n_h behaviour is to be ascribed to the enhanced thermal motions that lead to lower residence times of water molecules in the nearby hydration of the polymer, in spite of the rupture of a certain fraction of hydrogen bonds in water which rises the number of water molecules available for bonding with the polymer The hydration number value, $n_h \approx 31$, at T=30°C is in excellent agreement with the crossover value for $\omega_{D\text{-LAM}}$, $\gamma_{D\text{-LAM}}$.

As we have observed for viscosity measurements, also in this case, the hydration number increases with the increasing of polymerisation degree. This result can be rationalized by assuming that at low M_w only tightly bound water is associated with the PEO chain. Furthermore the number of bonded water molecules falls with temperature: in the case of PEG 600, for example, n_h^β decreases from 34.6 at T=5° C to 31.2 at T=25° C. As we can see from Table I, this is a general behaviour independent from M_w. In fact at low temperatures, it is reasonable to hypothesize that the interaction strength allows water to bond not only to the polymer backbone oxygens but also to the terminal groups, or to form a second coordination shell. As the temperature increases the water molecules weakly bonded to the polymer (such as second water shell) break their bonds [27].

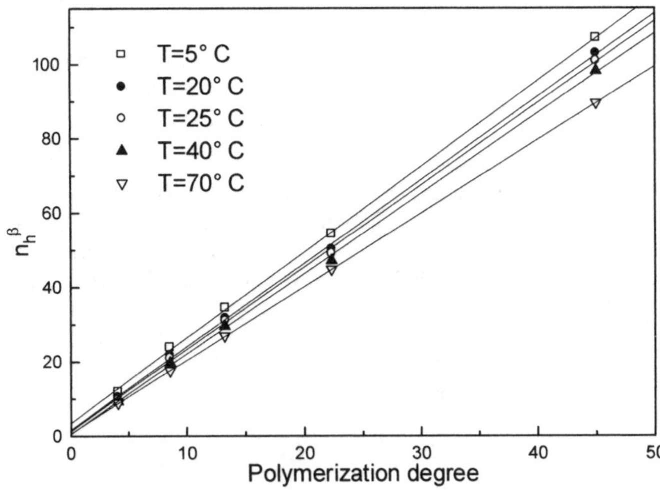

Fig.7 Hydration numbers for poly(ethylene glycol)s aqueous solutions at different temperatures deduced by acoustic data.

Tab.I Hydration numbers deduced by acoustic and viscosity data.

Molecular Weight	n_h^β (5° C)	n_h^β (25° C)	n_h^η (25° C)
200	12.0	7.4	9.9
400	23.9	21.0	24.0
600	34.6	31.2	38.0
1000	54.3	49.1	78.9
2000	107.0	101.0	228.0

From the comparison of viscosity and acoustic data, see Tab. I, it emerges that hydration numbers determined by viscometry are larger than those determined by acoustic measurements. In particular the n_h^η results higher than n_h^β for the investigated samples; this occurrence could be justified by the different spatial sensitivity of the probe. On this concern, the ultrasonic probe is more local in respect to viscosity technique and this occurrence can justify the observed discrepancy. From the same reason the obtained value of hydration number results different from that obtained by other probes, like neutron [28].

4. Concluding remarks

The present work shows how the data obtained by the joint employment of Raman scattering, density, viscosity and sound velocity measurements on PEO aqueous solutions at different temperature and concentration values.
The D-LAM Raman analysis as a function of polymerisation degree indicates the presence of an olygomer-polymer transition for m=13. Contrary to the case of pure samples, the most striking features revealed in the D-LAM spectra in PEO aqueous solutions, are the remarkable frequency increase, towards values corresponding to the crystal ones, and the significative sharpening of the spectral contribution. These findings indicate that the addition of water destroys the intermolecular interactions and stiffens the coil structure, promoting a more ordered conformation in respect to the melt phase. Also in this frame it is possible to evidence a crossover of the $\omega_{D\text{-}LAM}$ and $\gamma_{D\text{-}LAM}$ to plateau values in correspondence of a certain amount of water, which signals the full hydration of the polymer.
The evaluated hydration number, in excellent agreement with the crossover value for $\omega_{D\text{-}LAM}$ and $\gamma_{D\text{-}LAM}$, confirms the goodness of the interpretation. In aqueous solution, the temperature analyses indicate clearly that the solvent power of water increases up to 45°C and then decreases at higher temperatures. A calculation of solvation numbers from the change of solution viscosity with solute concentration and from ultrasonic measurements can be achieved. The evaluated hydration numbers both from viscosity and acoustic data increase remarkably with increasing polymerisation degree.
Furthermore, from acoustic data, we observe that the temperature increase lowers the polymer interaction strength giving rise to the loss of the water molecules not tightly bonded to the polymer. Finally, from a comparison of the hydration numbers evaluated from acoustic and viscosity data, differences have been evidenced. The apparent discrepancy in experimental determination of hydration number deduced by viscosity and acoustic measurements is intriguing and points toward different definitions of hydration number. Moreover attempts are in progress to clarify the probe dependence of this relevant physical parameter that plays an important role in H-bonded system.

References

[1] R. Kjellander and E. Florin, J. Chem. Soc. Faraday Trans., **77** (1981) 2053.

[2] F. E. Bailey and R. W. Callard, J. Appl. Polym. Sci.,**1** (1959) 56.

[3] A. C. Barnes, J. E. Enderby, J. Breen and J. C. Leyte, Chem. Phys. Lett., **142** (1987) 404.

[4] P. Molyneux, Water-Soluble Synthetic Polymers: Properties and Uses, CRC Press Boca Raton, 1983.

[5] S. Magazù, *Physica B Condensed Matter*, **92** (1996) 226.

[6] T. W. N. Bieze, A. C. Barnes, C. J. M. Huige, J. E. Enderby, and J. C. Leyte, *J. Phys.Chem.*, **98**, (1994) 6568.

[7] F. E. Bailey and J. V. Koleske, Poly(Ethylene Oxide), Academic Press, New York 1976.

[8] V. Crupi, M.P. Jannelli, S. Magazù, G. Maisano, D. Majolino, P. Migliardo and D. Sirna, *Mol. Phys.,* **84**, (1995) 645.

[9] J. F. Rabolt, K. W. Johnson and R. N. Zitter, *J. Chem. Phys.,* **61** (1974) 504.

[10] T. Miyazawa, *J. Chem. Phys.*, **35** (1961) 693.

[11] J. L. Koenig and A. C. Angood, *J. of Polymer Science: Part A-2,* **8**, (1970) 1787-1796.

[12] F. Volino, Spectroscopic Methods for the Study of Local Dynamics in Polyatomic Fluids, NATO ASI-series B vol.33, J. Dupuy and A. J. Dianoux editors, Plenum Press, New York, (1978).

[13] H. Nishide, M. Ohyanagi, O. Okada and E. Tsuchida, Macromolecules, 19, 496-498, (1986).

[14] I. Kim and S. Krimm, Macromolecules, 29, 7186-7192, (1996).

[15] R. G. Snyder and H. L. Strauss, J. Chem. Phys., 87, 3779, (1987).

[16] A. Y. Grosberg and A. R. Khokhlov, Giant Molecules, Academic Press, (1997).

[17] P. G. De Gennes, Scaling Concepts in Polymer Physics, Cornell University, Ithaca, New York (1979).

[18] C. H. Wang, Y. H. Lin and D. R. Jones, Mol. Phys., 37, 287, (1979).

[19] L.Börjesson, P. Jacobsson and L.M. Torell, J. of Non-Crystalline Solids, 131-133, 104-108, (1991).

[20] A. P. Sokolov, A. Kisliuk, D. Quitmann, E. Duval, Phys. Rev. B, 48, n°10, 7692-7695, (1993).

[21] K. J. Linow and B. Philipp 1984 Z. Phys. Chemic 265 321-329.

[22] R. B. Seymour, C.E. Carrher Jr., in : J.J. Lagowski (Ed) 1998 Polymer Chemistry.

[23] S. P. Moulik, S. Gupta, Can. 1989 J. Chem. 67 356- 363.

[24] H. Bahri, D. Guveli 1988 Colloid Polym. Sci. 266 141-144.

[25] R. Kjellander and E. Florin 1981 J. Chem. Soc. Faraday Trans., 77 2053.

[26] J. Stuher, E. Yeager 1965 Physical Acoustic, edited by P.W. Mason (Academic Press, New York), p. 351.

[27] C Branca, S Magazù, G Maisano, P Migliardo and V Villari 1998 J. Phys.: Condens. Matter 10 10141-10157.

[28] T.W.N. Bieze, A.C. Barnes, C.J.M. Huige, J.E. Enderby, and J.C. Leyte, 1994 J. Phys. Chem. 98 6568-6576.

Influence of low level nitrogenation on the structural properties of pulsed laser ablation deposited a-CN$_x$ films

E. Fazio, F. Barreca and F. Neri

Dipartimento di Fisica della Materia e Tecnologie Fisiche Avanzate and Istituto Nazionale per la Fisica della Materia, Salita Sperone 31, I-98166, Messina, Italy

S. Trusso

Istituto di Tecniche Spettroscopiche del CNR Via La Farina 237, I-98123, Messina, Italy

Abstract. The effect of the addition of up to 18% of nitrogen on the structure of diamond like carbon films deposited by laser ablation, has been investigated by means of visible micro-Raman spectroscopy and X-ray photoelectron spectroscopy. The Raman spectra of the films with nitrogen content up to 8%, are characterized by a broad asymmetric band near 1550 cm^{-1} (G band), with a shoulder in the 1300-1400 cm^{-1} (D band) typical of diamond like carbon systems. Increasing the nitrogen content the D band intensity increased and the G band position shifted towards higher frequencies while its width decreased. The C1s photoelectron peak was found to broaden and shift as the nitrogen content of the films increased. Such a behaviour is interpreted in terms of a transformation of sp^3 bonded carbon atoms into sp^2 ones, induced by the incorporation of nitrogen atoms, besides a clustering process of the sp^2 domains.

1. Introduction

Since the theoretical calculations of Liu and Cohen about the stability of the β-C$_3$N$_4$ phase which should present a hardness comparable with that of diamond [1], a continuing interest in carbon nitride materials persists. Most of the films deposited with very different preparation techniques, have shown an amorphous or disordered structure even if, in some cases, small crystallites were observed embedded in the amorphous phase [2]. Nevertheless, also non stoichiometric carbon nitride CN$_x$ films have shown promising technological application due to their peculiar mechanical, optical and electronic properties which can be tailored as a function of the nitrogen content x [3,4]. In a previous paper we estimated the relative concentrations of threefold and fourfold coordinated carbon atoms in amorphous CN$_x$ films as a function of the nitrogen content by means of reflection electron spectroscopy [5]. We found that the introduction of nitrogen produced a progressive graphitization of the films. In particular we found that nitrogen atoms induced a progressive decrease of the C-C sp^3 bonds concentration while the C-C sp^2 one was nearly unaffected. Such a behaviour was evident also for the lowest nitrogen content sample ($x =$ 14%)[1] investigated.

In this paper, we focused the attention on the influence of small amount of nitrogen atoms (from x=18% down to 3%), on the structure of laser ablation deposited CN_x films [6], by carrying out measurements of X-ray photoelectron spectroscopy (XPS) and micro-Raman scattering.

2. Experimental details

The films were deposited in a high vacuum chamber with a residual pressure better than 1.0×10^{-4} Pa. The graphite rotating target were ablated by the focused beam of a KrF excimer laser (λ=248nm, 150 mJ pulse energy, 10 Hz repetition rate and 25 ns pulse duration). The estimated laser fluence was 5.5 J/cm^2. The films were deposited at room temperature onto c-Si substrates. The target-substrate distance was 40mm. The nitrogen partial pressure P_{N2} was kept constant, during each ablation process, in the range 0.13-26.6 Pa by means of a mass flow controller. Films thickness, measured by an Alpha-step 500 surface profiler, ranged from 70 to 570 nm depending on the N_2 pressure. Room temperature X-ray photoemission spectra were collected by a VG Scientific spectrometer equipped with a conventional twin-anode Mg/Al Kα X-ray source and a concentric hemispherical analyzer CLAM 100. Raman scattering were performed by means of a U1000 Jobin-Yvon monochromator coupled with an Olympus BX-40 microscope. The 514.5 nm line of an Ar$^+$ laser was focused on the samples surface through the 100× objective of the microscope. The backscattered radiation was collected by the same microscope optics and dispersed by the monochromator, equipped with two holographic grating (1800 line/mm). The dispersed radiation was detected by means of a LN$_2$ cooled CCD sensor.

3. Results

The composition and the chemical bonding nature of carbon nitride films was investigated by studying the modifications induced, by the progressive nitrogen incorporation, on the C 1s and N 1s photoemission peaks lineshapes. In particular, the N/C ratio x for the films deposited at different nitrogen partial pressure P_{N2} was obtained from the integrated areas of the carbon and nitrogen peaks, weighted by the relative sensitivity factors. Increasing the nitrogen gas partial pressure up to 26.6 Pa, the nitrogen content reached a value close to x=18% with respect to the carbon one, showing also a systematic and progressive modifications of the C1s line [5,7]. Such results are illustrated in Fig.1 for three of the investigated samples, having x values of 2.6%, 11.9% and 17.5%, respectively.

The effects of nitrogenation are an asymmetric broadening (from 1.85 to 2.7eV FWHM values) and a shifting towards higher binding energies of the C1s peak (from 284.7 up to 285.0 eV). For the lowest nitrogen content sample (x=2.6%) the C1s peak is centred at 284.6 eV, very close to the position observed for pure carbon systems. Upon increasing nitrogenation, the presence of new components are evident [8,9] indicating a change of carbon chemical bonding structure due to the formation of new C-N bonds. Although an accurate determination of the different contributions is difficult to give, it is quite well established that the main contribution detected at 284.5 eV is attributed to pure carbon

---

[1] Concentrations are intended as the percentage of the number of nitrogen atoms with respect to carbon ones.

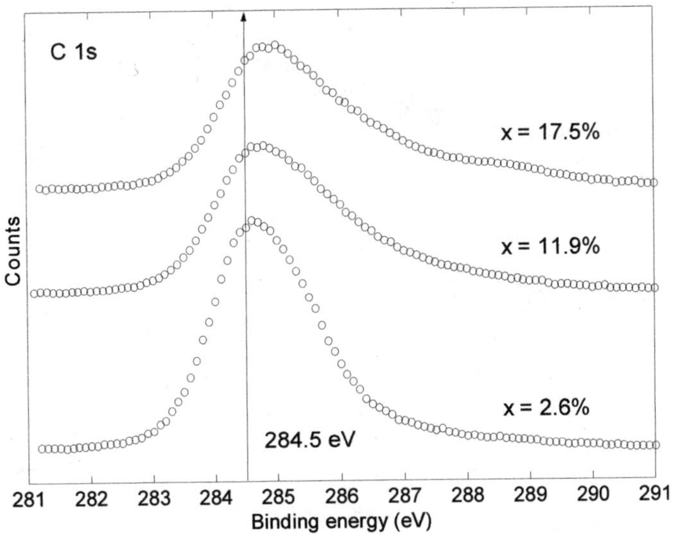

Fig.1 Experimental C1s XPS spectra of CN_x films having different nitrogen content. The arrow shows the position of C1s peak in pure carbon systems

while the asymmetry on the high binding energy side depends on the local environment of the carbon atoms and on the particular bonding structures [10,11,12]. The shoulder at 289 eV is attributed to C-O bonds due to some surface oxygen contamination.

The Raman spectra of the samples are shown in Fig. 2. In general, the Raman spectra of the films with very low nitrogen content, up to 8%, were characterized by a broad asymmetric
band positioned near 1580 cm^{-1} (referred as G band) while the films, with higher nitrogen content, showed also the presence of a second band (referred as D band) peaked near 1330 cm^{-1}.
Raman spectra were deconvoluted using a Breit-Wigner-Fano (BWF) line for the G peak plus an additional gaussian line for the D peak [13,14].
The Breit-Wigner-Fano line-shape is described by:

$$I(\omega) = I_0 \frac{\left[1 + 2(\omega - \omega_0)/Q\Gamma\right]^2}{1 + \left[2(\omega - \omega_0)/\Gamma\right]^2}$$

where I_0 is the peak intensity, ω_0 is the peak position, Γ is the full width half maximum (FWHM) and Q^{-1} is the BWF coupling coefficient. In the limit $Q^{-1} \to 0$ the Lorentzian line shape is recovered. Furthermore a BWF line, due to its asymmetric line-shape, is used to account for Raman contribution extending beyond 1300 cm^{-1} especially for films with low x in agreement with VDOS of graphite or amorphous carbon [13].
It is important to remember that the maximum of the BWF line lies at:

$$\omega_{max} = \omega_0 + \frac{\Gamma}{2Q}$$

Since Q is negative, the peak position ω_{max} is lower than the line position of the undamped mode peaked at ω_0 [15]. The positions, line-widths and intensities of the G and D bands were deduced from the fitting procedure. Moreover, we obtained the ratio I_D/I_G between

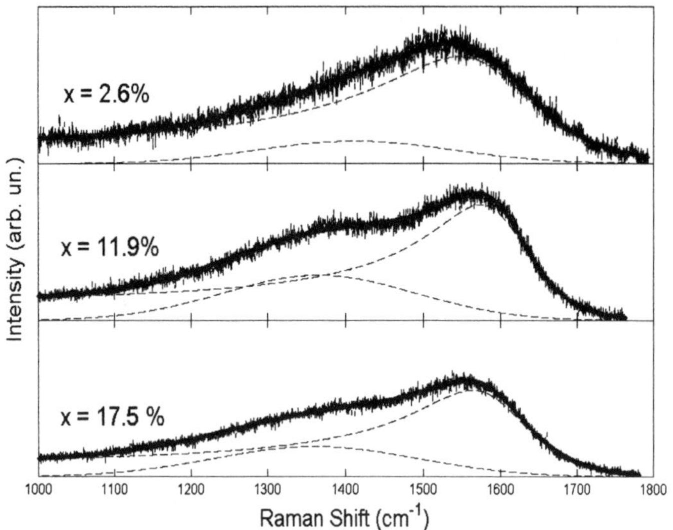

Fig.2 Raman spectra of CN_X films having different N/C ratio. The dashed lines represent the fitting of D and G peaks for the films with 2.6, 11.9, 17.5 at. %

the intensities of the D and G bands which gives information about the presence and the size of

graphitic domains in the films and can be used, even if with some caution, to evaluate the sp^3/sp^2 bonding ratio [13].

Fig.3 shows the G peak position ω_G, its width at half maximum (FWHM) and the relative intensity ratio I_D/I_G as a function of nitrogen content. The I_D/I_G ratio is always less than 1, increasing for sample with nitrogen content equal to 8%. In particular, I_D/I_G ranged from 0.2, for samples with low nitrogen content ($x < 8\%$), to about 0.40 for samples having $x=18\%$. Similarly, the G band position ω_G shifted gradually from 1540 cm^{-1} up to 1570 cm^{-1}. Γ_G showed an opposite trend: it decreases from 240 cm^{-1} down to 180 cm^{-1}.

4. Discussion

The atomic structure of amorphous carbon nitride films presents several aspects not yet clarified due to the lack of knowledge about the local bonding configuration and coordination involving C-C, C-N and N-N bonds. It has been also found that the structural and chemical properties of CN_x films prepared by different techniques such as ion-beam deposition [16], chemical vapour deposition [17,18] and laser deposition [19,20] differ in an appreciable way. XPS and Raman spectroscopy are very useful in the investigation of such a topic. In particular, change in the carbon microstructure can be well evidenced from the modifications in the C1s and N1s core level photoemission spectra as well as from the evolution of the G and D bands in vibrational Raman spectra. In fact, the G peak is connected with the relative motion of sp^2 carbon atoms while D peak is related to the breathing modes of the aromatic rings [13].

It is known that carbon atoms in amorphous carbon films hybridize s and p orbitals of carbon into sp^3 and sp^2 hybrids to form π and σ bonds. If the C1s core level binding

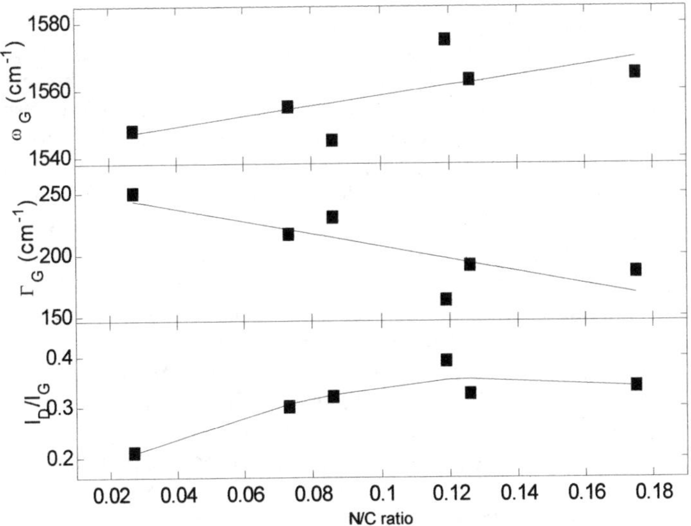

Fig.3 Variation of G peak position, G peak FWHM and I_D/I_G ratio of the Raman spectra vs N/C ratio, the lines are guide to the eye.

relative photoemission spectra are deconvoluted in two lines which are identified with the sp^3 and sp^2 hybrids. In the case of CN_x, the situation becomes more complex due to the presence of an extra atom, namely N, and the photoemission spectra are described as a convolution of different contributions attributed to different atomic bonding configurations: e.g. a single nitrogen atom linked to one, two or three carbon atoms with single and multiple bonds [21,22]. In particular Weich et al. [23] showed that approximately one-half of them are sp^3 hybridized with one filled lone-pair orbital, while the other half shows sp^2 hybridization with C-N-C bonds. In the literature, there is an almost general agreement upon the effect of increasing nitrogen incorporation: a slight increase in both the sp^2 coordinated and multiply bonded carbon fraction and a marked decrease in the amount of carbon species in sp^3 configuration. As it was shown in Fig.1 and described in the previous section, we observe that, for x=2.6%, the C1s photoemission position and FWHM is very close to that observed for pure carbon systems, while for x=17.5%, it is characterized by an asymmetric broadening indicating a change of carbon chemical bonding structure. Then, the nitrogen incorporation induce an increase of the fraction of sp^2 bonded carbon clusters and the contemporary decrease of sp^3 contributions. The Raman spectra of carbon based material can be interpreted starting from the characteristics of the graphite Raman spectra. A perfect graphite crystal shows a single line peaked at 1580 cm^{-1}, referred as the G band, and due to the E_{2g} vibrational mode. If the translational symmetry of the crystal is broken, as in the case of microcrystalline graphite, the Raman inactive A_{1g} mode becomes Raman active and a second band in the 1300-1400 cm^{-1} region appears, indicated as the D band [13-24]. The position and the relative intensity of these two bands depend strongly not only on the structural properties of the material but also on the exciting wavelength used in the Raman experiment. Visible Raman spectroscopy is, in fact, sensitive to the sp^2 phase due to the resonance with π-π^* transitions which occur in sp^2 domains. On the other hand σ-σ^* transitions, occurring in both sp^2 and sp^3 domains are in resonance with UV light. In this respect visible Raman spectroscopy can only indirectly probe the evolution of a sp^3 phase through the evolution of the sp^2 one, and the sp^3/sp^2 bonding ratio can be evaluated from the I_D/I_G ratio only in some cases. In fact, the increase of intensity of the D band can have different origins. Being originated from the breathing mode of sixfold aromatic rings, its evolution in microcrystalline graphite is indicative of a disordering process which produces smaller and smaller crystalline domains. On the contrary, when a diamond-like carbon system is concerned, i.e. a mainly sp^3 connected system, its appearing and evolution is related to the development and evolution of an sp^2 phase with the creation of sixfolds aromatic rings.

In this framework the low value of the I_D/I_G ratio for the samples having no or low nitrogen content (less than 8%) is indicative of a mainly sp^3 connected phase typical of DLC material [25]. The D intensity is low with respect to the G one, due to the absence or the very low content of aromatic rings, while the G band arising from the stretching of sp^2 C-C bonds does not need the presence of aromatic ring to be detectable. Conversely, the increasing of the I_D/I_G ratio, the shift towards higher frequencies of the G peak position and the shrinking of its linewidth can be explained in terms of an increase of sp^2 phase with the appearing of aromatic rings. Upon increasing the sp^2 fraction, the graphitic domains grow in size giving rise to a clustering process. Then, the G band position shifts towards the graphite position and the linewidth decreases, being inversely proportional to the graphitic domain size. As a consequence the graphitisation of the material, with increasing nitrogen content, is evidenced also from Raman scattering measurements. Moreover, from the behaviour of ω_G, Γ_G and of I_D/I_G it is evident that the graphitisation process starts as the nitrogen atom are incorporated, i.e. at nitrogen content as low as 2.6%. It can be argued, then, that nitrogen atoms play a role in the building up of a low density structure binding to a carbon atom with either sp^2 or sp^3 configuration, irrespective of the total nitrogen concentration. In this respect, the picture that emerged from the Raman scattering analysis and XPS spectroscopy data, can be interpreted in terms of a prevalence of the sp^2 hybridization for C-N bonds rather than an sp^3 one, at least for the highest nitrogen containing films.

5. Conclusions

Thin films of a-CN$_x$ have been deposited by means of pulsed laser ablation. Incorporation of nitrogen at different atomic percent were accomplished by performing the ablation process in a controlled nitrogen atmosphere. The influence of nitrogen introduction in the amorphous carbon matrix, starting with atomic percentage as low as 2.6%, on the structural and electronic properties of the films have been investigated by means of micro-Raman spectroscopy and X-ray photoelectron spectroscopy. A progressive asymmetric broadening and shift towards higher energies of the C1s photoelectron peak have been observed upon increasing the films nitrogen content. The Raman spectra were characterized by a broad band given by the contributions of the D and G peaks. An increasing trend of the I_D/I_G ratio, a shift towards higher frequency and a shrinking of the G peak were observed as the nitrogen content of the films increased. These findings are interpreted as a consequence of the progressive development of sp^2 carbon clusters containing aromatic rings promoted by the incorporation of nitrogen atoms. Such a process has been clearly observed also for the lowest nitrogen content sample.

References

[1]   A.Y. Liu, Cohen, Phys. Rev. B **41**, (1990) 10727.

[2] C. Niu, Y.Z. Lu, C.M. Lieber, Science **261**, (1993) 334.

[3] H. Sjostrom, S. Stafstrom, M. Borman, J.E. Sundgren, Phys. Rev. Lett.75, (1995) 1336.

[4] M.N. Semeria, J. Baylet, B. Montmayeul, C. Germain, B. Angleraud, A. Chaterinot, Diamond Rel. Mater. **8**, (1999) 801.

[5] A.M. Mezzasalma, G. Mondio, F. Neri, S. Trusso, Appl. Phys. Lett. **78**, (2001) 326.

[6] D.B. Chrisey and G.K. Hubler, Pulsed Laser Deposition of Thin Films, Wiley, New York (1994).

[7] F. Barreca, A. M. Mezzasalma, G. Mondio, F. Neri, S. Trusso, C. Vasi, Phys. Rev. B **62**, (2000) 16893.

[8] D. Marton, K.J. Boyd, A.H. Al-Bayati, S.S. Todorov, J.W. Rabalais, Phys. Rev. Lett. **73**, (1994) 118.

[9] S. Muhl, J.M. Mendez, Diamond Relat. Mater. **8**, (1999) 1809.

[10] M. Tabbal, P. Merel, S. Moisa, M. Chaker, A. Ricard, M. Moisan, Appl. Phys. Lett. **69**, (1996) 1698.

[11] S.Bhattacharyya, J. Hong, G. Turban, J. Appl. Phys. **83** (1998) 3917.

[12] Y.F. Lu, Z.M. Ren, W.D. Song, D.S.H. Chan, T.S. Low, J. Appl. Phys. **84**, (1998) 2909.

[13] A.C. Ferrari, J. Robertson, Phys. Rev. B **61**, (2000) 14095.

[14] D.G. Mc Culloch, S. Prawer, A. Hoffman, Phys. Rev. B **50**, (1994) 5905.

[15] M.V. Klein, in Light Scattering in Solids, edited by M.Cardona and G. Güntherodt, Topics in Applied Physics, Vol. **51** (Springer-Verlag, Berlin, 1982).

[16] K. Ogata, J.F. Diniz Chubaci, F. Fujimoto, J. Appl. Phys. **76**, (1994) 3791.

[17] S. Bhattacharyya, C. Cardinaud, G. Turban, J. Appl. Phys. **83**, (1998) 4491.

[18] A.G. Fitzgerald, Liudi Jiang, M.J. Rose, T.J. Dines, Appl. Surf. Sci. **175**, (2001) 525.

[19] M.L. De Giorgi, G. Leggieri A. Luches, M. Martino, A. Perrone, A. Zocco, G. Barucca, G. Majni, E. Gyorgy, I.N. Mihailescu, M. Popescu, Appl. Surf. Sci., **127**, (1998) 481.

[20] M. Jelínek, J. Zemek, M. Trchová, V. Vorlíček, J. Lančok, R. Tomov and M. Šimečková et al. Thin Solid Films **366**, (2000) 69.

[21] J.M. Ripalda, I. Montero, L. Galan, Diamond Relat. Mater. 7, (1998) 402.

[22] C. Ronning, H. Felderman, R. Merk, H. Hofsass, P. Reinke, J.U. Thile, Phys. Rev. B **58**, (1998) 2207.

[23] F. Weich, J. Widany, Th. Frauenheim, Phys. Rev. Lett. **78**, (1997) 3326.

[24] R.O. Dillon, J.A. Woollam, V. Katkanant, Phys. Rev. B **29**, (1984) 3482.

[25] J.D. Hunn, S.P. Withrow, C.W. White, D.M. Hembree, Phys. Rev. B **52** (1995) 8106.

GNSR 2001
G. Messina and S. Santangelo (Eds.)
IOS Press, 2002

Intensity and frequency vibrational spectroscopy: nonlinear optical response of polyconjugated materials

Mirella Del Zoppo and Giuseppe Zerbi

*Dipartimento di Chimica Industriale Politecnico di Milano, 20133
Milano, Italy*

Abstract. In this paper we wish to show the importance of vibrational spectroscopy in the study of the molecular electronic properties of polyconjugated molecules. in particular it will be shown how it is possible from the knowledge of vibrational absolute intensities (infrared, Raman and hyperRaman) to obtain the vibrational contribution to nonlinear optical responses. This contribution, in the case of polyconjugated systems, is very large (of the same order of magnitude or even larger than the electronic one) and in many instances it correlates well with its electronic counterpart. Moreover new criteria to relate intensity patterns to molecular structure of polyconjugated systems are discussed.

1. Introduction

In the recent past nonlinear optical processes have been increasingly exploited in a variety of optoelectronic and photonic applications. The basic phenomenon of an optically induced intensity dependent change of the refractive index is of fundamental importance in all-optical switching and computing. Optical switching allows much faster data processing than electronic switching, and parallel processing also becomes feasible. Of course the efficiency of these nonlinear optical processes relies on the material employed. Most of the materials currently used in the fabrication of photonic devices are inorganic crystals (e.g. potassium dideuterium phosphate KDP, lithium niobate $LiNbO_3$, barium titanate $BaTiO_3$). Although the technology for these materials is highly developed and their nonlinear optical susceptibilities are large enough for most current photonic applications, they have some drawbacks. A particularly severe limitation is the relatively slow optical switching time characteristic of photorefractive ferroelectric inorganic crystals. For this reason new nonlinear optical materials are needed to extend the range of photonic applications. Organic materials and inorganic semiconductors are the main candidates as new nonlinear optical media. Organic materials [1-3] are of major interest because of their relatively low cost, ease of fabrication and integration into devices, tailorability (which allows fine tuning of the chemical structure and properties for a given nonlinear optical process), high laser-damage thresholds, low dielectric constants, fast nonlinear optical response times and off resonance nonlinear optical susceptibilities comparable to or exceeding those of inorganic crystals. These features justify the enormous interest which has recently grown in the study and characterisation of these materials. Several techniques, both experimental and theoretical, have been employed in efforts to understand the mechanisms which lead to an improved performance of the material.

The basic relationship which rules the nonlinear optical response of a macroscopic sample under the action of an external, intense electromagnetic field ($\bar{E}$) is given in eq. 1

$$\bar{P} = \bar{P}_0 + \chi^{(1)} \cdot \bar{E} + \chi^{(2)} \cdot \bar{E} \cdot \bar{E} + \chi^{(3)} \cdot \bar{E} \cdot \bar{E} \cdot \bar{E} + \ldots\ldots \tag{1}$$

where $\chi^{(1)}$, $\chi^{(2)}$ and $\chi^{(3)}$ are the bulk susceptibilities and $\bar{P}$ is the volume polarization. The corresponding microscopic equation is:

$$\bar{\mu} = \bar{\mu}_0 + \alpha \cdot \bar{E} + \beta \cdot \bar{E} \cdot \bar{E} + \gamma \cdot \bar{E} \cdot \bar{E} \cdot \bar{E} + \ldots\ldots \tag{2}$$

In the case of molecular organic materials, where intermolecular interactions are relatively weak, the nonlinear optical behaviour is essentially determined by eq. 2. For this reason we will focus on molecular hyperpolarizabilities of the first and second order (respectively β and γ).

The main goal of the research in this field is the establishment of criteria for the choice or the design of the "best" material. In order to optimise the response of the material, it is important to obtain correlations between structural parameters and nonlinear optical behaviour. This can be satisfactorily done with the help of frequency and intensity vibrational spectroscopy.

Up to now frequency spectroscopy has been always preferred to intensity spectroscopy. Frequency seems a much more easily accessible datum, but from the experience gathered over the years, it appears that the intensity datum is much richer in information, especially when the electronic properties of the molecules are considered.

Infrared intensities have already been successfully used to obtain quantitative estimates of equilibrium fixed atomic charges $q_\alpha°$ and charge fluxes (or charge redistribution induced by the vibration) $\partial q_\alpha°/\partial Q_k$ [4,5]. Within the Equilibrium Charges and Charge Fluxes model (ECCF) [6,7] infrared intensities can be expressed as functions of (q_α, $\partial q_\alpha°/\partial R_t$ and $\partial R_t/\partial Q_k$.)

On the other hand, the vibrational spectra of polyconjugated molecules, i.e. the class of materials of interest for this work, are very peculiar. In particular their Raman spectra are relatively simple with only a few normal modes with extremely large scattering cross sections. The prototype is the well known Raman spectra of polyacetylene. Its Raman spectrum has been rationalised since a long time [8] in terms of a vibrational coordinate $\mathcal{R}$ which describes the geometry variation between the ground state and the first excited state structure and which corresponds to an in-phase stretching of double CC bonds and shrinking of single CC bonds.

In the case of push-pull systems, where the conjugated backbone joins a donor and an acceptor end group, the same kind of normal modes appear also in the infrared spectra with strong activity [9]. Still larger infrared intensities can be found in the case of conjugated charged systems (e.g. doped polyaceylene, cyanine dyes, etc.). Again the extreme intensity enhancement is obtained for normal modes with $\mathcal{R}$ character [10,11].

2. Molecular nonlinear optical responses

The peculiar spectroscopic behaviour just described, is related to the existence of a delocalized network of π electrons which are easily polarizable and which determine all the electronic properties of these materials including their nonlinear optical behaviour which is our main interest in this paper.

It is well known that nuclear relaxations give a contribution to the static nonlinear optical response.

We have shown [12] that in the double harmonic approximation and in the limit of an applied static field the vibrational contribution to polarisabilities are:

$$\alpha_{nm}^{v} = \frac{1}{4\pi^2 c^2} \sum_k \left(\frac{1}{v_k^2}\right)\left[\left(\frac{\partial \mu_n}{\partial Q_k}\right)\left(\frac{\partial \mu_m}{\partial Q_k}\right)\right] \tag{3}$$

$$\beta_{nmp}^{v} = \frac{1}{4\pi^2 c^2} \sum_k \left(\frac{1}{v_k^2}\right)\left[\left(\frac{\partial \mu_n}{\partial Q_k}\right)\left(\frac{\partial \alpha_{mp}}{\partial Q_k}\right)+\left(\frac{\partial \mu_m}{\partial Q_k}\right)\left(\frac{\partial \alpha_{np}}{\partial Q_k}\right)+\left(\frac{\partial \mu_p}{\partial Q_k}\right)\left(\frac{\partial \alpha_{nm}}{\partial Q_k}\right)\right] \tag{4}$$

$$\gamma_{nmps}^{v} = \frac{1}{4\pi^2 c^2} \sum_k \left(\frac{1}{v_k^2}\right)\left[I + II\right] \tag{5}$$

with $\quad I = \left(\frac{\partial \mu_n}{\partial Q_k}\right)\left(\frac{\partial \beta_{mps}}{\partial Q_k}\right)+\left(\frac{\partial \mu_m}{\partial Q_k}\right)\left(\frac{\partial \beta_{nps}}{\partial Q_k}\right)+\left(\frac{\partial \mu_p}{\partial Q_k}\right)\left(\frac{\partial \beta_{nms}}{\partial Q_k}\right)+\left(\frac{\partial \mu_s}{\partial Q_k}\right)\left(\frac{\partial \beta_{nmp}}{\partial Q_k}\right)$

and $\quad II = \left(\frac{\partial \alpha_{nm}}{\partial Q_k}\right)\left(\frac{\partial \alpha_{ps}}{\partial Q_k}\right)+\left(\frac{\partial \alpha_{np}}{\partial Q_k}\right)\left(\frac{\partial \alpha_{ms}}{\partial Q_k}\right)+\left(\frac{\partial \alpha_{ns}}{\partial Q_k}\right)\left(\frac{\partial \alpha_{mp}}{\partial Q_k}\right)$

where v_k is the vibrational frequency of the k-th normal mode Q_k and the quantities $\partial \mu_n / \partial Q_k$, $\partial \alpha_{nm} / \partial Q_k$, and $\partial \beta_{nmp} / \partial Q_k$ are derivatives of the molecular dipole moment, polarisability and hyperpolarisability with respect to normal coordinates. These parameters can be obtained from infrared intensities, Raman and hyperRaman cross sections, respectively.

Expressions relating the vibrational contribution to molecular polarizabilities to vibrational observables , and more precisely to vibrational intensities, were known since a long time [13-15], but it was not until a few years ago that we realised, for the first time, how important this contribution is when conjugated organic systems are considered [16].

The hyperpolarisability values measured with the vibrational method are quite large and reach a value, which makes them interesting for technological applications. Moreover, what is even more astonishing is that in most cases the vibrational contribution to β and γ is very similar to the corresponding electronic quantities. See for example the data reported in Table 1.

The close agreement between both calculated and experimental vibrational and calculated electronic values needed some theoretical justification. The rationale behind this observation is that whenever the vibration-induced charge redistribution reproduces the effect of the polarisation induced by an electronic excitation, the coincidence is not fortuitous but takes advantage of the strong correlation existing between nuclear relaxations and electronic charge redistribution in conjugated systems [17]. When the vibrational nuclear oscillation (ΔQ_k) reflects the geometry variation between the ground and the excited state and the Raman intensity can be expressed in the following semplified way:

$$\frac{\partial \alpha_{mn}}{\partial Q_k} = (8\pi^2 v_k^2 c^2)\frac{\Delta Q^{eg} \mid M_{mn}^{ge} \mid^2}{E_g^2} \tag{6}$$

where M_{mn}^{ge} is the transition dipole moment between ground and excited state, vibrational and electronic hyperpolarisabilities are ruled by the same quantities and behave in a similar

Table 1 "Ab initio" calculated (6-31 G) third order vibrational and electronic polarizabilities of various conjugated linear molecules. Units are esu.

	$\gamma^{v}_{yyyy}[\alpha^2]$	$\gamma^{v}_{yyyy}[\mu\beta]$	$\gamma^{v}_{yyyy}TOT$	γ^{e}_{yyyy}
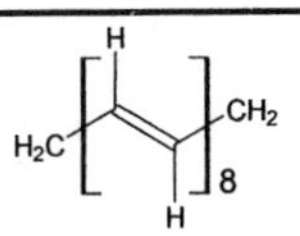	4.14E-33	9.64E-35	4.24E-33	3.16E-33
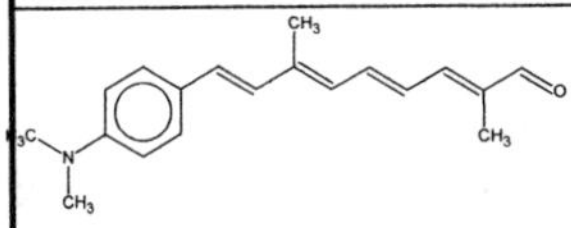	5.48E-34	1.96E-34	7.39E-34	1.21E-33
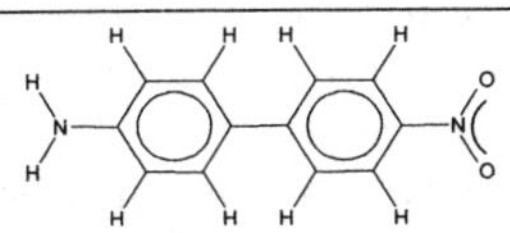	1.23E-34	1.01E-34	2.24E-34	2.54E-34
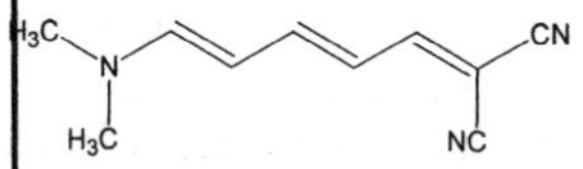	4.28E-34	-0.39E-34	3.89E-34	4.14E-33
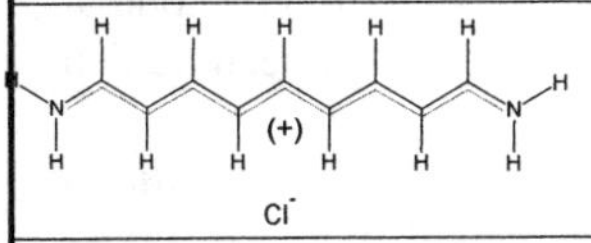	1.03E-33	-6.66E-33	-5.63E-33	-7.69E-34
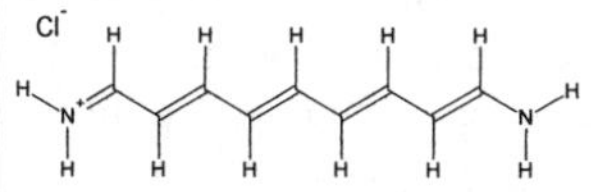	1.31E-33	3.56E-34	1.67E-33	5.90E-34

way. We will see that when this is not the case (e.g. in cyanine molecules of C_{2v} symmetry) the agreement between electronic and vibrational quantities is lost (see Table 1).

Another advantage of using the vibrational method is that by means of its use it is possible to understand which are the most important structural parameteres in determining the nonlinear optical behaviour. Vibrational intensities turn out to be very sensitive to the degree of bond length alternation (BLA) along the skeleton. This structural parameter is known to be related to the delocalization length within the molecule and heavily affects also electronic polarizabilities [18]. If one compares the dependence of electronic and vibrational polarizabilities on BLA, similar trends are found [19].

3. Second order polarisability β

The first observation that can be made by looking at eq. 4 is that in order to obtain large second order vibrational polarisabilities it is necessary to have vibrational normal modes simultaneously very active in both the infrared and the Raman spectra. This seems to be an apparent contradiction since it is quite accepted as a "rule of thumb" that what has a strong Raman activity is rather silent in the infrared and vice versa. However, there exists a special class of conjugated materials that escapes this sort of rule. These are the push-pull systems described in the Introduction. These molecules take advantage of the extremely intense Raman spectra of the conjugated bridge, where collective backbone vibrations can take place, and of the simultaneous polarisation of the skeleton induced by the polar end groups which makes the same normal modes extremely active also in the infrared. Whenever such infrared and Raman coincidence is found, one expects to measure large second order polarisability β^v as is shown for some systems in Fig. 1. This plot shows that β^v increases i) when the number of double bonds increases and ii) when the strength of the polar end groups increases. The plot shows also that given a pair of end groups β values cannot increase indefinitely just adding double bonds but there is a sort of maximum length beyond which saturation sets in and a linear relationship between β and N is established. This behaviour can be rationalised with the existence of a threshold length beyond which the charge transfer is no more effective and the infrared intensity does not increase any more. A greater threshold length can be obtained only if one moves to stronger end groups.

Another way to optimise molecular polarisabilities or, in other words, simultaneously maximise vibrational infrared and Raman intensities is through solvent effects. Dissolving push-pull molecules in solvents of increasing polarity favours the charge transfer and hence modulates the charge distribution and consequently the intensity pattern. The intramolecular charge transfer has a deep influence also on the molecular structure. It has been proven [18,20] that starting with a polyenic molecule, the continuous increase of intramolecular charge transfer changes the molecular structure from polyenic, through cyanine-like, to zwitterionic structure. This means that we obtain a continuous modulation of the bond length alternation (BLA) defined as the average difference between the length of double and single CC bonds, from negative to positive values.

Let us see how this is reflected on vibrational intensities. A decrease in BLA corresponds to diminishing ΔQ (see eq. 6) and hence implies a weaker Raman intensity. The decrease in ΔQ can be seen as a consequence of the fact that the new ground state contains some of the charge transfer character of the excited state and its equilibrium

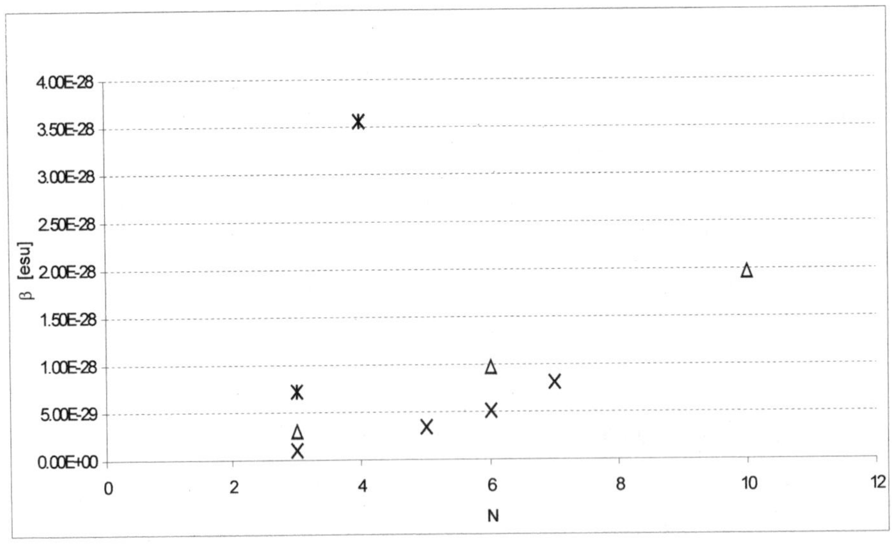

Fig.1 Experimental β^v values for molecules I (*); II (Δ) and III ($\times$) of Fig.2.

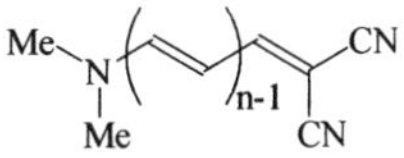

I.a : n = 3
I.b : n = 4

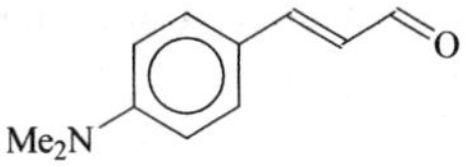

II.a

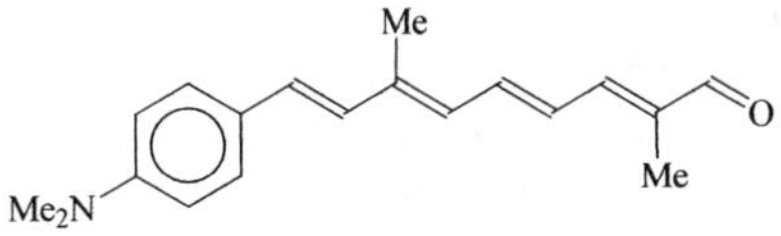

II.b

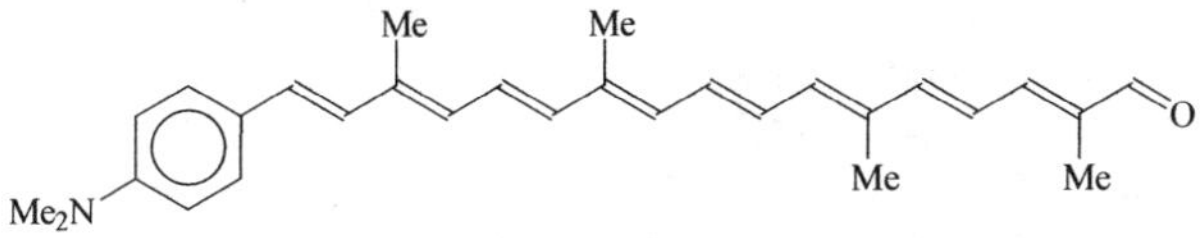

II.c

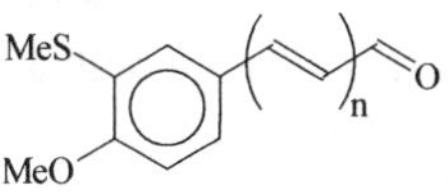

III.a : n = 1
III.b : n = 3
III.c : n = 4
III.d : n = 5

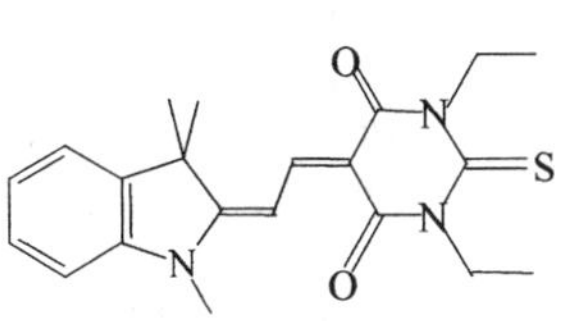

IV

Fig.2 Sketch of the molecules in the plot of Fig.1

geometry must be more similar to that of the excited state. On the contrary infrared intensities increase since the molecular vibration sustains the charge transfer and this builds up the dipole moment (larger $\partial\mu/\partial Q$) [21].

On the whole, even if infrared intensities become larger, the global effect is that of diminishing β. For example the Raman spectra of molecule (IV) dissolved in CCl_4 and $CHCl_3$ shows a decrease of one order of magnitude of the skeletal normal mode $I_k(CCl_4) = 3.16\times10^{-7}\ cm^4g^{-1}$, $I_k(CHCl_3) = 4.83\times10^{-8}\ cm^4g^{-1}$ (it must be noted that the remaining part of the spectrum remains practically unchanged). This corresponds to the zero crossing of β predicted theoretically and experimentally verified with Electric Field Induced Second Harmonic Generation (EFISH) [20]. A similar modulation of the structural parameter can be obtained with any other surrounding medium. In particular care must be taken in using evaluation of β in solution (the usual experimental condition) in order to predict the value of the molecular response of the solid.

4. Third order polarisability γ

In the case of third order molecular polarisability the analysis in terms of vibrational intensities is more complex. In eq. 5 two kinds of contributions appear. Let us consider first term II. Contributions of this kind contain only Raman intensities and can be identified as $[\alpha^2]$. It is to be expected that this term dominates whenever large Raman intensities occur together with a weak infrared spectrum. This is what happens in the case of conjugated systems with no strongly polarising groups (e.g. polyenes, aromatic and heteroaromatic compounds with a strongly alternated structure). In this case the vibrational contribution can be estimated neglecting term I and optimisation corresponds to maximising Raman intensities. Thus one can hope to increase the non linear optical response by increasing the number of conjugated units in the chain.

When the Raman intensity begins to decrease, in less alternated structures, term II becomes gradually less important and term I must be considered. If in addition, the presence of polarising groups induces large infrared intensities a delicate balance between the two terms must be considered. Actually also the hyperRaman spectrum comes into play. This is the major problem in the use of the vibrational method. No experimental data on hyperRaman spectra are available and it is still harder to think of obtaining absolute cross sections. One way to avoid this problem is to use quantum chemical calculations. Even so it is not easy to get the desired quantities ($\partial\beta/\partial Q$) as no quantum chemical program routinely calculates hyperRaman spectra. The easiest and quickest way to solve this problem is to evaluate $\partial\beta/\partial Q$ for the normal modes of interest by means of finite differences [22,23]. This is exactly the procedure we have followed in the calculations here discussed. It must be noticed that while term II can only be positive, term I ($[\mu\beta]$) can be both positive or negative thus allowing the possibility of having negative values of vibrational γ.

It has been shown that electronic γ can indeed be modulated by BLA in such a way that starting with a positive value for polyene systems, it reaches a maximum then starts to decrease till it becomes negative, reaches a deep minimum and finally increases again in a symmetric way [20,24]. If the vibrational γ is to be related to the electronic γ the same trend must be followed. As matter of fact the $[\mu\beta]$ term is rather small and adds to $[\alpha^2]$ when the system is strongly alternated (e.g. molecule II.b). As BLA diminishes $[\mu\beta]$ becomes negative and always larger (e.g. molecules I.a. and charged polyenes). This is shown in Table 1 where we have compared the two contributions in different classes of molecules with smaller and smaller BLA.

If one wishes large negative vibrational γ, systems with small Raman intensities and large infrared activity must be considered. From the above observations this means that almost equalised systems (BLA~0 $\Rightarrow$ vanishing Raman) with polarising groups come into play. For this reason we have focused our attention on charged odd polyene systems and cyanine-like molecules.

As an example of our approach we consider a cyanine-like dye: NH_2-$(CH)_7$-NH_2^+. This molecule is expected to have an equalised backbone structure, a large infrared intensity and correspondingly negative electronic and vibrational γ values. Indeed the "ab initio" calculations (at the RHF/6-31G level) give a mean BLA of 0.0175 Å which is indeed a small value (to be compared with <BLA> = 0.129 Å calculated for C_6H_{14}). The infrared spectrum shows only two bands which are much more intense than all the other active normal modes (the band at 1254 cm^{-1} has an intensity of $\approx$ 10200 Km/mol which is orders of magnitude higher than the strongest infrared band of regular polyenes of similar length). The Raman spectrum, on the contrary, shows many more bands and the skeletal normal modes have intensities comparable with those of NH stretchings. The geometry displacement forced in order to calculate $\Delta\beta/\Delta Q$ is chosen along the normal mode

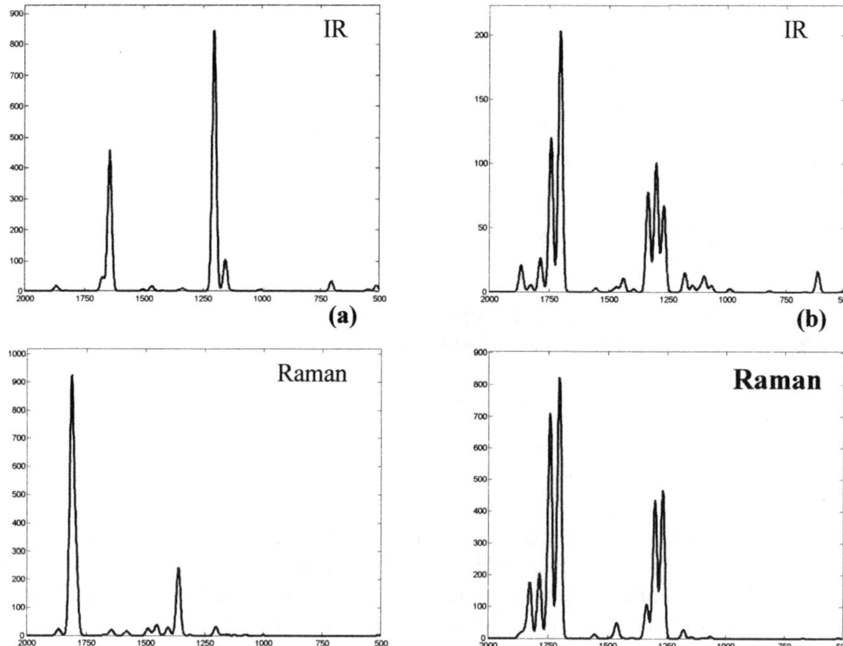

Fig.3 Comparison of the "ab initio" calculated (6-31 G) infrared and Raman spectra of the cyanine with n=4 in the configurations (a) and (b) of Tab. 1

associated with the strongest IR band. This normal mode corresponds, as already discussed, to a collective stretching of the double bonds and shrinking of the single bonds. When, in the calculation, we force such a displacement we induce in addition to a large $\Delta\mu_y/\Delta Q$ also a large $\Delta\beta_{yyy}/\Delta Q$ so that a large, negative γ_{yyyy} value is obtained (y is the chain axis direction). Similar results can be obtained for $C_{13}N_2H_{17}^{+}$. The $[\mu\beta]$ term dominates over the $[\alpha^2]$ term and determines the sign of the total vibrational contribution. Indeed the negative contribution is so large that the whole vibrational contribution turns out to be much larger than the electronic one.

It is interesting to note that the same polymethine chain reported in the last row of Table 1 shows two completely different behaviours in terms of the nonlinear optical response depending on the position of the counterion. Since the polymethine molecule is a charged species, in most situations, a counterion must be present. According to the relative position of molecular cation and anion different properties can be found. In particular we have focused on the two extreme situations reported in Table 1. When the counterion preserves the molecular symmetry (C_{2v}) the chain is indeed equalised and we find a negative γ and a vanishing β as can be seen by looking at the calculated spectra. In Fig. 3, where the molecule with 9 Carbon atom is considered, it is apparent that the normal modes strongly active in the infrared have different frequencies from those appearing in the Raman spectrum. On the contrary when the symmetry of the complex is lowered both electronic and vibrational γ become positive and what is even more striking, the diagonal component of the β tensor along the chain axis becomes extremely large (see Table 2). If one looks at the calculated spectra it appears that in this case the same normal modes are strongly active in both spectra. This behaviour is consistent with an alternated chain as is indeed confirmed

Tab.2 : "Ab initio" calculated (6-31G basis set) β_{yyy} values (y is the direction of chain axis). Units are esu

		β^e	β^v
$C_9N_2H_{13}{}^+$	β_{yyy}	-3.5907E-037	-1.5114e-036
$C_9N_2H_{13}{}^+ + Cl^-$ (C$_{2V}$)	β_{yyy}	6.0314e-036	4.5716e-035
$C_9N_2H_{13}{}^+ + Cl^-$ (asymmetric)	β_{yyy}	9.94E-29	3.69E-28

by the calculated BLA. In this case the stabilisation of the intramolecular charge transfer operated by the counterion makes the molecule behave as a push-pull system.

5. Conclusions

We have presented a few selected cases in which it has been shown how it is possible to use the vibrational approach in the characterisation of molecular nonlinear optical properties. This gives promptly available data to be used as a feedback in the continuous interplay between the synthetic effort and the physical characterisation of the molecular compounds to be optimised.

This is possible because of the extreme sensitivity of vibrational intensities to structural changes. On the other hand, vibrational intensities are indeed electronic quantities, ruled by transition dipole moments and excitation energies exactly in the same way as electronic polarisabilities.

The vibrational method can help in elucidating general criteria for the choice of the most suitable material or for the design of improved compounds. The optimisation of infrared, Raman and hyperRaman intensities leads to the optimisation of α^v, β^v and γ^v through eqs. (3)-(5).

The vibrational spectra are helpful not only as a diagnostic tool but they also give a measure of the vibrational contribution to hyperpolarizabilities. Provided that accurate intensities are known γ^v and β^v can be estimated.

The theoretical (calculated) γ^v values (also because of the lack of hyperRaman spectra) of cyanines turn out to be much larger than the corresponding electronic quantities. This class of polyconjugated molecules behaves differently from the others previously considered, in that there seems not to be a strict correlation between electronic and vibrational quantities (in the asymmetric case the agreement is much better). The explanation of the existence of this discrepancy, must be searched in the structure of the electronic excited states. Preliminary results [25] indicate that while in the asymmetric case there is mainly only one excited state involved (for which the trajectory of the geometry variation is similar to that described by the strongest normal modes), in the asymmetric case the situation is more complex. First of all, there is at least another electronic excited state which can not be neglected since it has a large transition dipole moment with the first excited state. Secondly, the most relevant Raman normal modes (for which the nuclei oscillate following the geometry variation between the electronic states involved) are not responsible for the dominating contributions to γ^v.

With this work we think we have given further evidence of the importance of the results that can be obtained with the vibrational spectroscopy of polyconjugated materials in the field of non linear optics.

References

[1] P.N. Prasad, D.J. Williams, Introduction to Nonlinear Otical Effects in Molecules and Polymers, Wiley, New York, 1991

[2] S.R. Marder, J.E. Sohn, G.D. Stucky (eds.), Materials for Nonlinear Optics: Chemical Perspectives, ACS Symp. Ser. Vol. 455 American Chemical Society, Washington, DC, 1991

[3] H.S. Nalwa, S. Miyata (eds.), Nonlinear Optics of Organic Molecules and Polymers CRC Press, New York, 1997

[4] C. Castiglioni, M. Gussoni, G. Zerbi, J. Mol.Struct., 198 (1989) 475

[5] M. Gussoni, C. Castiglioni, M.N. Ramos, M. Rui, G. Zerbi, J. Mol.Struct., 224 (1990) 445

[6] J.C. Decius, J. Mol. Spectrosc., 57 (1975) 348

[7] A.J. van Straten, W.M.A. Smit, J.Mol. Spectrosc. 62 (1976) 297

[8] M. Gussoni, C. Castiglioni, G.Zerbi, in "Advances in Material Science Spectroscopy" (ed. R.J.H. Clark, R.E. Hester) Wiley, New York (1991), 251

[9] M. Del Zoppo, C. Castiglioni, P. Zuliani, G. Zerbi in Handbook of Conducting Polymers. II Edition, Dekker, New York, 1998, p. 765.

[10] A. Bianco, M. Del Zoppo, G. Zerbi, Synth. Met., in press

[11] M. Del Zoppo, C. Castiglioni, M. Tommasini, P. Mondini, C. Magnoni, G. Zerbi, Synth. Met. 102 , (1999) 1582

[12] C. Castiglioni, M. Gussoni, M. Del Zoppo and G. Zerbi, Solid State Comm., 82 (1992) 13.

[13] C. Flytzanis, Phys. Rev. B, 6 (1972) 1264

[14] R. Hellwarth, J. Cherlow and T.-T. Yang, Phys. Rev. B, 11 (1975) 964

[15] D.M. Bishop, Rev. Mod. Phys., 62 (1990) 343

[16] M. Del Zoppo, C. Castiglioni, M. Veronelli, G. Zerbi, Synth. Metals 55 (1993) 3919

[17] C. Castiglioni, M. Del Zoppo, G. Zerbi, Phys. Rev. B 53 (1996) 13319

[18] F. Meyers, S.R. Marder, B.M. Pierce, J.L. Brédas Chem. Phys. Lett., 228 (1994) 171

[19] M. Del Zoppo, C. Castiglioni, P. Zuliani, A. Razelli, M. Tommasini, G. Zerbi, M. Blanchard-Desce, J. Appl. Polymer Sci., 70 (1998) 1311

[20] G. Bourhill, J.L. Brédas, L.-T. Cheng, S.R. Marder, F. Meyers, J.W. Perry and B.G. Tiemann, J. Am. Chem. Soc., 116 (1994) 2619.

[21] P. Zuliani, M. Del Zoppo, C. Castiglioni, G. Zerbi, S.R. Marder, J.W. Perry, J. Chem. Phys. 103 (1995) 9935

[22] B. Kirtman, B, Champagne and J.-M. André, J. Chem. Phys.,104 (1996) 4125.

[23] M.C. Magnoni, P. Mondini, M. Del Zoppo, C. Castiglioni and G. Zerbi, J. Chem. Soc. Perkin Trans. 2, (1999) 1765

[24] F. Meyers, S.R. Marder, J.W. Perry, G. Bourhill, S. Gilmour, L.-T. Cheng, B.M. Pierce and J.L. Brédas, Nonlin. Opt., 9 (1995) 59.

[25] A. Bianco Thesis in Materials Engineering, Politecnico di Milano (2000)

GNSR 2001
G. Messina and S. Santangelo (Eds.)
IOS Press, 2002

Local bonding-nature investigation in hydrogenated carbon nitrides deposited by reactive sputtering of graphite

G. Fanchini*, A. Tagliaferro

INFM, Dipartimento di Fisica, Politecnico di Torino, corso Duca degli Abruzzi 24, 10129 Torino, Italy
** E-mail: fanchini@polito.it tel.: +39.(0)11.5647348; fax:+39.(0)11.5647399*

G. Messina, S. Santangelo

INFM, Dipartimento di Meccanica e Materiali, Facoltà di Ingegneria, Università "Mediterranea", località Feo di Vito, 89060 Reggio Calabria, Italy

Abstract An accurate investigation of the vibrational and structural properties of a set of low-gap disordered hydrogenated carbon nitride (a-C:N:H) thin films, grown by reactive sputtering, is performed. The nature of the local bonding, the nitrogen- and hydrogen- atoms are engaged-in within the carbon network, is clarified through the comparative analysis of the results of the infrared (IR) characterisation and the Raman analysis. The disorder-induced features of both IR and Raman spectra are discussed in light of the current assessment on the vibrational properties of carbon-based materials. The effects of charge redistribution and bond polarisation, due to the presence of nitrogen, are shown to be responsible for the differences therewith evidenced between spectra of a-C:N:H and non-nitrogenated a-C:H materials.

1. Introduction

In spite of large efforts, devoted to acquire a better understanding of the properties of carbon-nitride based materials, recently carried out, many concerned questions are still open. The prediction of super-hard β-C_3N_4 crystalline phase *[1]*, confirmed by other theoretical works *[2,3]*, has given great impulse to the research aimed at the development of growth techniques apt to prepare sp^3 coordinated carbon-nitride based thin films. Nevertheless, the most suitable deposition process to obtain films endowed with attractive characteristics in view of mechanical applications is still an object of controversy, since the most of coatings has been shown to be sp^2 (and sp) rich and paracyanogenic *[4]*. In this scenery, theoretical models and experimental techniques related to the structural characterisation of carbon-nitride based films play a key role. Furthermore, since, owing to the large amount of research work carried out, the applications of these materials are, today, extended to several diverse fields of relevant technological-interest *[5,6]*, the full characterisation of the carbon-nitrides based films has become a fundamental step for the realisation of highly-performing applications.

The aim of this paper is, hence, to achieve a deeper comprehension of the vibrational properties of hydrogenated carbon nitrides. An accurate Raman and IR investigation is carried out on a set of *a*-C:N:H thin films, deposited by reactive sputtering of graphite, without deliberate external bias of the substrate. The nature of the local bonding, the nitrogen- and hydrogen- atoms are engaged-in within the carbon network, is clarified. The results of the IR characterisation and Raman analysis are comparatively discussed and the relationships evidenced understood in terms of the charge-redistribution and bond-polarisation effects of nitrogenated groups on the back-bonded carbon phase. Finally, the mutual influence-factors between the IR-active and the Raman-active vibrational modes, the optical properties and the microstructures are emphasised.

2. Experimental

Thin a-C:N:H films were simultaneously deposited on glass and undoped (100) *c*-Si substrates in a conventional radio-frequency (13.56 MHz) diode sputtering system, using a (99.999% purity) graphite cathode, in an $Ar+He+N_2+H_2$ reactive atmosphere. The He/Ar flow ratio (3/7) and the substrate temperature (100 °C) were kept fixed during deposition. The other deposition conditions were varied as reported in detail in Tab.1. In all cases no external bias voltage was applied. One additional sample (#1) was prepared under very low self-bias voltage conditions, previously found to produce sp^2 richer and more conductive materials. The measured self-bias voltages, as well as the film thickness, as achieved by the analysis of interference fringes in the IR photon energy range, are reported in Tab.2.

The N content, evaluated, with a relative uncertainty of ~30 % both by Rutherford back-scattering (RBS) and nuclear reaction analysis, was always less than ~8 %. The H content, evaluated by elastic recoil detection analysis (ERDA), ranged from 14% (#2) to 20% (#7).

Tab.1 Growth conditions of the investigated *a*-C:N:H samples: W_{rf} and p_{tot} respectively indicate the r.f. power and the total pressure. Φ_{N_2} and Φ_{H_2} denote the flows of the reactive gases (N_2 and H_2).

Tab.2 Values of self-bias voltage (V_{sb}), thickness (*t*), Tauc gap (E_{gap}) and E_{04} gap measured in the investigated *a*-C:N:H films.

Film	W_{rf} (W)	p_{tot} (mTorr)	Φ_{N_2} (sccm	Φ_{H_2} (sccm
#1	220	38	4	3
#2	300	25	10	7
#3	300	20	10	5
#4	300	20	7	5
#5	300	20	10	7
#6	400	20	10	7
#7	300	20	20	7

Film	V_{sb} (V)	*t* (nm)	E_{04} (eV)	E_{gap} (eV)
#1	-350	250	0.95	0.58
#2	-890	370	1.20	0.68
#3	-890	370	1.34	0.71
#4	-890	520	1.18	0.73
#5	-890	440	1.46	0.84
#6	-1000	460	1.09	0.88
#7	-890	300	1.90	1.14

Of course, the quoted compositions include both bonded and free N and H.

Preliminary results of small angle X-ray scattering (XRD), obtained on samples deposited under similar conditions, signalled the films to be amorphous or, at least, with graphite cluster islands below the minimum (1.2÷1.5 nm) detectable size.

The values of the Tauc gap, E_{gap}, and of the E_{04} gap (i.e. the photon energy at which the absorption coefficient value is 10^4 cm^{-1}), reported in Tab.2, were extracted from the optical measurements performed in the visible-near ultraviolet (VIS-NUV) energy range.

The IR spectra were recorded, in the 450-4000 cm^{-1} range, on a Perkin-Elmer FTIR-2000 spectrometer (1 cm^{-1} resolution) by using a bare substrate as a reference. Up to 64 interferograms were recorded and averaged to improve the S/N ratio.

The Raman spectra were recorded, at room temperature, in the 200 to 3600 cm^{-1} region, by using a Jobin Yvon Ramanor U-1000 double monochromator equipped with an electrically cooled Hamamatsu R943-02 photo-multiplier as a detector, and photon counting electronics. The S/N ratio has been improved by recording multiple scans. A power density of ~20 W/mm^2, at the sample surface, was utilised in order to prevent sample annealing.

3. Results and discussion

3.1 *Raman spectroscopy*

Figure 1 shows the evolution of the Raman spectra of the investigated samples. The spectra are dominated by the amorphous sp^2 carbon related features, the so-called D- and G- bands. As well-known, D- and G- bands are related to the enhancement and the modifications, in presence of disorder, of the graphite modes exhibiting A_{1g} and E_{2g} symmetries [7]. The corresponding broad nearly-featureless second-order structure, detected approximately at double frequencies [8], is superimposed to a relevant photo-luminescent (PL) background. In the framework of this study, the main interest in Raman analysis originates

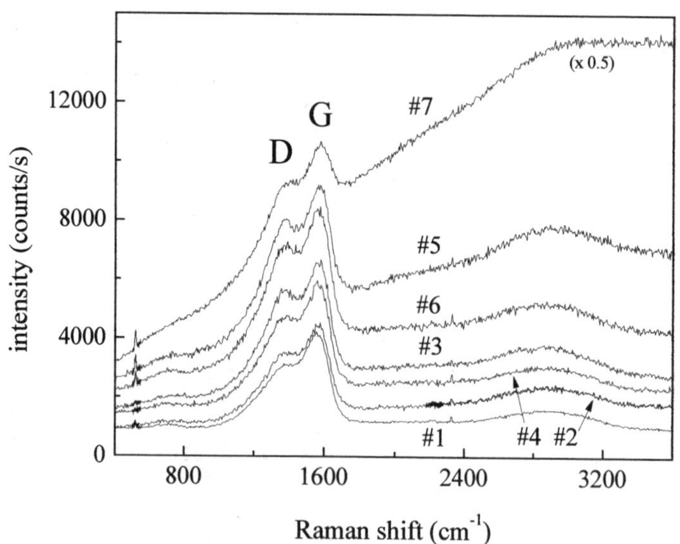

Fig.1 Evolution of the as-measured Raman spectra of the investigated *a*-C:N:H films.

Tab.3 Frequency positions (ω_D and ω_G) and widths (σ_D and σ_G) of the D- and G- bands detected in the Raman spectra of the investigated a-C:N:H samples. The D/G intensity ratio is also indicated.

Film	ω_D (cm^{-1})	σ_D (cm^{-1})	ω_G (cm^{-1})	σ_G (cm^{-1})	I_D/I_G (ad.u.)
#1	1393	383	1570	128	3.17
#2	1376	293	1574	146	1.82
#3	1372	318	1575	148	2.22
#4	1375	312	1572	147	2.01
#5	1365	240	1574	143	1.61
#6	1371	258	1573	148	1.59
#7	1368	260	1575	140	1.66

from its ability in estimating the degree of clusterisation in the sp^2 phase. At this purpose, the spectra were reproduced by using Gaussian shapes *[8-10]*, found to fit the observed spectral features much more satisfactorily than the often used (although mainly for sp^3 rich materials *[11-12]*) Lorentzian (D-mode) and Breit-Wigner-Fano (G-mode) shapes. The frequency positions, widths (FWHM) and intensities were chosen by a least-square best-fit method. The results relative to the D- and G- bands are reported in Tab.3.

Recently, the G mode has been shown to represent a general feature of all sp^2 structures, both olefinic and aromatic, while the D mode being necessarily connected to the presence of aromatic sixfold rings *[12]*. In the presently investigated a-C:N:H films, the PL background (usually associated to undistorted and floppy networks *[13]*) tends to become more intense in samples (such as, #5 and #7), whose Raman spectra exhibit the D-band shifted at lower frequency-positions. This suggests that the quoted Raman shift cannot be related *[14]* to higher amounts of non-optimal bond-angles (i.e. to five- and seven- fold rings, chains or distorted sixfold rings) as far as the network floppiness prevents the formation of local distortions. Hence, the sp^2 aromatic clusters can be assumed mainly organised in ordered sixfold rings.

Moreover, two competing phenomena have been shown to relate the D/G intensity ratio, I_D/I_G, with the average graphitic-cluster size, L_C: the disorder-induced enhancement of the D-mode cross-section when L_C decreases (Tuinstra-Koenig effect *[7]*) and the (decreasing with L_C) probability to find some of the cluster-forming sp^2 atoms organised in a D-mode vibrating sixfold ring *[12]*. Nevertheless, as a critical D/G intensity ratio $(I_D/I_G)_{crit}=2.2$ [corresponding to a critical cluster-dimension $(L_C)_{crit}=2$ nm] is reached, the two effects compensate each other. It is, however, worthwhile noticing that, since the exact $(I_D/I_G)_{crit}$ and $(L_C)_{crit}$ values are expected to depend on the shape of the fitting curves (Gaussian or Lorentzian plus Breit-Wigner-Fano), the Raman D/G intensity ratio can provide only qualitative information about the average dimensions of the graphitic domains. In order to quantitatively estimate L_C values, the support from other techniques is required.

Actually, if the mean cluster size of the samples under investigation is estimated via the Tuinstra-Koenig relationship *[7]*,

$$L_C = \frac{4.4}{I_D/I_G}, \qquad \text{for } L_C \geq 2 \text{ nm,}$$

L_C values obtained (ranging from 2.0 to 2.8 nm), too large, are not consistent with the results of the XRD preliminary measurements (*Sect.2*). Also the Ferrari-Robertson relationship (utilised, as an alternative, for small clusters) *[12]*,

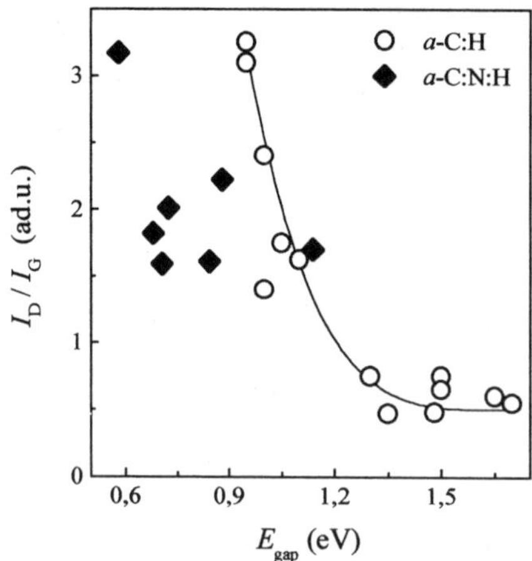

Fig.2 Correlation between the D/G intensity ratio, I_D/I_G, and the Tauc gap, E_{gap}, in a-C:H films [9]. No well-defined relationship contrarily links I_D/I_G and E_{gap} in nitrogenated materials.

$$L_C = \sqrt{\frac{I_D/I_G}{0.55}}, \qquad \text{for } L_C \leq 2 \text{ nm,}$$

(giving values ranging from 1.7 to 2.4 nm) clearly overestimates the mean cluster sizes. It has to be pointed out that the latter equation, worked out for non-nitrogenated carbon materials [12], derives the cluster size with the additional support of optical measurements and empirical Robertson-O'Reilly formula [13],

$$E_{gap} \approx \frac{6}{\sqrt{M}} \approx \frac{6\sqrt{3}a}{L_C},$$

correlating the Tauc gap (measured in eV), E_{gap}, and the number, M, of sixfold rings involved in a L_C sized sp^2 cluster ($a \approx 0.14$ nm being the lattice bond-length), when the effect of matching with the sp^3 phase or the presence of impurity atoms is disregarded [15]. Nevertheless, the cluster dimensions attained from the Tauc gap (ranging from 0.7 to 1.4 nm) do not agree with those deduced from the D/G intensity ratio. Such a disagreement suggests that in hydrogenated carbon-nitrides, probably due to the peculiarity of the material optical and electronic properties, the cluster size might be not univocally related to the Tauc gap. This is actually demonstrated, in terms of D/G intensity ratio, in Fig.2, where the clear correlation existing between I_D/I_G and E_{gap} in a-C:H films [9] is set against the lack of a well-defined relationship in nitrogenated materials.

3.2 FT-IR spectroscopy

Figure 3 shows the IR absorption spectra of the investigated a-C:N:H films. The spectrum of a hydrogen-free film [16], used below for some comparative remarks, is also reported.

The main spectral features analysed to gain qualitative information about the film vibrational-properties are listed below:

i) peaks arising from several stretching- and bending- modes with very different characteristics, contributing in the 1000-1900 cm^{-1} ('CC') region;

ii) peaks due to the stretching of several terminal and non-terminal groups involving sp-hybridised N and C, giving their contribution in the 2000-2250 cm^{-1} ('CN') region. Some bending peaks, probably related with groups that stretch in this region, are further detected in the lower (500-950 cm^{-1}) energy region;

iii) peaks originating from the stretching modes of the Csp^3H_x (2850-2950 cm^{-1} range), Csp^2H_x (2950-3080 cm^{-1} range), and $CspH_x$ bonds (3150-3250 cm^{-1} range), vibrating in the 'CH' region;

iv) peaks, around 3200 cm^{-1}, originating from the stretching of the N_xH_y bonds ('NH' region), superimposed to the intense peaks, around 3500 cm^{-1}, arising from the modes of groups (such as O–H, O–H···N, O–H···O,... [4]) related to the oxygen contamination, indicative of the porosity and softness of the films (i.e. of their relatively-low

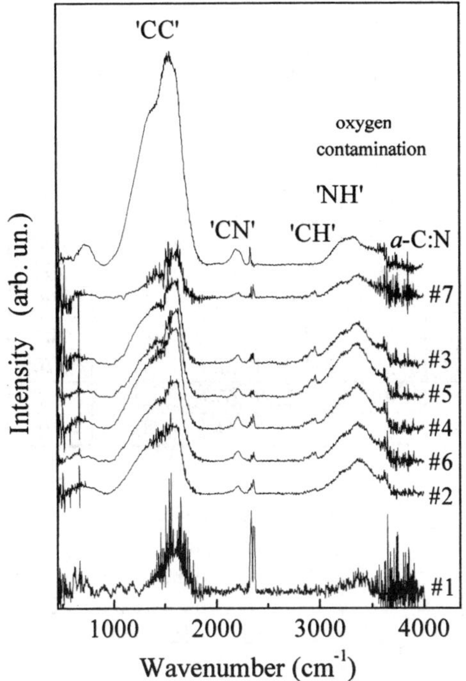

Fig.3 IR absorption spectra of the investigated *a*-C:N:H films after subtraction of the signal due to the interference fringes. The spectrum of an hydrogen-free sample is also shown.

structural disorder).

The IR spectra were fitted assuming Gaussian shapes for each contribution to the film absorption. The very poor S/N ratio, due to the very limited film absorption in the IR region, caused no reasonable fitting of the IR spectrum to be achieved for sample #1.

The effective oscillator density, $N_{\text{eff,i}}$, (proportional to the group concentration) and the optical oscillator density, N_i, for each bond-type can be evaluated through the relationship *[17]*

$$N_{\text{eff,i}} = C_{\text{IR,i}} \cdot N_i = \int_0^\infty \frac{\alpha_i(\omega)}{\omega} d\omega,$$

where the inverse of the IR cross-section, $C_{\text{IR,i}}$, written as

$$C_{\text{IR,i}} = \frac{9 \cdot N_A \sqrt{\varepsilon_1}}{f_i(\varepsilon_1+2)^2}$$

(N_A being the Avogadro's number) if vibrating dipoles in an hosting matrix are considered *[17]*, strongly depends on the integrated bond-strength, f_i, and on the real part, ε_1, of the complex dielectric constant. The bond strengths (not directly measured in disordered materials) are usually adjusted referring to their actual values in gases or organic molecules. However, due to the strong local fluctuations (undergone by ε_1 in inhomogeneous media, like carbon-nitrides) and to the different bond strengths, only very-similar vibrating-groups have comparable $C_{\text{IR,i}}$ values (i.e. comparable oscillator densities N_i). Instead, comparing different contributions and achieving information on their relative concentration generally requires knowledge on $C_{\text{IR,i}}$. This is an important point since all the considerations below will be based on the optical oscillator densities N_i rather than on $N_{\text{eff,i}}$.

3.2.1 The 'CC' region

Since the decomposition of the spectra shows that all features (excepted the weak one located around 1550 cm^{-1}) are present in both *a*-C:N and *a*-C:N:H films (top and bottom of Fig.4, respectively), the contribution of H related (Csp^3H_x and/or N_xH_y) bending modes in the 1000-1900 cm^{-1} region is concluded to be rather-marginal. The intense bands detected in this region are, hence, here simply referred as C=C stretching modes ('CC').

Three major peaks, centred approximately at 1150 cm^{-1}, 1385 cm^{-1} and 1600 cm^{-1}, are detected in the 'CC' region. Their frequency positions, widths and fractional oscillator-strength densities are reported in Tab.4.

No definite hypothesis can be made on the lowest-energy contribution, here called X-

Tab.4 Frequency positions ($^{IR}\omega_X$, $^{IR}\omega_D$ and $^{IR}\omega_G$), widths ($^{IR}\sigma_X$, $^{IR}\sigma_D$ and $^{IR}\sigma_G$) and fractional optical oscillator-strength densities (N_X/N_{tot}, N_D/N_{tot} and N_G/N_{tot}) of the X-, D- and G- bands detected in the IR spectra of the investigated a-C:N:H samples. The total optical oscillator-strength density (N_{tot}) in the region 1000-1900 cm^{-1} is also reported. Due to the low S/N ratio, no fitting of the single IR-active features can be achieved on sample #1.

Film	$^{IR}\omega_X$ (cm^{-1})	$^{IR}\sigma_X$ (cm^{-1})	$^{IR}\omega_D$ (cm^{-1})	$^{IR}\sigma_D$ (cm^{-1})	$^{IR}\omega_G$ (cm^{-1})	$^{IR}\sigma_G$ (cm^{-1})	N_X/N_{tot} (%)	N_D/N_{tot} (%)	N_G/N_{tot} (%)	N_{tot} (cm^{-1})
#1	==	==	==	==	==	==	==	==	==	105
#2	1163	52	1380	107	1599	50	5.8	68.2	26.0	304
#3	1175	42	1393	107	1605	50	5.0	64.7	30.3	299
#4	1172	50	1389	105	1600	51	7.7	65.0	27.2	309
#5	1176	50	1386	101	1604	52	4.2	58.6	37.2	270
#6	1175	54	1393	107	1605	45	5.2	67.9	26.9	266
#7	1175	51	1385	70	1603	51	7.2	44.2	48.6	173

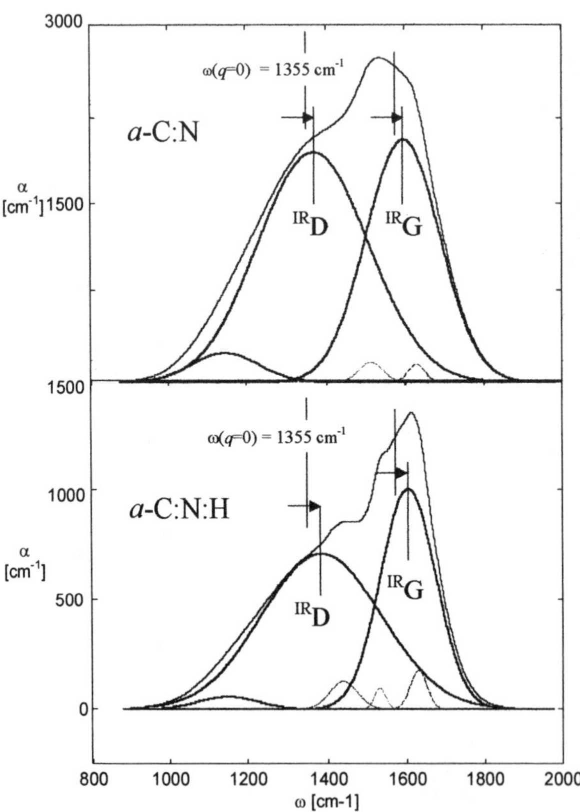

Fig.4 Comparison between the IR absorption of a-CN and a-C:N:H in the 1000-1900 cm^{-1} region. The shifts of the peak frequency-positions are evidenced.

peak, since no clear correlation is evidenced between its relative optical oscillator strength density and optical gap. On the basis of its frequency-position, the X-peak is here supposed to be related with disorder-activated single carbon-carbon bonds (whose vibrations, in presence of disorder, are not necessarily associated to the sp^3 phase), even if the connection of such mode with some asymmetric vibration of sp^2 chain or ring cannot be excluded at all. The IR-absorption peaks at 1385 cm^{-1} and 1600 cm^{-1}, although not straightforwardly assimilable (as, f.i., in [18]) to the Raman-active vibrations (D- and G- bands), since:

i) photon-scattering (Raman) and absorption (IR) processes exhibit very different cross-sections;

ii) the lineshape of the IR

features, then, turns out to be Gaussian, while, as already mentioned (*Sect.3.1*), depending on the film characteristics, it is not always the same for Raman D- and G-bands;

iii) moreover, the [IR]D- and [IR]G- peaks are usually shifted in position (see f.i. *[19]*);

iv) finally, the Raman D/G intensity ratio, I_D/I_G, and the IR oscillator density ratio, N_D/N_G, are different;

are currently improperly named, for the sake of simplicity, [IR]D- and [IR]G- peaks.

In the carbon-nitride based materials, the Csp^2 islands are currently hypothesised *[20]* to be directly activated by the presence of N atoms inside such olefinic structures or graphitic rings. Contrarily, below (*Sect.3.3*), demonstration is given that the modes vibrating in the 'CC' region, in the investigated *a*-C:N:H films, are, rather, indirectly activated by the long-range polarisation originated into the back-bonded carbon phase by the *sp*-bonded nitrogenated groups vibrating into the 'CN' region.

3.2.2 The 'CN' region

The comparison between hydrogen-free and hydrogenated films provides useful information. In *a*-C:N films (top of Fig.5) the 'CN' region is reproduced by three Gaussian contributions centred around $\omega_1 = 2130$ cm^{-1}, $\omega_2 = 2165$ cm^{-1} and $\omega_3 = 2210$ cm^{-1}, respectively. In *a*-C:N:H samples (bottom of Fig.5), the first contribution is absent, while the others are found at similar energy-positions. The higher-energy contributions have been suggested *[4]* to originate from (terminal) Csp_xNsp_y groups (e.g. nitriles -C≡N), while many non-terminal groups involving Csp and Nsp (such as carbodiimides) stretch in the 2050-2150 cm^{-1} region *[18]*.

The comparison between *a*-C:N and the *a*-C:N:H films, thus, evidences that the formation of terminal nitrogenated groups is promoted by the presence of hydrogen during deposition.

In addition, since the numerous Gaussian-shaped contributions, typical of the stretching modes of $C_{1-x-y}N_xH_y$ groups, vibrating in the 'CN' region *[17]*, are not observed, these groups are argued to be not present in the investigated films.

3.2.3 The 'CH' region

Additional information is deduced from the comparison with the spectra of graphite-like *a*-C:H thin films (top of Fig.6) grown by

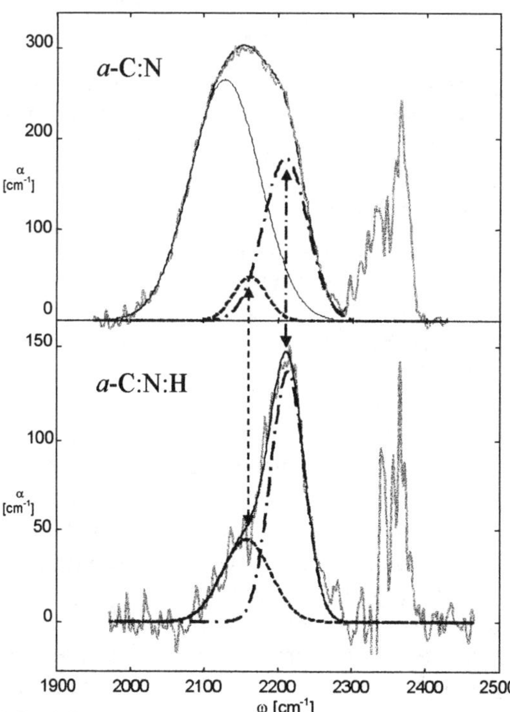

Fig.5 Comparison between the IR absorption of *a*-C:N (top) and *a*-C:N:H (bottom) in the 'CN' region. The two common peaks at energy-position ω_2 and ω_3 (see text) are evidenced. The absorption features in the range 2280-2400 cm^{-1} are due to the environmental CO_2.

reactive sputtering *[22,23]*. The absence, in the analysed *a*-C:N:H samples (bottom of Fig.6), of any contribution associated to the stretching modes of Csp^2H_x (and/or $CspH$) groups is noticed. Contrarily, a relevant contribution is found at the energy positions, where the Csp^3H_x stretching modes are detected in *a*-C:H *[21]*.

3.2.4 The 'NH' region

Finally, the very-low oscillator-strength density of N_xH_y groups is indicative of the rare occurrence of such groups in the investigated films.

The latter finding, together with the absence of $C_{1-x-y}N_xH_y$ groups, suggests that bonded H and N atoms are located in different regions of the C skeleton in *a*-C:N:H films: H is present chiefly in the Csp^3 phase, while N binds mainly with sp (and/or sp^2) groups, located inside a very poorly-hydrogenated Csp^2 phase.

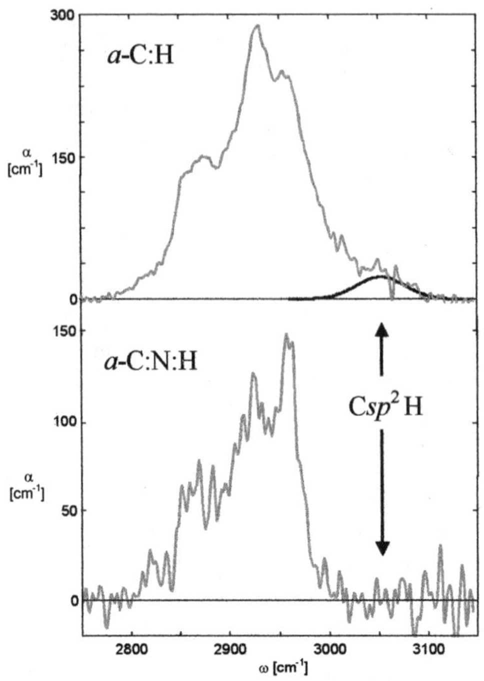

Fig.6 Comparison between the IR absorption of graphite-like *a*-C:H (top) and *a*-C:N:H (bottom) in the 'CH' region.

3.3 Indirect activation of IR absorption modes of Csp^2 islands

Figure 7 demonstrates the existence of a clear correlation between the oscillator densities, $N_{'CC'}$, of the three major IR peaks in the 1000-1900 cm⁻¹ 'CC' region and the oscillator densities, $N_{'CN'}$, of the (Nsp-related) peaks in the 2000-2250 cm⁻¹ 'CN' region.

In order to explain such an experimental evidence, the polarisation of the sp^2 phase generated by the terminal Csp_xNsp_y groups, which the graphitic clusters are back-bonded to, is assumed to be responsible for the activation of the Csp^2 island related features dominating the 1000-1900 cm⁻¹ region of the IR absorption spectra. If the field, locally generated by nitrogenated sp^1 groups (behaving like dipoles *[4]*), is thought to polarise the π-electrons of Csp^2 clusters, consequently activating the 'CC' IR absorption modes in the clusters, $N_{'CC'}$ is proportional to the total polarisation of the sp^2 phase and the saturating trend of the $N_{'CC'}$ vs. $N_{'CN'}$ curve can be interpreted as indicative of a full polarisation of the sp^2 phase (in other words, all the clusters are back-bonded to Csp_xNsp_y groups) *[24,25]*.

To show how such an effect could arise, it has to be noted that the nitrogenated (sp) impurities vibrating in the 'CN' region surely behave as sharp lattice perturbations, since they are both characterised by very strong triple bonds (or by also rigid double-bond chains) and by the presence of an host atom (N) forming a local inhomogeneity in the π-charge distribution. Such effects are well known in crystalline solids, where impurities are able to modulate the electronic charge density by Lindhard waves *[26]*, due to electronic screening effects, that polarise the material and create sequences of electric dipoles. Hence, the nitrogenated groups are expected to play an important disordering role. Such a disorder

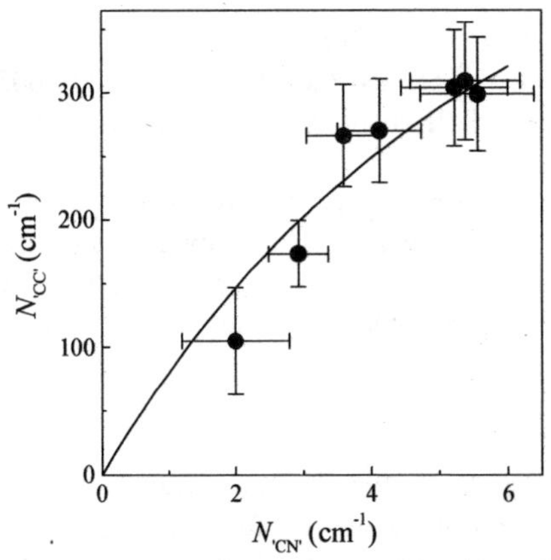

Fig.7 Correlation between the (total) optical oscillator strength densities in the 'CC' and 'CN' regions. The continuous line represents a saturation fit: $N_{'CC'} = 485[1 - \exp(-0.18\,N_{'CN'})]$

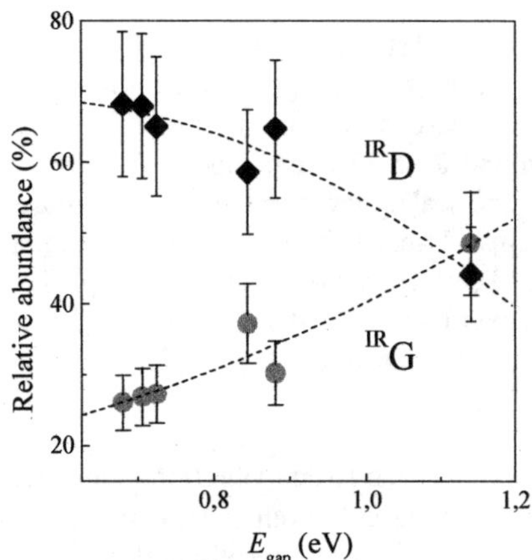

Fig.8 Importance of the two chief (^{IR}G and ^{IR}D) contributions, relative to the total optical oscillator density in the 1000-1900 cm^{-1} 'CC' region.

adds to that typical *[13]* of nitrogen-free carbon-based materials.

In presence of disorder, both electronic and vibrational states tend to spatially localise, since their wave-functions are featured by decay lengths related to the local amount of disorder. As, in a given position, the disorder amount is the same for electronic and vibrational states, their decay lengths are correlated. However, for the sake of simplicity, in the following, the same decay length, r_0, is assumed to describe both electronic states and lattice vibrations *[*]*. The nitrogen-induced polarisation affects electronic and vibrational states localised in a sphere of radius r_0 centred on the nitrogenated group.

Dealing with heterogeneous materials, it is important to remind that electronic and vibrational states with different degrees of localisation may be present. For instance, the graphite-related G-mode can even be highly localised, as it can even degenerate to the stretching mode of a single ($a \approx 0.14$ nm long) Csp^2-Csp^2 bond, when order can be found at such a short-range, that only the very border-zone characters of the graphite band-structure survive *[12]*.

In presence of a bond polarisation due to a back-bonded perturbation, the symmetry with respect to the mid-bond plane, featuring the G (Raman-active) stretching mode of a single bond, vanishes and is replaced by anti-symmetry with respect to the same mid-bond plane. The as-generated IR-active bond stretching mode (^{IR}G) will not be Raman-active and will most probably move to higher frequencies, since heteropolar bonds have generally higher bond-strength than their homopolar counterparts. In fact, in the investigated a-C:N:H samples, the ^{IR}G-peak is detected at frequency positions, $^{IR}\omega_G$, up to 1600-1605 cm^{-1}, while the Raman-shift, ω_G, is always around 1570-1575 cm^{-1}. As the ^{IR}G-mode could be highly localised, it can be activated even by small polarisation-radii ($r_0 \sim a$), so that it would be even detectable where the permanently-polarised electronic states are localised too.

As the vibrating modes of sixfold rings back-bonded to nitrogenated groups are concerned, the picture is more complicate and several conjectures can be made. Different

---

[*] It is worthwhile noticing that, even if their actual values are different, as far as they are affected in the same way by disorder, the discussion still holds.

charge redistribution can give rise to different IR-active modes in a single sixfold ring type. However, the strong IR mode around 1380-1395 cm^{-1} (^{IR}D) and the ^{IR}G mode (1600-1605 cm^{-1}) reasonably scale, in their relative positions, with the Raman D- and G- modes located around 1365-1395 cm^{-1} and 1570-1580 cm^{-1}, respectively.

The scaling law turns to be even more accurate, if one notes that it is the lowest possible Raman D-band frequency position (i.e. 1355 cm^{-1} *[12]*) that has to be compared with the candidate IR-active breathing modes. To justify such a claim, it has to be reminded that the Raman D-mode is disorder-activated thanks to a strong enhancement of its cross section by the quasi-selection rule *[10,12]*

$$q = \Delta k = L^{-1}$$

where q is the lattice-vibration momentum and Δk represents the electronic momentum in a super-lattice approximation of the system (i.e. Δk represents the disorder-induced loss of coherence *[27]* of a localised-electron due to its finite mean free-path, L, being of the order of the cluster size, L_C). Of course, this rule does not apply to an IR-absorption process where, as no external mode polarisation is required, it is replaced by $q = 0$). Therefore, it is the super-lattice value $\omega(q=0) = 1355$ cm^{-1}, and not the $\omega(q=\Delta k) = \omega_D = 1365$-$1395$ cm^{-1} ones, that should scale with $^{IR}\omega_D$ value when a comparison of Raman and IR modes is attempted.

As previously anticipated the rigid shift of both the IR modes, with respect to the Raman modes, is easily justifiable in terms of different bond-strengths (i.e. slightly perturbed super-lattices) in *a*-C:N:H films.

Below, the existence of a clear correlation between IR oscillator-density ratio, N_D/N_G, and optical gap, E_{gap}, (set against the much less clear trend that I_G/I_D bears vs. E_{gap}) is tentatively justified, on the same basis. This is possible in light of the above hypotheses of:

    i) indirect activation by the (sp) nitrogenated groups (vibrating in the 'CN' region) of the three major contributions vibrating in the 'CC' regions; and

    ii) rigid energy-shifts of the so-called ^{IR}D- and ^{IR}G- peaks, relative to the positions the same vibrating modes assume on the band-structure of an unperturbed graphite super-lattice.

At this purpose, it has to be noticed that the breathing mode of single ($b \approx 0.26$ nm large) carbon sixfold-ring is anti-symmetric (IR-active) only when it is uniformly-polarised (i.e. dominated by fully anti-symmetric π-electron wave-functions). This only occurs where a polarisation radius $r_0 > b$ is present. Such an actual r_0 value is larger than that ($r_0 \sim a$) required to produce an IR-active stretching mode of single bond. The N_D/N_G ratio is then expected to be related to the r_0-size of the aromatic clusters back-bonded to nitrogenated groups. If the conventional optical gap is controlled by the nitrogen back-bonded clusters (for instance, because they are usually the largest ones) a well-definite N_D/N_G vs. E_{gap} trend is present. Contrarily, the I_D/I_G ratio is related with the average size of the Raman-active clusters, that are not necessarily sitting near the nitrogenated sites. Hence, *a*-C:N:H materials are expected to follow the same clear I_D/I_G vs. E_{gap} trend of *a*-C:H materials, only if the nitrogen inclusions not markedly affect neither the conventional optical-gap, nor the average cluster size, nor the spread of the cluster sizes around their average value. The absence (Fig.2) of a clear I_D/I_G vs. E_{gap} trend in *a*-C:N:H demonstrates that it is not so, while the well-defined trends of both N_D and N_G vs. E_{gap} (Fig.8) support the hypotheses that the conventional optical-gap is controlled by the fraction of aromatic clusters monitored by the IR analysis. Such clusters present the largest sizes and they are back-bonded to the nitrogenated inclusions. This suggests that nitrogenated groups promotes the nucleation of the sp^2 phase around them.

4. Conclusions

The results are presented of an accurate investigation of the vibrational and structural properties of a set of low-gap a-C:N:H thin films, deposited by reactive sputtering of graphite. The different bonding topologies of nitrogen- and hydrogen- atoms within the carbon network, is clarified through the comparative analysis of the indications emerging from the complementary film-characterisation by IR and Raman spectroscopy. In particular, hydrogen is found to bind almost exclusively with sp^3 hybridised C atoms, meanwhile nitrogen is suggested to be mainly located in Csp_xNsp_y groups, back-bonding the aromatic carbon-clusters.

The hydrogen addition during deposition is then supposed to play an indirect role in promoting the formation of nitrogenated terminal groups, acting as nucleation centres for the Csp^2 phase.

The existence is further demonstrated of a clear correlation between the oscillator densities of the Csp^2 island related features vibrating in the 1000-1900 cm^{-1} region and the oscillator densities of the Nsp-related peaks in the 2000-2250 cm^{-1} region of the IR spectra. Such an experimental evidence is understood by assuming the long-range polarisation of the sp^2 phase, generated by the terminal Csp_xNsp_y groups, to be responsible for the indirect activation of the IR absorption modes in the graphitic clusters.

Finally, the mutual influence-factors between the IR-active and the Raman-active vibrational modes, the optical properties and the microstructures are emphasised.

Acknowledgements

We would like to acknowledge G.Mina (Politecnico di Torino) for technical support in sample deposition, E.Tresso and C.Vanzini (Politecnico di Torino) for X-ray small-angle scattering measurements, G.Ottaviani, C.E.Nobili and F.Gambetta (Università di Modena) for compositional analysis. This study was carried out with financial support from the MADESS II project of CNR (Italian National Research Council).

References

[1]   A.Y.Liu, M.L.Cohen, *Science* **245** (1989) 841

[2]   T.Hughbanks, Y.Tian, *Solid State Commun.* **96** (1995) 321,

[3]   D.M.Teter, *Mater. Res. Soc. Bullet.*, **23** (1998) 22

[4]   S.Muhl, J.M.Mendez, *Diamond Relat. Mater.* **8** (1999) 1809

[5]   A.Grill, V.Patel, *Diamond Films Technol.* **2** (1992) 25

[6]   F.L.Freire jr, *Jap. J. Appl. Phys.*, **36** (1997) 4886

[7]   F.Tuinistra, J.L.Koenig, *J. Chem. Phys.*, **53** (1970) 1126

[8]   G.Messina, S. Santangelo, G.Fanchini, A. Tagliaferro, "High-frequency features in Raman spectra of reactively sputtered a-CN:H thin films", *present volume*

[9]   M.A.Tamor, W.C.Vassel, *J. Appl. Phys.* **76** (1994) 3823

[10] I.Pocsik, M.Hundhausen, M.Koos, L.Ley, *J. Non-Cryst. Solids* **227-230** (1998) 1083

[11] D.G.Mc Culloch, S.Prawer, A.Hoffman, *Phys. Rev.* B **50** (1994) 5905

[12] A.C.Ferrari, J.Robertson, *Phys. Rev.* B **61** (2000) 14095

[13] J.Robertson, *Diamond Rela.d Mater.* **4** (1995), 297

[14] F.Parmigiani, E. Kaym and A. Seki, *J. Appl. Phys.* **64** (1998) 3031

[15] G.Fanchini, A.Tagliaferro, *Diamond Relat. Mater.* (in press)

[16] The *a*-C:N film, whose IR absorption spectrum is comparatively shown in Fig.3, was deposited, using the same sputtering system, under the following conditions: 100 °C temperature, 3/7 He/Ar flow ratio, 300 W r.f. power, 20 mTorr chamber pressure, 20 sccm N_2 flow rate. No hydrogen was added during deposition.

[17] G.A.N.Connell, J.R.Pawlik, *Phys. Rev.* B **13** (1976) 787

[18] J.H.Kaufman, S. Metin, D.D.Saperstein, *Phys. Rev.* B **39** (1989) 13053

[19] Y.Taki, T.Kitawaga, O.Takai, *Thin Solid Films* **304** (1997) 183

[20] C.De Martino, F.De Michelis, A.Tagliaferro, *Diamond Relat. Mater.* **4** (1995) 1210

[21] G.Sokrates, *Infrared characteristic group frequencies*, Wiley, Chichester (1980)

[22] N.Mutsukura, K.Akita, *Diamond Relat. Mater.* **8** (1999) 1720

[23] B.Dischler, in *Europen Mater. Res. Soc. Symposia Proc.* Vol. XVII (P.Koidl Ed.), Les Editions de Physique, Les Ulis (1987) p.189

[24] G.Fanchini, G.Messina, A.Paoletti, C.S.Ray, S.Santangelo, A.Tagliaferro, A.Tucciarone, *Surf. Coat. Technol.* (2001) in press

[25] G.Messina, S. Santangelo, G.Fanchini, A. Tagliaferro, "The G-band frequency-position in Raman spectra of amorphous carbon-nitride based materials: correlation with the chemical composition", *present volume*

[26] N.Ashcroft, N.D.Mermin, *Solid State Physics*, Saunders College, Philadelphia (1975) p.343

[27] N.F.Mott, E.A.Davis, *Electronic processes in non crystalline materials*, Clarendon, Oxford (1979) p.19.

Luminescence properties of point defects in silica

Marco Cannas

INFM, Dipartimento di Scienze Fisiche ed Astronomiche, via Archirafi 36 I-90123 Palermo
Phone: +39-091-6234220; Fax: +39-091-6162461; E-mail:
cannas@fisica.unipa.it

Abstract. The optical properties of point defects in as-grown natural silica are reviewed. Two emissions peaked at 4.2 eV (α_E band) and at 3.1 eV (β band), related to an absorption band at 5.1 eV ($B_{2\beta}$), have been experimentally investigated on the basis of their temperature dependence and their kinetic decay. Our results allow to characterize the excitation pathway of these luminescence bands and to make clear the competition between the radiative relaxation rates and the phonon assisted intersystem crossing process linking the singlet and the triplet excited states from which α_E and β, respectively, originate. Finally, we discuss the role played by the disordered vitreous matrix in influencing the optical features of defects.

1. Introduction

The study of vitreous silica, the amorphous silicon dioxide (a-SiO$_2$), is currently an attractive research field in solid state physics and material science [1-3]. The physical properties of high transparency in a wide spectral region (visible, UV, vacuum-UV) and low conductivity, in combination with favorable mechanical characteristics and low manufacturing costs, have led to the widespread utilization of silica-based materials in many technological applications, like manufacturing of optical fibers, lenses and optoelectronic devices.

These exceptional features depend critically on the maintenance of defect free band gap [4]. For this reason the understanding of the nature and the formation mechanisms of defects in a-SiO$_2$ plays a fundamental role in both the technological and basic research. To this end, the combined use of several spectroscopic techniques and silica materials different for their manufacturing processes or for external treatments (irradiation, heating) could provide a powerful method to improve the knowledge of the properties of defects. In this paper we focus our attention on the optical absorption (OA) and photoluminescence (PL) bands detected in as grown natural silica in a wide spectral region from visible to UV. Aim of this work is to exemplify the whole optical activity in terms of an energy level diagram with the radiative and non radiative transitions accounting for the spectral and kinetics features, also as a function of temperature. Moreover, we investigate the role played by the vitreous matrix in determining the observed optical activities. It is worth noting that the study of defects in silica holds a more general significance in obtaining new insights on the relationship between the properties of an optically active defect and the structural and dynamic properties of its environment in other disordered materials.

The present paper is so structured: in the section 2 we outline the theoretical background on the optical properties of a generic point defect in silica, in section 3 we describe the experimental method and finally in the section 4 we review and discuss our experimental results.

2. Theoretical background

2.1 Point defect in silica

The point defect is usually defined in the contest of a crystalline network, if the lattice site is occupied differently than in the perfect crystal [5]. The defects may be classified as intrinsic and extrinsic. The first type includes unoccupied sites (vacancies) and occupied sites that in the perfect crystal are unoccupied (interstitial). The second type consists of impurities at sites that in the crystal lattice either are occupied by atoms of the pure material (substitutional impurities) or are unoccupied (interstitial impurities). An overview of the different kinds of defects is shown in Fig. 1.

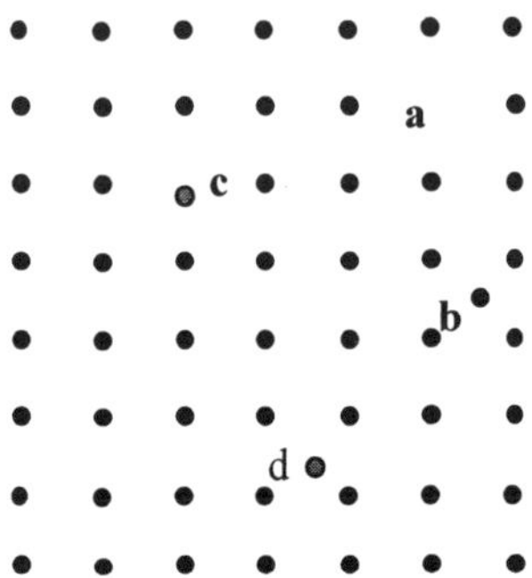

Fig.1 Crystal lattice with different types of defects: intrinsic vacancy (a) and interstitial (b); extrinsic substitutional (c) and interstitial (d).

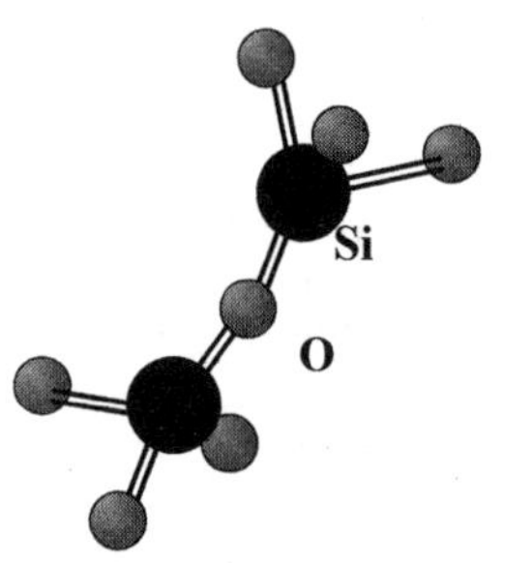

Fig.2 Fragment of a regular silica network

The concept of defect can also be extended to amorphous materials, like silica, whose structure matches the crystalline α-quartz only in a short-range order [2]. As depicted in Fig. 2, the structural unit of silica is the SiO_4 tetrahedron where the Si atom is bonded to four O atoms with O-Si-O angle of 109.5°. The lack of long-range atomic order is due to the large spread of tetrahedral linkage angle Si-O-Si, statistically distributed between 120° and 180°. In this framework, a point defect is present when the array of Si and O atoms of the ideal silica network is broken down by an imperfection.

The presence of defects in the vitreous matrix may drastically modify the optical properties of the host material [4, 6, 7]. Indeed, defects exist in different electronic states that can cause optical transitions as absorption and luminescence with lower energies than the fundamental absorption edge of the silica material, approximately 9 eV, from valence to conduction band. For this reason, a point defect is also defined a color center or chromophore.

Even if these transitions are localized at the point defect, the optical spectra are influenced by its environment. While in an isolated center, the energy transferred in an optical transition has to match the difference between the electronic levels whose spread is limited only by the excited state lifetime, for a chromophore embedded in a matrix, it can be shared between a local electronic contribution and a wide variety of phonon excitations of the vibrational modes of the matrix. Moreover, owing to the amorphous nature, each defect can exist in different local rearrangements of the surrounding matrix (conformational inhomogeneity) and the energy associated to transitions between the electronic levels is largely distributed. Then, a wide range of photon energies can be involved in the transition and the spectrum of absorption or luminescence consists of broad bands.

2.2 Optical absorption transition

We consider now the transition occurring between two electronic states of a point defect: the ground state (0) and the excited one (1) [5, 8, 9]. In Fig. 3, we sketch the potential energy curves ε_0 and ε_1 as a function of the generalized normal coordinate Q_f. In the same figure are also depicted the vibrational levels associated to the quantum numbers n_f and m_f in the states (0) and (1), respectively, due to the nuclear oscillations. If the equilibrium positions of atoms in the ground and excited state are different, the set of normal coordinates changes from Q_f (in the ground state) to $Q'_f = Q_f - \Delta_f$ (in the excited state). Then, the total energies associated to the (0) and (1) states can be expressed by:

$$E(0)_{Tot} = \varepsilon_0^{eq} + \sum_{f=1}^{N_f} h\nu_f\left(n_f + 1/2\right) \qquad n_f = 0, 1, 2, \ldots \qquad (1)$$

$$E(1)_{Tot} = \varepsilon_1^{eq} + \sum_{f=1}^{N_f} h\nu_f\left(m_f + 1/2\right) \qquad m_f = 0, 1, 2, \ldots \qquad (2)$$

where ε_0^{eq} and ε_1^{eq} are the energies of the two electronic states when all the nuclei are in their equilibrium position. In the above Eqs., we have assumed that the frequency ν_f of normal modes is the same in the two states (0) and (1) (linear coupling approximation).

According to the *Franck-Condon* approximation, that is, the electronic transitions occurring in a time much faster than the nuclear motion, the absorption is described as a vertical transition with respect to the energies associated to the two states in the configuration coordinates diagram. If $E_{1,0} = E_1 - E_0 = h\nu_{1,0}$ is the energy value matching the quantum transition between the two states (0) and (1), for a dl path length of a sample having N_0 identical non-interacting absorbers per unit volume in the ground state, the differential energy loss by the electromagnetic field is given by [8]:

$$-dI(E) = I(E)N_0 \frac{4\pi^2}{3\hbar^2} \frac{1}{4\pi\varepsilon_0} \frac{E}{c} \cdot \delta\left(E - E_{1,0}\right)\left|R_{1,0}\right|^2 dl \qquad (3)$$

where $I(E)$ is the intensity of the electromagnetic field flowing through the sample and $R_{1,0}$ is the quantum-mechanical matrix element of the electric dipole moment $\mathbf{M}$ between the total eigenfunctions ψ_0 and ψ_1 of the two states (transition moment).

$$\left|R_{1,0}\right|^2 = \left|\left\langle\psi_1\left|\mathbf{M}\right|\psi_0\right\rangle\right|^2 \qquad (4)$$

By integration over a unitary path of absorbing material, one obtains the following expression for the absorption coefficient:

$$\alpha(E) = Ln\frac{I_0(E)}{I(E)} = \frac{N_0}{3\hbar^2\varepsilon_0 c}\frac{\pi E}{}\left|R_{1,0}\right|^2 \delta\left(E - E_{1,0}\right) \qquad (5)$$

where $I_0(E)$ is the intensity of the electromagnetic field incident on the sample at energy E. Taking into account the finite lifetime τ of the excited state, the δ function can be replaced by a *Lorentzian* shape function (with $2\Gamma = 1/\tau$):

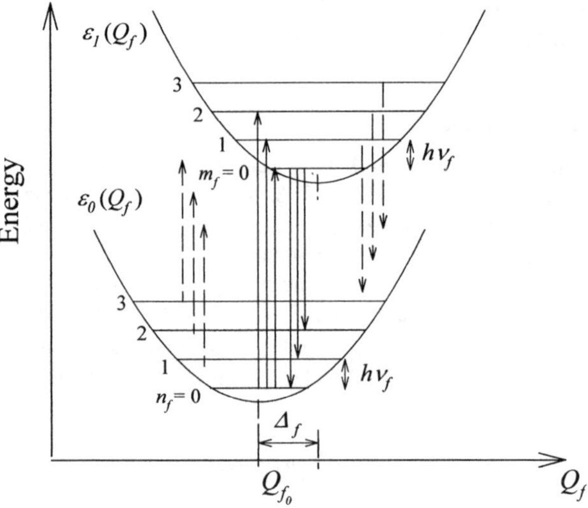

Fig.3 Optical absorption and luminescence transitions between the ground (ε_0) and excited (ε_1) electronic states in a configuration coordinate diagram

$$\alpha(E) = \frac{N_0 \, \pi \, E}{3\hbar^2 \varepsilon_0 c} |R_{1,0}|^2 \frac{\Gamma}{(E - E_{1,0})^2 + \Gamma^2} \tag{6}$$

In order to obtain a suitable expression of the transition moment $R_{1,0}$, we make use of the *Born-Oppenheimer* approximation and write the eigenfunction ψ of the total system by the product of the electronic wavefunction ϕ and the nuclear one φ [5, 8]:

$$\psi_0(r,Q) = \phi_0(r,Q) \cdot \varphi_0(Q) \tag{7}$$

$$\psi_1(r,Q') = \phi_1(r,Q') \cdot \varphi_1(Q') \tag{8}$$

where r is the electronic coordinate and Q and Q' are the set of normal nuclear coordinates in (0) and (1), respectively, which allow to write φ_0 and φ_1 as a product of harmonic oscillator functions:

$$\varphi_0(Q) = \prod_{f=1}^{N_f} \varphi(Q_f); \quad \varphi_1(Q') = \prod_{f=1}^{N_f} \varphi(Q_f - \Delta_f) \tag{9}$$

We can consider the electronic dipole moment **M** as composed by an electronic term, $\mathbf{M}_{el.}(r)$, and nuclear one, $\mathbf{M}_{nucl.}(Q)$ [8]. So, we can write:

$$\begin{aligned} R_{1,0} = & \iint \psi_1^*(r,Q') \mathbf{M} \psi_0(r,Q) d\tau_e d\tau_v = \\ & \iint \phi_1^*(r,Q') \varphi_1^*(Q') \mathbf{M}_{el.}(r) \phi_0(r,Q) \varphi_0(Q) d\tau_e d\tau_v + \\ & + \iint \phi_1^*(r,Q') \varphi_1^*(Q') \mathbf{M}_{nucl.}(Q) \phi_0(r,Q) \varphi_0(Q) d\tau_e d\tau_v \end{aligned} \tag{10}$$

where the integral is carried out on the space of the electronic τ_e and vibronic τ_v coordinates. Since during the transition from (0) to (1) the nuclei remain almost stationary in the equilibrium position of the ground state, the electronic eigenfunction ϕ can be assumed to depend upon $Q=Q_0$. Besides, we note that $\mathbf{M}_{nucl.}(Q)$ does not depend on the electronic coordinates but only upon vibrational coordinates Q. In this way, the expression for $R_{1,0}$ can be readjusted as follows:

$$R_{1,0} = \int \phi_1^*(r,Q_0)\mathbf{M}_{el.}(r)\phi_0(r,Q_0)d\tau_e \cdot \int \varphi_1^*(Q')\varphi_0(Q)d\tau_v +$$

$$+ \int \phi_1^*(r,Q_0)\phi_0(r,Q_0)d\tau_e \cdot \int \varphi_1^*(Q')\mathbf{M}_{nucl.}(Q)\varphi_0(Q)d\tau_v \qquad (11)$$

Owing to the orthogonality of electronic eigenfunctions, the second term vanishes and the transition moment $R_{1,0}$ is given by:

$$R_{1,0} = R_e \cdot \int \varphi_1^*(Q')\varphi_0(Q)d\tau_v \qquad (12)$$

where R_e indicates the transition moment associated to the electronic states ϕ_0 and ϕ_1:

$$R_e = \int \phi_1^*(r,Q_0)\mathbf{M}_{el.}(r)\phi_0(r,Q_0)d\tau_e \qquad (13)$$

The term $\int \varphi_1^*(Q')\varphi_0(Q)d\tau_v$, known as the *Franck-Condon* integral, measures the overlap between the vibrational functions φ_0 and φ_1. If there is no coupling between the electronic transition and the vibrational modes, the set Q' coincides with Q ($\Delta_f = 0$) and, owing to the orthogonality of harmonic oscillator wavefunctions, only terms with the same vibrational quantum numbers (i.e. $m_f = n_f$) contribute to the transition. In the presence of coupling, $Q' \neq Q$ and also terms with $m_f \neq n_f$ will contribute to the transition.

Taking into account the Eqs. (9) and (12), the absorption coefficient is given by:

$$\alpha(E) = M_0 \; E \left| \prod_{f=1}^{N_f} \int \varphi_f^*(Q - \Delta_f)\varphi_f(Q)d\tau_v \right|^2 \frac{\Gamma}{(E - E_{1,0})^2 + \Gamma^2} \qquad (14)$$

with $M_0 = \dfrac{N_0 \pi}{3\hbar^2 \varepsilon_0 c}|R_{1,0}|^2$.

At T=0 K, only the vibrational level with $n_f=0$ in the ground electronic state (0) is populated and Eq. (14) can be rewritten as:

$$\alpha(E,T=0) = M_0 \; E \cdot \sum_{\{m_f\}} \left[\left(\prod_{f=1}^{N_f} e^{-S_f} \frac{S_f^{m_f}}{m_f!} \right) \times \frac{\Gamma}{\left(E - E_{00} - h\sum_{f=1}^{N_f} m_f v_f \right)^2 + \Gamma^2} \right] \qquad (15)$$

where $E_{1,0}$ has been expressed in agreement with Eqs. (1) and (2) to take into account the energies associated to the transitions from $n_f = 0$ in the state (0) to various m_f in the state (1), as shown by the upwards arrows (continuous line) in Fig.3. In particular, E_{00} represents the energy difference between the $m_f = 0$ level in (1) and $n_f = 0$ in (0), i.e. the energy of the purely electronic transition. The dimensionless *Huang-Rhys* factor S_f [10] (linear coupling constant) is defined by:

$$S_f = \frac{h v_f}{2} \Delta_f^2 \qquad (16)$$

and it measures the coupling strength between the electronic transition $0 \to 1$ and the f–th vibrational mode that determines the rearrangement of the nuclei from their equilibrium position by Δ_f. Then, Eq. (15) shows that the absorption band at T=0 K results from the superposition of a series of *Lorentzians* whose intensity is modulated by the *Poissonian* distributions product:

$$\prod_{f=1}^{N_f} e^{-S_f} \frac{S_f^{m_f}}{m_f!} \tag{17}$$

At $T \neq 0$, the vibrational levels with $n_f \neq 0$ in the ground state (0) can be thermally populated according to the *Boltzmann* law and contribute to the transitions toward the excited state (1), as shown by the dashed line arrows in Fig.3. Generally, if $T_{max.}$ is the upper limit of the temperature range investigated, it is possible to distinguish between vibrational modes with high (ν_h) and low (ν_l) frequency [11]. The ν_h modes have frequency such that $h\nu_h >> K_B T_{max}$, where K_B is the *Boltzmann* constant, so that only the vibrational level with $n_h = 0$ is occupied in the temperature range up to T_{max}. This fact implies that, assuming a set of N_h high frequency modes, only transitions from $n_h = 0$ in the state (0) to $m_h = 0, 1, 2,...$ in the state (1) occur. At variance, the ν_l modes can change their population on varying the temperature. Therefore, assuming a set of N_l low frequency modes, transitions from $n_l = 0, 1, 2,...$ in (0) to $m_l = 0, 1, 2,...$ in (1) take place and are relevant in changing the shape of the absorption spectrum. If we consider the contribution of these N_l modes as a single mode (*Einstein* oscillator) with mean frequency value $\langle \nu_l \rangle$ and mean linear coupling constant S_l, the expression of the absorption at the temperature T is given by:

$$\alpha(E,T) = M_0 \, E \left[\sum_{\{m_f\}} \left(\prod_{f=1}^{N_h} e^{-S_f} \frac{S_f^{m_f}}{m_f!} \right) \right] \times$$

$$\times \frac{\Gamma}{\left(E - E_{00} - h\sum_{f=1}^{N_f} m_f \nu_f \right)^2 + \Gamma^2} \otimes \frac{1}{W(T)} \cdot e^{-\frac{1}{2} \frac{E^2}{W^2(T)}} \tag{18}$$

where $\otimes$ represents the convolution operator, $f(E) \otimes g(E) = \int f(E - E') \cdot g(E') dE'$, and where

$$W^2(T) = N_l S_l h^2 \langle \nu_l \rangle^2 \coth \frac{h\langle \nu_l \rangle}{2 K_B T} \tag{19}$$

According to Eqs. (17), (18) and (19), the absorption profile is given by the superposition of a series of *Voigtians* (*Gaussian* convolutions of *Lorentzians*) whose width increases as the temperature increases while the energy peak remains constant. We recall that the Eq. (18) for the absorption spectrum has been obtained in the linear coupling approximation, i.e. assuming the same vibrational frequency ν_f of normal modes in the states (0) and (1). At variance, if the transition from (0) to (1) changes the vibrational frequencies ν_f (non linear coupling), temperature effects on the peak position of the absorption band are also present and must be taken into account [11, 12].

Further contributions to the absorption lineshape are the inhomogeneous effects arising from the different local environments surrounding the point defects in amorphous materials. Generally, this site-to-site non-equivalence results in a spectral distribution of the purely electronic transition energies E_{00}. If the mapping between the conformational and the spectral heterogeneity is linear [11], the distribution function for E_{00} is given by a *Gaussian* function:

$$g_{abs}(E_{00}) = \frac{1}{\sqrt{2\pi}\sigma_{inh}} e^{-\frac{(E_{00}-E'_{00})^2}{2\cdot\sigma^2_{inh}}} \tag{20}$$

peaked at the mean energy E'_{00} whose width σ_{inh} does not depend on the temperature. So, the whole optical absorption spectrum is given by the convolution of Eqs. (18) and (20) and its total width is determined by the different weights of the broadening mechanisms: lifetime, electron-phonon interaction, inhomogeneous broadening.

2.3 Photoluminescence activity

Following light absorption, the inverse transition from the excited state (1) to the ground state (0), shown in Fig. 3, causes a spontaneous emission of light also called photoluminescence. In particular, for two states having the same spin multiplicity, the electronic transition is allowed and in this case it is called fluorescence. As in the excited state the nuclei relax towards the minimum energy configuration at $Q_f = Q_{f0} + \Delta_f$ in a much shorter time (10^{-12} s) than the fluorescence lifetime (10^{-8} s), the light emission occurs after that the state (1) has reached the thermal equilibrium [9]. Because the excitation of phonons reduces the energy available to the photon, the luminescence emission occurs at lower energies than the absorption.

The expression of the intensity of light emitted from the excited state (1) is obtained by the relation between the *Einstein* coefficients for absorption and spontaneous emission [13]. If $E_{0,1} = E_1 - E_0 = h\nu_{0,1}$ is the energy value matching the quantum transition $(1)\rightarrow(0)$, the luminescence intensity of a sample having N_{lum} identical non-interacting centers per unit volume in the excited state is given by:

$$I_{PL}(E) = N_{lum} \frac{1}{3\hbar^4 \pi\varepsilon_0 c^4} |R_{0,1}|^2 E^4 \frac{\Gamma}{(E - E_{0,1})^2 + \Gamma^2} \tag{21}$$

where $R_{0,1}$ is the quantum-mechanical matrix element of the electric dipole moment **M** for the transition between the states (1) and (0):

$$|R_{0,1}|^2 = |\langle\psi_0|\mathbf{M}|\psi_1\rangle|^2 \tag{22}$$

By comparing the Eqs. (21) and (22) with the Eqs. (6) and (4), respectively, it is possible to see that the emission band profile is mirror like to the absorption one apart from a shift of the peak position toward lower energies. Because the excited state (1) is at thermal equilibrium during the luminescence emission, its vibrational levels are populated according to the *Boltzmann* law. At T=0 K, only transitions from the vibrational level with $m_f=0$ in the state (1) to different vibrational levels with various n_f of the state (0) can occur, as depicted by downwards arrows (continuous line) in Fig.3. So, the luminescence spectrum assumes a shape which depends on the linear coupling constant S_f as in Eq. (15). At T $\neq$ 0, the coupling between the electronic transitions and the low frequency ν_l modes induces changes in the luminescence spectrum like those reported for the absorption. In this case, the width of the emission profile as a function of the temperature can be expressed by Eq. (18), but $\langle v_I \rangle$ indicates the mean frequency value of the v_I modes in the state (1). Moreover, as for the absorption, the conformational heterogeneity in amorphous materials causes a spread of the emission energies associated with the transition from (1) to (0), which results in the inhomogeneous broadening of the photoluminescence spectrum.

Generally, following light absorption, different excitation pathways can occur in a point defect. In Fig. 4, we depict a typical energetic level scheme consisting in a singlet ground state S_0, two singlet excited state S_1 and S_2 and the first triplet excited state T_1 [14]. The absorption $S_0 \rightarrow S_1$ and the fluorescence $S_1 \rightarrow S_0$ has been already discussed through the Fig. 3. At variance, if the system is excited to an electronic state S_2, it rapidly relaxes to the lowest vibrational levels of S_1. This process is called internal conversion and it occurs in $\sim 10^{-12}$ sec. As the fluorescence lifetime is $\sim 10^{-8}$ sec, the internal conversion is complete before the radiative emission from S_2 to a lower singlet state so that the transition $S_0 \rightarrow S_2$ is able to excite the emission $S_1 \rightarrow S_0$. Finally, when the system is in the S_1 state, it can undergo a radiation-less transition to the first triplet state T_1 of lower energy. This non radiative conversion mechanism, known as intersystem crossing process, can be thermally activated by the interaction of luminescent defect with the lattice (phonon assisted process) [8]. The relaxation of T_1 towards the ground state causes a light emission (phosphorescence) at energies lower than the fluorescence. As the transition $T_1 \rightarrow S_0$ is forbidden, the phosphorescence lifetimes $(10^{-1}\text{-}10^{-5}$ sec) are several orders of magnitude longer than those of fluorescence. Therefore, the light absorption due to the transitions from the ground state S_0 to higher excited states S_1 and/or S_2 can excite at least two luminescence bands: one associated to the $S_1 \rightarrow S_0$ transition (fluorescence) and the other associated to the $T_1 \rightarrow S_0$ transition (phosphorescence).

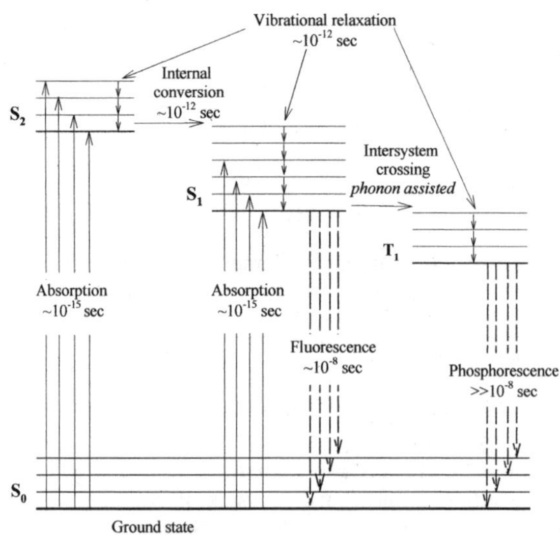

Fig.4 Excitation and relaxation pathways involving the ground and the first excited states localized on a point defect

3. Experimental method

3.1 Samples

In this work, we investigated a set of silica specimens of commercial origin, chosen so as to cover a wide spectrum of preparation techniques. These samples can be grouped in four standard silica types according to the early *Hetherington* classification [15]:

Type I natural dry silica is obtained by fusion of powder of quartz crystal via electric melting in vacuum or in an inert gas at low pressure. It contains negligible OH amount but about the same metallic impurities such as Al, Ge or alkali, totally of the order of 10 part per millions (ppm) by weight as the unfused raw material.

Type II natural wet is prepared by fusion of quartz crystal in a flame. This material has higher chemical purity than the type I because some impurities are volatilized in the flame but as it is prepared in a water-vapor atmosphere it contains nearly 150 ppm of OH groups.

Type III synthetic wet is made by the vapor-phase hydrolysis of pure silicon compounds such as $SiCl_4$. It contains the highest OH content (up to 1000 ppm) but it is virtually free from metallic impurities.

Type IV synthetic dry is obtained by the reaction of O_2 with $SiCl_4$ in a water-vapor-free plasma. In this way, the concentration of OH is reduced to less than 1 ppm but excess of Cl and oxygen in the form of -O-O- linkages are present.

In Table 1 we list the investigated silica types with their name and OH content. The materials Infrasil (I), Herasil (H), Homosil (HM), and Suprasil (S) were supplied by Heraeus [16]; silica EQ (EQ) were supplied by Quartz&Silice [17]; Vitreosil (VTS) was supplied by TSL [18]. All samples used in our measurements have sizes of $5\times5\times1$ mm^3 with the major surfaces optically polished.

3.2 Experimental techniques

Absorption measurements at a wavelength between 190 to 340 nm, corresponding to 4.0-6.5 eV, were performed at room temperature using a JASCO V-570 double-beam spectrometer.

Photoluminescence emission (PL) and excitation (PLE) spectra, measured in steady state regime, were obtained by a Jasco PF-770 instrument, mounting a Xenon lamp

Table 1 Sample list: sample name, silica type, and nominal OH content

Sample Name	Type	OH (ppm)
Infrasil 301 (I301) by Heraeus	Natural dry (I)	≤ 8
Puropsil QS (QPA) by Quartz & Silice	Natural dry (I)	15
Silica EQ906 by Quartz & Silice	Natural dry (I)	20
Silica EQ912 by Quartz & Silice	Natural dry (I)	15
Vitreosil (VTS) by TSL	Natural wet (II)	150
Homosil (HM) by Heraeus	Natural wet (II)	150
Herasil 1 (H1) by Heraeus	Natural wet (II)	150
Herasil 3 (H3) by Heraeus	Natural wet (II)	150
Suprasil 1 (S1) by Heraeus	Synthetic wet (III)	1000
Suprasil 311 (S311) by Heraeus	Synthetic wet (III)	200
Suprasil 300 (S300) by Heraeus	Synthetic dry (IV)	<1

of 150 W as light source. The samples were placed with the major faces at 45° with respect to the exciting beam and the PL light collected in the direction opposite to the reflected beam (45°-backscattering-geometry). The obtained excitation and emission spectra are corrected for wavelength dependent effects as spectral density of the source and the spectral response of detecting system.

PL and PLE measurements as a function of the temperature were also carried out by mounting the samples in a continuous flow helium cryostat (Oxford Optistat^{CF}), equipped with four optical windows and a temperature control (Oxford ITC503). The temperature could be varied from 4.2 up to 350 K and at the required temperature; the spectra were recorded after 10 min for thermal equilibrium.

Lifetime measurements were performed by using the synchrotron radiation (SR) at the SUPERLUMI station on the I-beamline of HASYLAB at DESY (Hamburg, Germany. The time decay of the transient PL was measured using 512 channels for scanning the time interval of 192 ns between adjacent SR pulses, pulse width 0.5 ns. These measurements were carried out at T=300 K and at T= 10 K.

4. Results

4.1 PL activity in natural silica excited in the UV range: emissions at 3.1 eV and 4.2 eV

In this section we review the experimental results concerning the optical absorption and the photoluminescence activity excited in the UV range in our natural silica [19-22] aiming to characterize the two PL bands centered at ~3.1 eV and ~4.2 eV and to make clear some aspects regarding their excitation mechanisms and their temperature dependence.

In Fig.5 is reported the UV absorption spectrum detected in the I301 sample. The main absorption band ($B_{2\beta}$ band [23]) is characterized by a peak energy value E_0=5.15±0.01 eV, by a full width at half maximum FWHM=0.46±0.02 eV and by an amplitude at maximum α_{max}=0.46±0.02 cm^{-1}, as obtained by a best fit procedure in *Gaussian* components. We note that the spectral components in the blue side of the spectrum are attributed to the tail

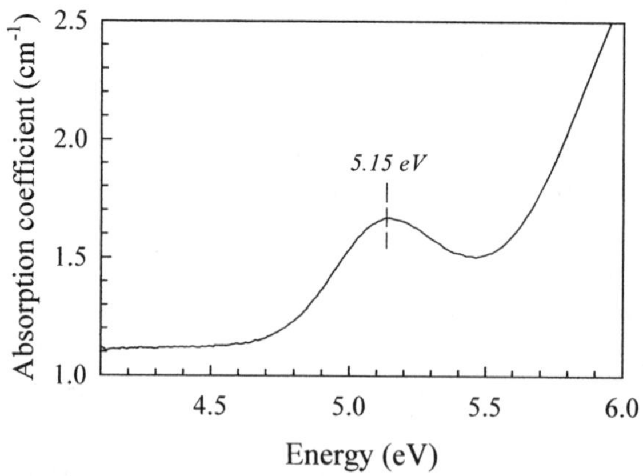

Fig.5 UV absorption profile for the natural silica I301. A band peaked at 5.15 eV ($B_{2\beta}$ band) is recognizable in the spectrum. Taken from *J. Non-Cryst. Solids* **245** (1999) 190-195, copyright 1999 by Elsevier Science.

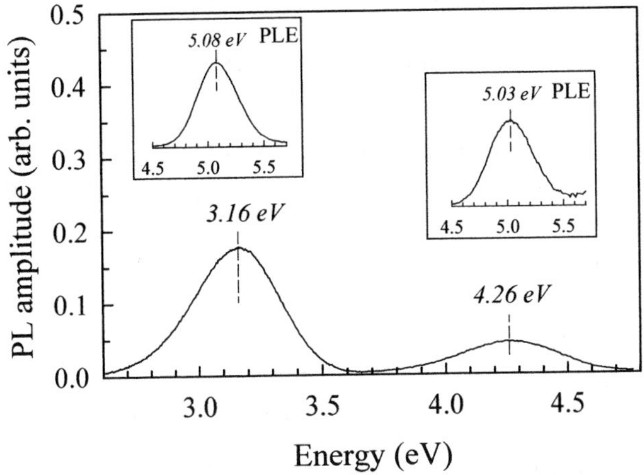

Fig.6 Photoluminescence spectrum excited at 5.0 eV for the I301 sample. Two main emissions are detected, centered at 3.16 eV (β) and 4.26 eV (α_E), respectively. The insets show the corresponding excitation profiles of β and α_E.

of other OA bands located at higher energies and are taken into account in the fit procedure.

The PL stationary activity excited at 5.0 eV at room temperature for the same sample is reported in Fig.6. This spectrum exhibits two PL emissions: the first, at low energies, is centered at E_0=3.16±0.01 eV (with a FWHM=0.43±0.02 eV), the second is peaked at E_0=4.26±0.01 eV (with a FWHM=0.48±0.02 eV). These two emissions are labeled as the β and α_E bands, respectively [24]. In the same figure, the PLE profiles for the emissions at 3.16 eV and 4.26 eV are also displayed. Apart from the small energy shift (0.05 eV), the excitation spectra of these two PL bands are quite similar to each other and have nearly the same FWHM (0.46 eV) as the $B_{2\beta}$ absorption band. We stress that the difference between the peak energies of the PLE spectra of β and α_E can be related to the inhomogeneous distribution of the relaxation rates of the excited states [7, 22, 25, 26] in disordered systems such as the amorphous SiO_2.

Table 2 Relevant spectral parameters (peak energy E_0, full width at half maximum FWHM and intensity I) of the absorption and luminescence bands as detected in natural silica samples. E_0 and FWHM are affected by an error of 0.01 and 0.02 eV, respectively, while M_0 is measured within an uncertainty of 5%.

	OA band $B_{2\beta}$			PL band β			PL band α_E		
Sample	E_0 (eV)	FWHM (eV)	I (cm^{-1}·eV)	E_0 (eV)	FWHM (eV)	I (a.u.)	E_0 (eV)	FWHM (eV)	I (a.u.)
I301	5.15	0.46	0.24	3.16	0.42	0.12	4.26	0.46	0.028
EQ906	5.14	0.42	0.20	3.16	0.41	0.095	4.27	0.44	0.023
EQ912	5.13	0.40	0.06	3.17	0.41	0.028	4.26	0.45	0.0059
QPA	5.13	0.41	0.07	3.16	0.41	0.036	4.27	0.44	0.0080
VTS	5.15	0.44	0.09	3.16	0.41	0.034	4.26	0.43	0.0077
H1	5.16	0.46	0.21	3.16	0.42	0.10	4.27	0.44	0.024
H3	5.12	0.43	0.08	3.16	0.41	0.035	4.26	0.44	0.0072
HM	5.15	0.47	0.19	3.16	0.41	0.079	4.27	0.44	0.018

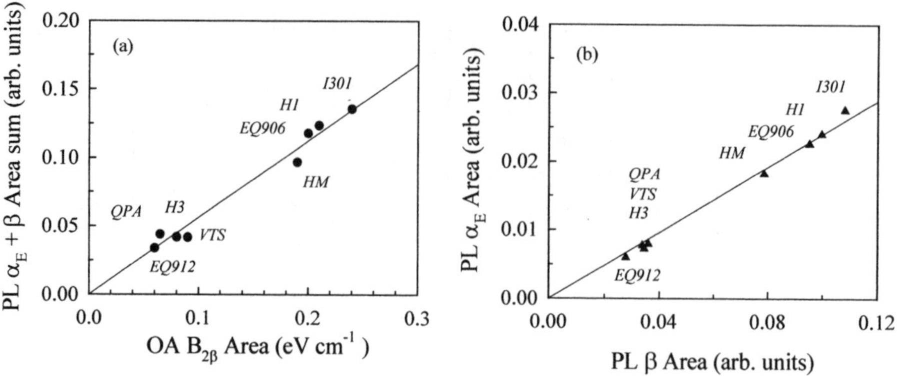

Fig.7 Correlation between the absorption and luminescence bands in our natural silica samples. The areas sum of the emissions α_E and β versus the area of the absorption $B_{2\beta}$ (a). Intensity of the α_E band versus the intensity of the β band. Taken from *Phys. Rev. B* **60** (1999) 11475-11481, copyright 1999 by the American Physical Society.

The optical activity observed in the I301 sample, characterized by the OA band $B_{2\beta}$ and the two PL emissions β and α_E, is common to all natural silica samples. We stress that these optical bands are not detected in synthetic silica types [27, 28].

In table 2, we list the experimental values of the relevant quantities of these optical transitions. The integrated intensity I or area of the bands is measured as the zero-th (M_0) energy moment of their spectral distribution $f(E)$ according to the definition:

$$I = M_0 = \int_{-\infty}^{\infty} f(E)dE.$$ It is worth to note that the spectral parameters (peak position E_0 and

FWHM) of the bands $B_{2\beta}$, β and α_E are the same in all investigated samples, within the experimental uncertainty. At variance, the integrated intensities depend on the specific sample, but they keep a strict correlation. This fact is evidenced in Fig.7 (a) where the sum of the areas of the two PL bands β and α_E is reported as a function of the $B_{2\beta}$ area, as

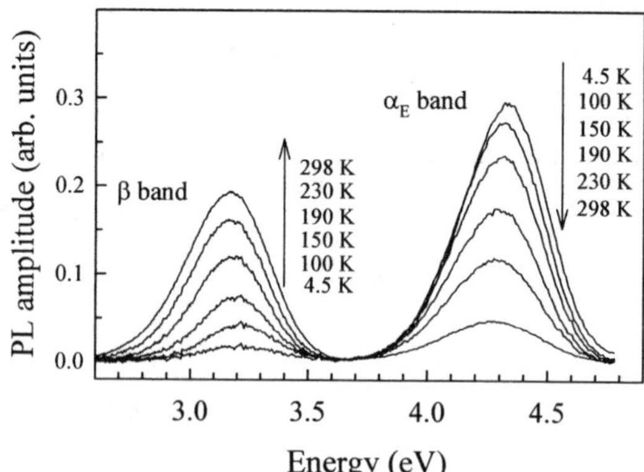

Fig.8 Emission spectra detected at various temperature in the sample I301 under excitation at 5.0 eV. Arrows indicate the changes induced on increasing the temperature. Taken from *Phys. Rev. B* 60 (1999) 11475-11481, copyright 1999 by the American Physical Society.

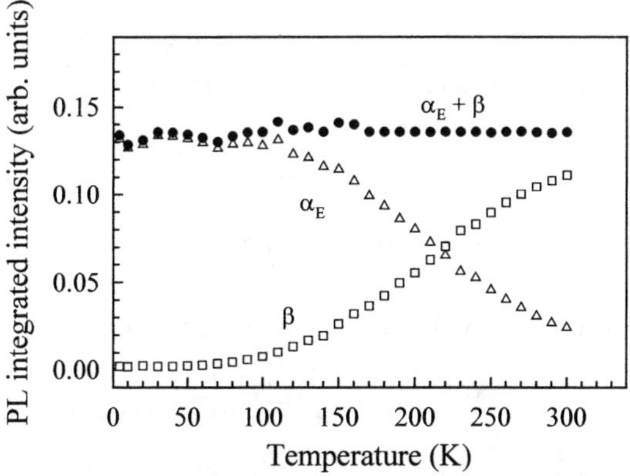

Fig. 9 Integrated intensities of the α_E and β PL bands as a function of the temperature. The sum of the two integrated intensities is also shown

measured in the eight silica samples considered here. Moreover, in Fig.7 (b) the β band area is plotted versus the α_E band one. The best linear fit is shown in both Figs.7 (a) and (b) (line curves) and indicates that both the area ratio between the whole PL activity and the OA band $B_{2\beta}$ and between the two emissions β and α_E areas is independent on the silica type. On the basis of these evidences, we can assert that the whole optical activity, here referred to as B type, originates from the same defect.

4.2 Temperature dependence: intersystem crossing process

The study of temperature effects on the PL activity is a useful tool to obtain information on the fine details of the processes involved in the excitation pathways. With this aim, we have investigated the temperature dependence of the β and α_E emissions in the range 4.5-300 K. Fig.8 shows the PL spectra excited at 5.0 eV, at various temperatures, for the I301 sample. The two PL bands exhibit an opposite behavior: on increasing the temperature, the α_E band decreases while the β band increases starting from a near-to-zero value at low temperature.

The anticorrelated dependence on T of the α_E and β emissions is summarized in Fig.9. We observe that the sum of integrated intensities of the two PL bands is constant in the temperature range investigated.

The above reported data can be taken into account by the energy level scheme shown in Fig.10 consisting in a singlet ground state S_0 and in two excited states of singlet S_1 and triplet T_1. For the sake of clarity, the radiative and non radiative processes with the relative rates are also indicated. We note that this quite general diagram was proposed by *L.N. Skuja* [25] to take into account the energetic levels of a defect consisting of a twofold coordinated Ge. According to the reported scheme, the α_E band is associated to the allowed $S_1 \rightarrow S_0$ fluorescence transition at a rate K_r^F, in agreement with its lifetime of the order of ns [21, 22]. On the other hand, the β emission is ascribed to the spin-rule forbidden $T_1 \rightarrow S_0$ phosphorescence transition at a rate K_r^P, in agreement with its longer decay time (~110 μs) [25]. Both emissions can be excited by the OA band $B_{2\beta}$ related to the $S_0 \rightarrow S_1$ transition: the α_E band directly and the β band via an intersystem crossing process between S_1 and T_1.

In this scheme, the intensities of the two emissions are proportional to the populations N^{S_1} and N^{T_1} of the states S_1 and T_1, and to the radiative rate K_r^F and K_r^P, respectively, according to:

$$I^F \propto K_r^F \cdot N^{S_1} \tag{23}$$

$$I^P \propto K_r^P \cdot N^{T_1} \tag{24}$$

Moreover, the rate equations of N^{S_1} and N^{T_1} can be written as:

$$\frac{dN^{S_1}}{dt} = I_0 \cdot \left[1 - e^{-\alpha \cdot d}\right] - \left[K_r^F + K_{nr}^F + K_{ISC}\right] \cdot N^{S_1} \tag{25}$$

$$\frac{dN^{T_1}}{dt} = K_{ISC} \cdot N^{S_1} - \left[K_r^P + K_{nr}^P\right] \cdot N^{T_1} \tag{26}$$

where $I_0 \cdot \left[1 - e^{-\alpha \cdot d}\right]$ represents the intensity of the light absorbed by ground state defects. In stationary (time-independent) condition, as obtained in our experiments under continuous light excitation, the first member in Eqs. (25) and (26) is reduced to zero and, combining with Eqs. (23) and (24), we get the steady-state (ss) solutions:

$$I_{ss}^F \propto K_r^F \cdot N_{ss}^{S_1} = Q^F \cdot I_0 \cdot \left[1 - e^{-\alpha d}\right] \tag{27}$$

$$I_{ss}^P \propto K_r^P \cdot N_{ss}^{T_1} = Q^P \cdot N_{ss}^{S_1} \cdot K_{ISC} \tag{28}$$

where Q^F and Q^P are the quantum yields of fluorescence and phosphorescence, respectively. They represent the fraction of chromophores that decay through radiative emission and are expressed by the ratio of the kinetic reaction constant of radiative decay and the sum of the kinetic reaction constant of all the decay processes:

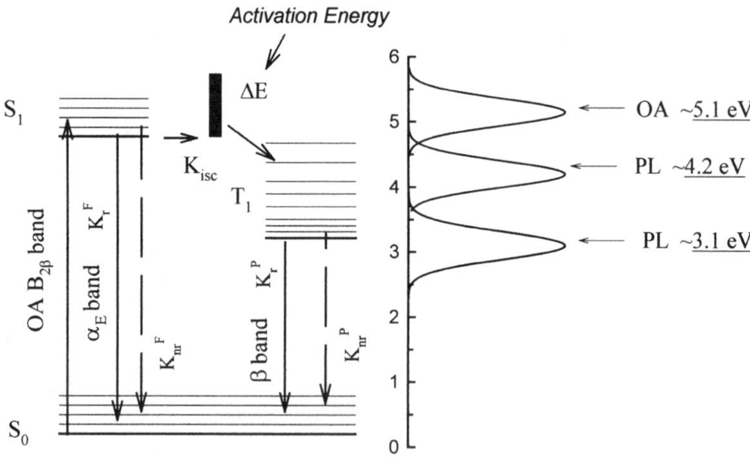

Fig.10 Outline of the electronic levels and of the related radiative and non radiative transitions accounting for the absorption at ~5.1 eV ($B_{2\beta}$) and the emissions at ~4.2 eV (α_E) and ~3.1 eV (β).

$$Q^F = \frac{K_r^F}{K_r^F + K_{nr}^F + K_{ISC}} \tag{29}$$

$$Q^P = \frac{K_r^P}{K_r^P + K_{nr}^P} \tag{30}$$

Now, we wish to address to the thermal evolution of the two PL bands evidenced in Figs.8 and 9. According to the energetic level diagram outlined in Fig.10, the interconversion between the two excited states S_1 and T_1 is related to the intersystem crossing process. As reported in Section 2, the efficiency of this non-radiative mechanism is expected to depend upon the temperature as it arises from the interaction with lattice dynamics (phonon assisted process). In this way, changes induced by the temperature on K_{ISC} could imply changes in the $S_1 \rightarrow S_0$ fluorescence quantum yield Q^F and, according to Eqs. (27) and (28), opposite changes in the $T_1 \rightarrow S_0$ phosphorescence intensity.

We can suppose, as a first approximation, that the intersystem crossing process is governed by the presence of an activation barrier, so that the rate K_{ISC} depends upon temperature according to the *Arrhenius* law [9]:

$$K_{ISC} = K_0 \cdot \exp(-\Delta E / K_B T) \tag{31}$$

where ΔE is the activation energy.

Moreover, since the sum of the integrated intensities of the two PL bands is constant in the temperature range 4.5-300 K (Fig. 9), we can assume that the temperature effects due to K_{nr}^F and K_{nr}^P are negligible in comparison with K_{ISC}. To corroborate this statement we also note that the lifetime of the β emission is temperature independent, as reported in [25], so evidencing that $K_{nr}^P \ll K_r^P$.

In this simplified scheme, the ratio η between the integrated intensities (zero-th moment M_0) of α_E and β is expected to be:

$$\eta = \frac{[M_0]_{\alpha_E}}{[M_0]_\beta} \cong \frac{K_r^F}{K_{ISC}} = A_0 \cdot \exp\left(\frac{\Delta E}{K_B T}\right) \tag{32}$$

with $A_0 = K_r^F / K_0$.

To make this interpretation quantitative, we report in Fig. 11 the temperature dependence of η. The scales used in the graph make easier the comparison with the Eq. (32). As shown in the inset, a simple *Arrhenius* law is obeyed only at high temperature (T>150 K) with $\Delta E = 0.078 \pm 0.003$ eV and $A_0 = (1.4 \pm 0.1) \times 10^{-2}$. At variance, below 120 K, the ratio between the integrated intensities of α_E and β tends to a constant value and the intersystem crossing process becomes temperature independent.

We stress that the knowledge of the η behavior allows us to take into account the temperature dependence of the PL bands trough the ratio between K_{ISC} and K_r^F. In particular, we get that the K_{ISC} value is ~4$\cdot K_r^F$ at T=300 K whereas it is ~0.02$\cdot K_r^F$ for T<120 K. Then, in agreement with the data of Fig.9, the α_E fluorescence quantum yield Q^f, expressed by Eq. (29) without the term K_{nr}^F, decreases by a factor ~5 on increasing the temperature from 4.5 to 300 K.

A final remark concerns the freezing of the intersystem crossing process at T<120 K. We observe that deviations from the *Arrhenius* law of K_{ISC} are known to occur in other

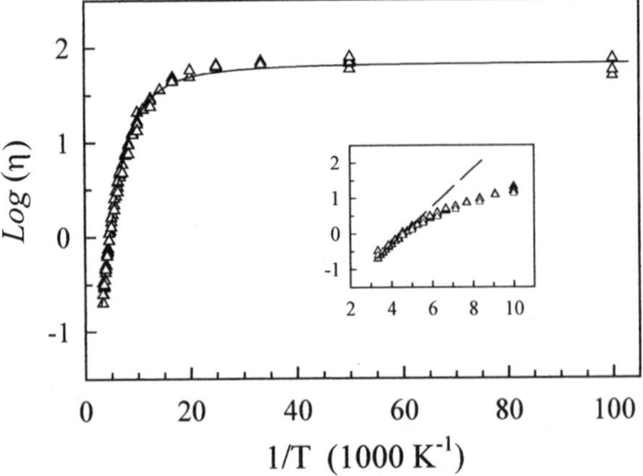

Fig.11 Temperature dependence of the ratio η between the integrated intensities of the α_E and β bands. The continuous line is a guide to the eye. The inset shows the initial part of the curve, in the temperature range 100 to 300 K. In this case, the dashed line results from a fit of Eq. (32) for T >150 K.

systems, like e.g. proteins [29, 30], and interpreted in terms of a distribution of the energy activation barriers ΔE. In this frame, we can qualitatively interpret our results by hypothesizing that among the site-to-site non equivalent defects in natural silica, there is a small fraction of B-active centers with ΔE values sufficient low to populate the T_1 state also at very low temperature.

4.3 Lifetimes measurements

A further experimental analysis of the luminescence activity of B-centers was performed by detecting the kinetic behavior of the α_E fluorescence under pulsed excitation. According to Eqs. (25) and (23), if the exciting light is abruptly switched off (I_0=0) we get:

$$\frac{dN^{S_1}}{dt} = -\left[K_r^F + K_{nr}^F(T) + K_{ISC}(T)\right]\cdot N^{S_1} \tag{33}$$

and the PL signal decays by following the single exponential low:

$$I^F \propto K_r^F \cdot N^{S_1}(0)\cdot e^{-\frac{t}{\tau^F}} \tag{34}$$

where $N^{S_1}(0)$ is the initial population in the excited state and $\tau^F = \left[K_r^F + K_{ISC}\right]^{-1}$ is the lifetime, assuming $K_{nr}^F \ll K_r^F$ and $K_{nr}^F \ll K_{ISC}$. Therefore, lifetime measurements can be useful to find out the temperature influence on the competition between radiative and the intersystem crossing rates arising from the S_1 state.

As shown in Fig.12, the decay of α_E under excitation at 5.0 eV evidences relevant variation on changing the temperature. At T=10 K (a) the time decay is well described by a single exponential law with a lifetime τ=7.3±0.1 ns. We note that at this temperature the relaxation of the excited state S_1 is essentially a radiative process and the lifetime tends to the value $\tau \approx \left(K_r^F\right)^{-1}$, so we can derive $K_r^F \sim 1.4\times10^8$ s^{-1}. On the other hand, the faster

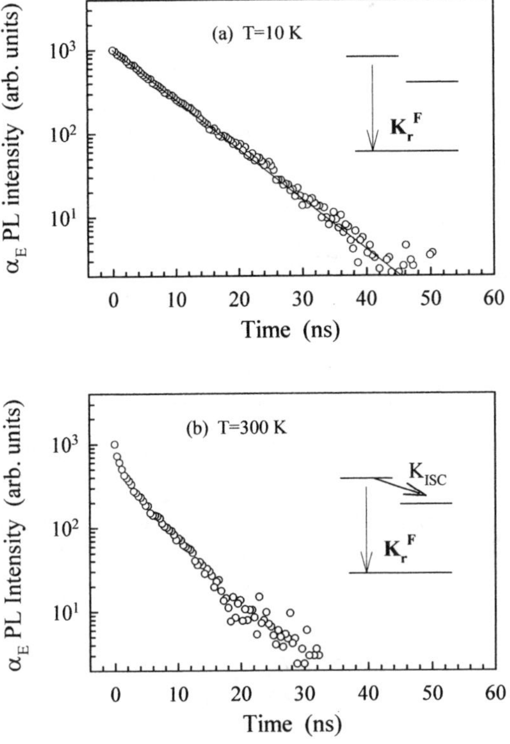

Fig. 12 Time decay of the α_E emission at 4.2 eV observed under excitation at 5.0 eV at T=10 K (a) and T=300 K (b). Taken from *Phys. Rev. B* **60** (1999) 11475-11481, copyright 1999 by the American Physical Society.

decay of curve (b), evidences the contribution of the intersystem crossing process to the S_1 relaxation and the lifetime is expected to decrease down to $\tau = \left(K_r^F + K_{ISC} \right)^{-1}$. We observe that at this temperature, the decay is not a single exponential. Then, it is intuitive to ascribe such a peculiar relaxation kinetic to the distribution of K_{ISC} rates, which are effective at room temperature.

4.4 Temperature dependence of spectral moments: dynamic properties of the silica matrix

To complete the study of the electronic properties of the optically B-active defects and investigate the influence of their environments, we have analyzed the energy moments of the spectral distributions, $f(E)$, relative to the α_E, as a function of temperature [31]. Indeed, as outlined in section 2, the change of the optical band profile on varying the temperature allows to study the local dynamic properties and the heterogeneity of the matrix surrounding the chromophores [11, 12].

Fig.13 shows the PL α_E band excited at 5.0 eV on varying the temperature, from room temperature to 4.5 K, in the I301 sample. In the paragraph *4.2* we have discussed about the marked temperature dependence of the α_E intensity (zero-th moment M_0), which is related to the effectiveness of the phonon, assisted intersystem crossing process. Here we focus

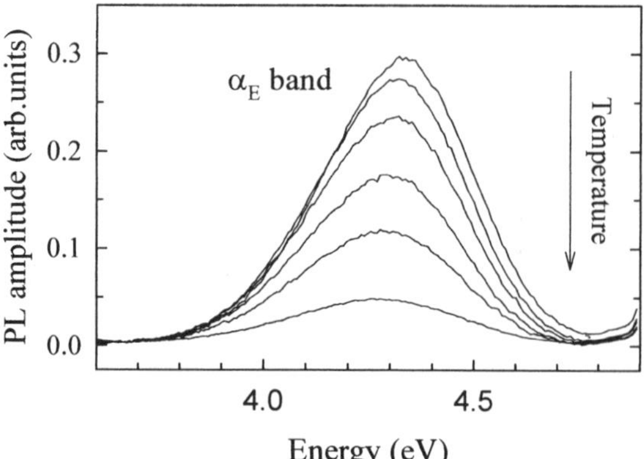

Fig. 13 α_E band profile detected at various temperature. The arrow indicates the changes induced by increasing the temperature, curves refer to 4.5, 100, 150, 190, 230 and 298 K, respectively, from top downwards.

our attention on the first (M_1) and second (M_2) spectral moments, which are calculated from the experimental data according to the following definitions:

$$M_1 = \frac{\int\limits_{-\infty}^{\infty} E \cdot f(E)\,dE}{M_0} \tag{35}$$

$$M_2 = \frac{\int\limits_{-\infty}^{\infty} E^2 \cdot f(E)\,dE}{M_0} - M_1^2 \tag{36}$$

We recall that M_1 is the mean value of the emission energy and measures the position of the PL band and M_2 is the mean square deviation of the spectral distribution $f(E)$ and measures its width.

In Fig.14, we report the thermal behavior of M_1 (a) and M_2 (b). Fig. (a) evidences the red-shift of the α_E emission on increasing the temperature, M_1 decreases from 4.32±0.01 eV at T=4.5 K to 4.26±0.01 eV at room temperature. We note that the temperature dependence of M_1 suggests the occurrence of a strong interaction of the electronic states with the surroundings. In this limit, the transition changes the vibrational frequencies of the ground and excited states (non linear coupling) resulting in temperature effects on the peak position of the optical bands [11, 12].

Fig. (b) shows the broadening of the PL band on increasing the temperature. At low temperature M_2 assumes the value of 0.028±0.001 $(eV)^2$, while it increases up to 0.031±0.001 $(eV)^2$ at room temperature. As already reported in section 2, we can take into account the thermal behavior of M_2, by assuming that the broadening of the PL emission of a single chromophore is dominated by its interaction with the local vibrational modes of the surrounding nuclei. The increase of the temperature causes an increase of the width of the band profile due to population of vibrational levels associated to the low frequency modes in their first excited states S_1. This effect adds to the coupling with the vibrational modes at high frequencies (temperature independent) whose energy distribution determines the shape of the PL bands. We stress that the broadening due to the electron-

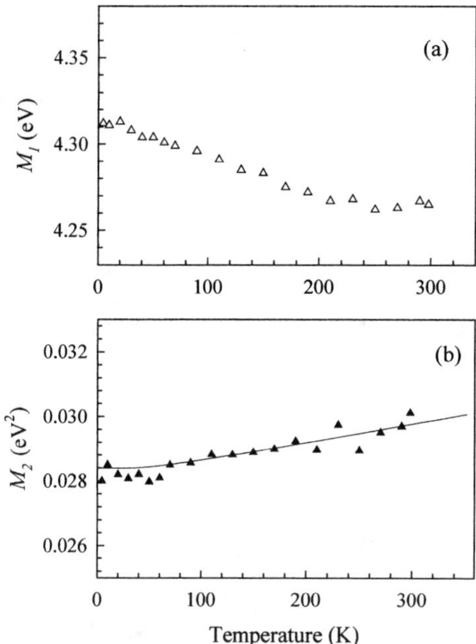

Fig. 14 Thermal behavior of the M_1 (a) and M_2 (b) for the PL α_E band. The continuous line in (b) represents the fitting of Eq. (37). Taken from *J. Non Cryst. Solids* **232-234** (1998) 514-519, copyright 1998 by Elservier Sciences.

phonon coupling is a property of the individual center and it appears to be the same for any center, so we can call it homogeneous contribution. Moreover, in an amorphous material such as the silica, the site-to-site non-equivalence among the point defects embedded in the silica matrix results in an additional inhomogeneous broadening of the PL spectra. We note that the values of the kinetic decay constants, that are of the order of $2 \cdot 10^8$ s^{-1} ($\sim 10^{-6}$ eV), allow us to exclude that the radiative decay rate contributes in a significant way to the observed bandwidth, $\sqrt{M_2} \approx 0.17$ eV. Therefore, we can take into account the total broadening of the α_E emission band by the following equation:

$$M_2(T) = N_l \cdot S_l \cdot h^2 \langle \nu_l \rangle^2 \coth\left(\frac{h\langle \nu_l \rangle}{2K_B T}\right) + \sigma_{stat.}^2 \tag{37}$$

where $\langle \nu_l \rangle$, N_l and S_l are the mean frequency value, the number and the mean linear coupling constant of the bath of vibrational modes at low frequency (*Einstein* oscillator model), and the static width $\sigma_{stat.}$ accounts for the temperature independent broadening due to the *Poissonian* distribution of the high vibrational structures of a single center smeared by the broadening due to the inhomogeneous distribution of the transition energy of the luminescent centers. The continuous lines in Fig. 14 (b) represent the fit of Eq. (37) to the experimental data, the relative parameters being reported in Table 3.

On the basis of these results, we can infer that, in the interaction with the surrounding silica matrix, the B-active defect has access to a low-frequency dynamics ($h\langle \nu_l \rangle \sim 12$ meV). Moreover, we stress that the bandwidth of the α_E band is mainly related to $\sigma_{stat} \sim 160$ meV. In this respect, a direct trial to distinguish between the coupling with high frequency modes and the conformational heterogeneity of the PL centers can be

Table 3 Values of the parameters obtained by fitting the Eq. (37) to the thermal behavior of M_2 for the α_E band. Taken from *J. Non Cryst. Solids* **232-234** (1998) 514-519, copyright 1998 by Elservier Sciences

$N_l \cdot S_l$	$h\langle v_l \rangle$ (meV)	$\sigma_{stat.}$ (meV)
6±1	12±2	160±5

achieved by using site-selective luminescence or spectral hole-burning techniques to suppress the inhomogeneous broadening effect [9]. These site-selective techniques have proven in detecting vibrational structures in non-bridging oxygen hole centers (NBOHC) in silica [32], whereas no such structures have been reported so far for other defects in silica.

5. Conclusions

In this paper, we have reviewed the optical properties of point defects in silica. In particular, our investigation has been focused on the study of PL emissions via their relationship with the absorption, their temperature dependence and their excitation pathways in the UV region. In this respect, the experimental evidences reported in section 4 contribute to outlining a clear picture of the optical features observed in a wide variety of natural silica samples and can be summarized as follows:

i)   Two PL bands α_E at ~4.2 eV and β at ~3.1 eV, related to the OA band at ~5.1 eV ($B_{2\beta}$) characterize the optical activity B associated to native defects present in natural silica.

ii)   The temperature dependence of the PL intensities is governed by a phonon assisted intersystem crossing process occurring at the rate K_{ISC} between the singlet and the triplet excited states from which the emissions α_E and β originate, respectively.

iii)   The departure from the simple *Arrenhius* law of the intersystem crossing thermal behavior and the no-single exponential time decay of α_E at room temperature indicate a distribution of K_{ISC} rates, probably related to different environments of the centers in the amorphous structure of the silica.

iv)   The analysis of the thermal behavior of the α_E emission profile (first and second spectral moment) enables to evaluate quantitatively the parameters of the electron-phonon coupling between the optically B active center and its environment.

As a final remark, the experiments and the related results described in this work prove that the parallel use of different luminescence techniques is a powerful tool to improve the understanding of the optical properties of defects in amorphous solids.

Acknowledgments

The author thanks S. Agnello, R. Boscaino, F.M. Gelardi and M. Leone for their collaboration and useful discussions during the experimental work and the preparation of this manuscript. This work is a part of a National Research Project supported by the Ministero della Ricerca Scientifica e Tecnologica, Roma, Italy.

References

[1]   H. Back, N. Neuroth, The properties of optical glass. Springer-Verlag, Berlin, 1995.

[2]   Devine RAB, Duraud JP and Dooryheé, Structure and imperfections in amorphous and crystalline silicon dioxide. John Wiley & Sons, New York 2000.

[3]   H.S. Nalwa, Silicon-based materials and devices. Academic Press, San Diego 2001.

[4]   G. Pacchioni, L.N. Skuja and D.L. Griscom, Defects in SiO_2 and related dielectrics: Science and technology. Kluwer Academic Publishers, Dordrecht 2000.

[5]   R.K. Watts, Point defects in crystals. J. Wiley & Sons, New York 1997.

[6]   D.L. Griscom, Optical properties and structure of defects in silica glass, *J. Ceramic Society of Japan* **99** (1991) 899916.

[7]   L.N. Skuja, Optically active oxygen-deficiency-related-centers in amorphous silicon dioxide, *J. Non Cryst. Solids* **239** (1998) 16-48.

[8]   G. Herzberg, Molecular spectra and molecular structure: III. Electronic spectra and electronic structure of polyatomic molecules. Van Nostrand Reinhold Company, New York, 1966.

[9]   D.R. Vij, Luminescence of Solids. Plenum Press, New York, 1998.

[10]   B.K. Huang, A. Rhys, Proc. R. Soc. Ser. A **204** 406 (1950)

[11]   A. Cupane, M. Leone, E. Vitrano and L. Cordone, Low temperature optical absorption spectroscopy: an approach to the study of stereodynamic properties of hemeproteins, *Eur. Biophys. J.* **23** (1995) 385-398.

[12]   G. Baldini, E. Mulazzi and N. Terzi, Isotope effects induced by local modes in the U band, *Phys. Rev.* **140** (1965) 2094-2101.

[13]   D. Curie, Luminescence in Crystals. Wiley, New York, 1963.

[14]   M. Weissbluth, Atoms and Molecules. Academic Press, New York 1978.

[15]   G. Hetherington, K.H. Jack, M.W. Ramsay, Phys. Chem. Glasses **6** (1965) 6

[16]   Heraeus Quartzglas, Hanau, Germany, catalogue POL-0/102/E.

[17]   Quartz and Silice, Nemours, France, catalogue OPT-91-3.

[18]   TSL Group PLC, Wallsend, England, Optical Products Catalogue (1996).

[19]   M. Cannas, R. Boscaino, F. M. Gelardi and M. Leone, Stationary and time dependent PL emission of v-SiO_2 in the UV range, *J. Non-Cryst. Solids* **216** (1997) 99-104.

[20]   M. Leone, R. Boscaino, M. Cannas and F. M. Gelardi, Low temperature photoluminescence spectroscopy: Relationship between 3.1 and 4.2 eV bands in vitreous silica, *J. Non-Cryst. Solids* **216** (1997) 105-110.

[21]   M. Cannas, M. Barbera, R. Boscaino, A. Collura F. M. Gelardi and S. Varisco, Photoluminescence activity in natural silica excited in the vacuum-UV range, *J. Non-Cryst. Solids,* **245** (1999) 190-195.

[22]   M. Leone, S. Agnello, R. Boscaino, M. Cannas and F. M. Gelardi, Conformational disorder in vitreous systems probed by the photoluminescence activity in SiO_2, *Phys. Rev. B* **60** (1999) 11475-11481.

[23]   R.Tohmon, H.Mizuno, Y.Ohki, K.Sasagane, K.Nagasawa and Y.Hama, Correlation of the 5.0- and 7.6-eV absorption bands in SiO_2 with oxygen vacancy, *Phys.Rev. B* **39** (1989) 1337-1345.

[24]   A. Anedda, F. Congiu, F. Raga, A. Corazza, M. Martini, G. Spinolo and A. Vedda, Time resolved photoluminescence of α centers in neutron irradiated SiO_2, *Nucl. Instr. Meth. Phys. Res. B* **91** (1994) 405-409.

[25]   L.N.Skuja, Isoelectronic series of twofold coordinated Si, Ge, and Sn atoms in glassy SiO_2: a luminescence study, *J. Non-Cryst. Solids* **149** (1992) 77-95.

[26]   M. Leone, R. Boscaino, M. Cannas and F. M. Gelardi, The landscape of the excitation profiles of the α_E and β emissions bands in silica, *J. Non-Cryst. Solids* **245** (1999) 196-202.

[27]   R. Boscaino, M. Cannas, F. M. Gelardi and M. Leone, Spectral and kinetic properties of the 4.4 eV photoluminescence band in SiO_2: Effects of γ-irradiation, *Phys. Rev. B* **54** (1996) 6194-6199.

[28]   R. Boscaino, M. Cannas, F. M. Gelardi and M. Leone, Experimental evidence of different contributions to the photoluminescence at 4.4 eV in synthetic silica, *J. Phys. Condens. Matter* **11** (1999) 721-731.

[29]   H. Frauenfelder, F. Parak and R.D. Young, Conformational substates in proteins, *Annu. Rev. Biophys. Chem.* **17** (1988) 451-479.

[30]   H. Gilch, R. Schweitzer-Stenner, W. Dreybrodt, M. Leone, A. Cupane and L. Cordone, Conformational substates of the Fe^{2+}-His F8 linakage in deoxymyoglobin and hemoglobin probed in parallel by the Raman line of the Fe-His stretching vibration and near infrared absorption of band III *J. Quantum Chem* **59** (1996) 301-313.

[31]   M. Leone, M. Cannas and F. M. Gelardi, Local dynamical properties of vitreous silica proped by photoluminescence spectroscopy in the temperature range 300-4.5 K, *J. Non-Cryst. Solids* **232-234** (1998) 514-519.

[32] L.N. Skuja, T. Suzuki and K. Tanimura, Site-selective laser spectroscopy studies of the intrinsic 1.9 eV luminescence center in glassy SiO_2, *Phys. Rev. B* **52** (1995) 15208-15215.

GNSR 2001
G. Messina and S. Santangelo (Eds.)
IOS Press, 2002

Micro-Raman characterisation of μc-Si:H film deposited by PECVD, μc-SiC:H deposited by ECR-CVD and 6H-SiC wafer

S. Ferrero[*], F. Giorgis, C.F. Pirri, P. Mandracci, G. Cicero, C. Ricciardi

*INFM and Physics Department, Polytechnic of Torino,
C.so Duca degli Abruzzi 24, 10129 Torino (ITALY)*
[*]*Phone +39 011 5647305 Fax +39 011 5647399 E-mail: sergio.ferrero@polito.it*

Abstract. Micro-Raman measurements have been performed on μc-Si:H films deposited by Plasma Enhanced Chemical Vapour Deposition (PECVD), μc-SiC:H deposited by Electron Cyclotron Resonance (ECR-CVD) and 6H-SiC wafer irradiated with Ar and Kr ions before and after thermal annealing. The experimental data are presented and discussed evidencing as this spectroscopy is a powerful tool for the structural characterization of both thin films and massive semiconductors.

1. INTRODUCTION

Micro-Raman spectroscopy was used in order to investigate on the phase transition from amorphous to microcrystalline silicon in films growth by rf PECVD induced by laser annealing performed by a frequency doubled Q-switched Nd:YAG laser at 532 nm.
Moreover we characterized the vibrational spectra of polycrystalline SiC films growth on silicon substrates by ECR-CVD. The films are grown in a plasma of silane/methane diluted in hydrogen under different conditions of temperature and power. The Raman spectra have shown the appearance of a peak at 780 cm^{-1} due to Si-C bonds and at the same time the disappearance of the clusterization of C-C bonds at 1500 cm^{-1} for samples deposited under optimised conditions.
Finally Micro-Raman technique was used to investigate recrystallization process in 6H-SiC wafer. In detail this wafers was damaged with different ions (Kr He) and later treated with thermal annealing in a hot furnace, for T_a=1000°C they shown the recovering of the damage, valuable through the Raman spectra, which evidences a structure similar to undamaged samples.

2. EXPERIMENTAL DETAILS

The measurements described in this paper were conducted on three different set of samples: μc-Si:H film grown on Pyrex substrates, μc-SiC:H growth on crystalline silicon substrate and SiC wafers supported by Cree Research Inc. All the samples were characterized with a Micro-Raman spectrophotometer [Renishaw system] with Ar$^+$ laser source (λ=514.5 nm) and a hot cell, which allow in situ measurement during annealing process (rt up 1500 °C). The spectral resolution of the system is 1 cm^{-1} and the lateral spatial resolution is 1 μm, with a depth of field of about 2 μm.

3. RESULTS AND DISCUSSION

Hydrogenated microcrystalline silicon (μc-Si:H) has received attention in the last years because of its superior properties such as high stability under light exposure and high carrier mobility compared to hydrogenated amorphous silicon. For these reasons μc-Si:H materials have been used in thin film device technology as active layer in thin film transistors and solar cells.

μc-Si:H films have been deposited by rf PECVD in a plasma of silane diluted in hydrogen. The silane concentration has been varied from 1% to 10%.

Hydrogenated silicon layers have been irradiated by a frequency doubled Q-switched Nd:YAG laser at 532 nm (FWHM = 7 ns), repetition rate of 10 Hz and excitation energy density from 0.15 to 1.5 mJ/cm^2. They show a change to amorphous vs crystalline phase.

In fig.1 it is represented the spectra of a sample before and after damaging. We can observe that, before damaging, Raman spectra show the typical broad band due to amorphous Si phase centered at 480 cm^{-1} with a full width at half maximum $\Delta\omega$ =90 cm^{-1} (TO-LO phonon branch). After laser damage, the spectra show the shift of the peak at 520 cm^{-1} with a full width at half maximum $\Delta\omega$ = 8 cm^{-1} (TO-LO phonons at the center of the Brillouin zone). We have also represented the spectra of crystalline silicon and we can observe as c-Si and μc-Si:H have a similar spectrum.

The other material investigated consists of polycrystalline silicon carbide. Silicon carbide is a wide band gap semiconductor very interesting, which has been intensively studied, in the recent years, due to its physical properties such as high breakdown field, high

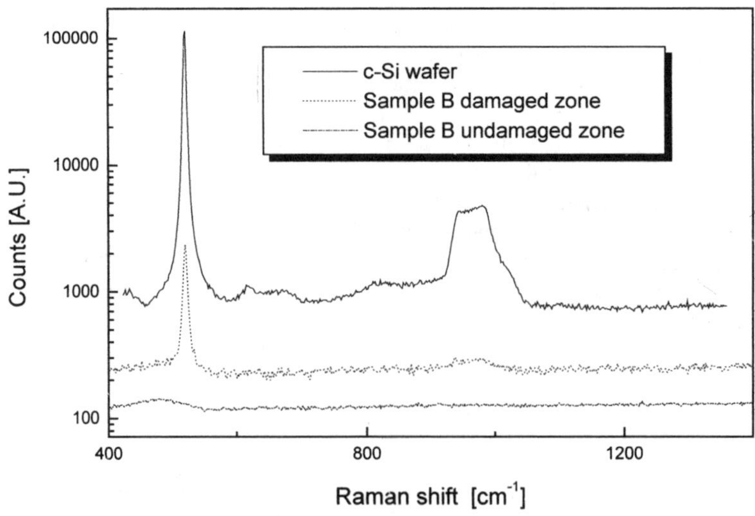

Fig.1 Comparison between m-Si:H sample before and after annealing and crystalline silicon.

saturated drift velocity and high thermal conductivity. This characteristics make SiC a very good candidate for the applications in which high temperature, high radiation intensity, high voltage or high power dissipation are involved. Typical applications of SiC are temperature sensors, nuclear radiation detectors, UV detectors, microwave devices and power devices.

Although of the high potential of this material for the use in the electronic industry, the SiC technology shows some limitations and requires further study in order to obtain electronic devices of the same quality standards as Si technology.

Both in its microcrystalline and polycristalline structure a large variety of applications can be planned. Among the several SiC polytipes, 3C-SiC is interesting since it can be grown on Si substrates.

In this work we report on microcrystalline SiC films grown by ECR-CVD in $H_2+SiH_4+CH_4$ gas mixtures at substrate temperature in the range of 850-1000°C over 4-inch crystalline silicon wafers. The evolution of the resonance yielded by carbon clusterization in graphitic phase of approximately 1400-1500 cm^{-1} is shown in Fig.2a: by decreasing the methane-silane ratio (R), a decrease of carbon clusters can be inferred, due to the stoichiometry approaching (Si/C=1). No Si clusterization was detected through the typical Raman signal at 480 cm^{-1}. On the other side, Fig. 2b shows as by

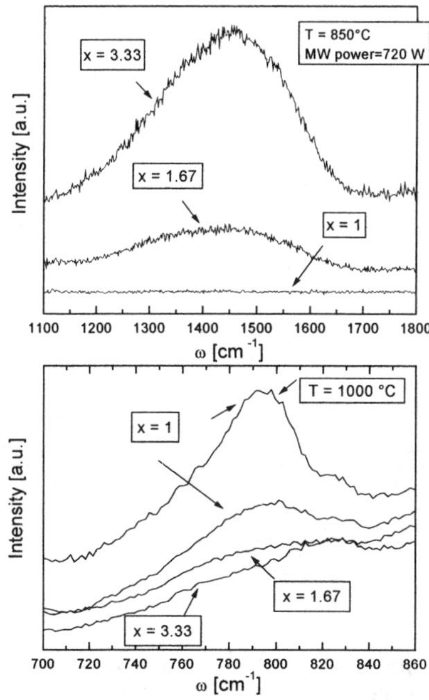

Fig.2a Evolution of the resonance yielded by carbon clusterization in graphitic phase of approximately 1400-1500 cm^{-1} where x is the methane-silane ratio

Fig.2b Evidence of the peak due to transversal optic phonons of policrystalline SiC matrix of approximately 800 cm^{-1}, by decreasing R and by and by enhancing the substrate temperature up to 1000°C.decreasing

R and by enhancing the substrate temperature up to 1000°C, the peak due to transversal optic phonons of policrystalline SiC matrix of approximately 800 cm^{-1} becomes evident. Raman analysis was also used to study the effect of recrystallization in Silicon Carbide wafer supported by Cree Research Inc.

One limitation for the industrial use of silicon carbide is that selective doping of SiC layers requires the use of ion implantation, since the diffusion coefficients for the doping species are too small for practical applications. Thus it is important to understand the evolution of the material after thermal annealing, necessary after ion implantation aimed to reach a defect recovering.

We have implanted SiC wafers with Ar ions and Kr ions, the sample after the implantion were annealed and it was observed a good recrystallization of the sample which evidences a structure similar to undamaged samples.

Two spectra of a sample of Silicon Carbide wafer are represented in figure 3, the wafer was implanted with Kr^{+} ions and we can observe the damage of the surface where a broad spectrum ascribed to the included amorphization is superimposed to the typical features of the crystalline phase. The same sample was then annealed at1000°C in an hot furnace for 1 h, the relative spectrum shows the recoverage of the damage.

MicroRaman analysis gives also information with different depth of the materials as it is possible to observer in fig. 4 where there are different spectra of the same sample recorded at different depths, by focusing the laser beam in different planes, so probing the sample structure from the surface to bulk regions.

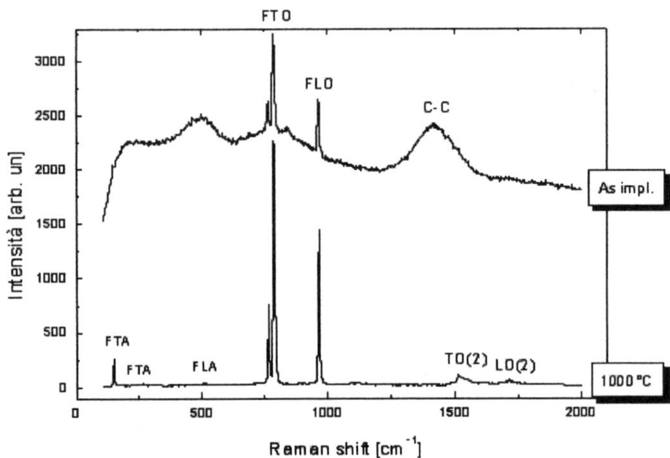

Fig.3 Evolution of a sample of Silicon Carbide wafer implanted with Kr^{+} ions and SiC implanted with Kr^{+} and annealed at 1000°C for 1 h.

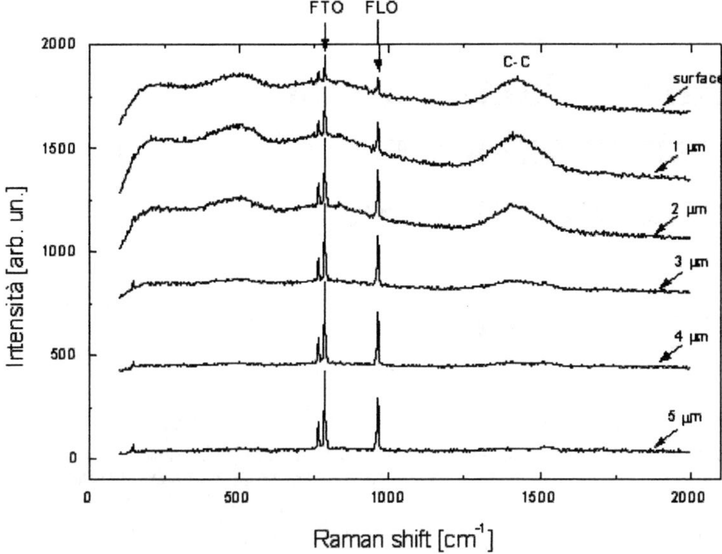

Fig.4 Raman spectra of unannelead implanted SiC at different focusing.

4.   SUMMARY AND CONCLUSIONS

MicroRaman spectroscopy gives important information in structural properties of semiconductors.

In particular such a technique is able to eliminate the contribute of the substrate in thin films analysis, as shown in Raman spectra of μc-SiC:H films growth on silicon substrates and in spectra of Kr implanted SiC wafers.

References

[1] "Confocal micro-Raman characterization of SiC epilayers"
    Ran-Liu
    Optical Microstructural Characterization of Semiconductors. Symposium (Materials Research Society Symposium Proceedings Vol. 588). Mater. Res. Soc, Warrendale, PA, USA; 2000; xi+333 pp.,p.233-8, 2000

[2] "Confocal micro-Raman characterization of lattice damage in high energy aluminum implanted 6H-SiC"
    Campos-FJ; Mestres-N; Pascual-J; Morvan-E; Godignon-P; Millan-J
    Journal-of-Applied-Physics. vol.85, no.1; 1 Jan. 1999; p.99-104, 1999

[3] "Micro-Raman characterization of crystallinity of laser-recrystallized silicon films on SiO/sub 2/ insulators"
    Mizoguchi-K; Nakashima-S; Sugiura-Y; Harima-H
    Journal-of-Applied-Physics. vol.85, no.9; 1 May 1999; p.6758-62, 1999

GNSR 2001
G. Messina and S. Santangelo (Eds.)
IOS Press, 2002

Micro-Raman investigation in mixed oxide films TiO₂-V₂O₅ grown by sol-gel method

E. Cazzanelli[a], S. Capoleoni[a], L. Papalino[a],
R. Ceccato[b] and G. Carturan[b]

[a]*Dipartimento di Fisica, Università della Calabria and
Unità INFM Cosenza; 87036- Arcavacata di Rende, Cosenza, ITALY*
e-mail:cazzanelli@fis.unical.it phone:+39-984-496114 Fax: +39-984-494401

[b]*Dipartimento di Ingegneria dei Materiali, Università di Trento,
38050- Mesiano di Povo , Trento, ITALY*

Abstract. Transparent thin films of mixed oxides, TiO_2 -V_2O_5 exhibiting both ionic and electronic conduction, have been grown by following the sol-gel route, in different compositions, and under different synthesis conditions. Micro-Raman mapping experiments, performed on these samples, show different degrees of structural and chemical homogeneity, which could be related to the synthesis procedures. In the more inhomogeneous samples, laser-induced crystallization from polyvanadate-containing gel to crystalline vanadium pentoxide can be observed.

1. Introduction

Mixed systems TiO_2-V_2O_5 are promising candidates for the improvement of counter-electrodes in electro-chromic cells. In principle, they must give a good compromise between the performances of the separate components, in terms of transparency, chemical and electrochemical stability and charge capacity [1-4].

Furthermore, a new application has been tested: after observations in cells of Nematic Liquid Crystal (NLC), containing a layer of WO_3 deposited on a ITO electrode [5-7] we expect a similar rectifying effect for others oxide films, having mixed conductivity (ionic+electronic), on the electro-optic response of the NLC cell between two crossed polarisers. In fact mixed titania-vanadia inserted films into a Nematic Liquid Crystal (NLC) modified cell, exhibit such modification with respect to the standard NLC cells, containing only purely electronic conductors[4].

The degree of crystalline order of the films is quite important to determine ionic and electronic conductivities, relevant for the electro-chromic applications, as well as for the insertion in NLC cells. In general, for electro-chromic materials, better ionic diffusion is obtained for the amorphous system, while, on the contrary, the electronic conductivity has been found to increase after thermal treatments promoting the size increase of the crystal-like ordered domains [8].

The use of different deposition methods and the further application of thermal treatments to the as-grown films allows a great variation of the structural order in the films, affecting their electrical, chemical and optical properties, quite relevant for the different kind of applications.

Among the various deposition techniques, the sol-gel method assures a better homogeneity of the starting precursors, thus a greater stability of the deposited films. In the present study we want to investigate as the changes of some chemical parameter, like the pH of hydrolysis solution, can generate films with different structural configurations, on microscopic as well as on nanoscopic scale.

The vibrational spectroscopy performed on a "local" scale by using a proper microscope interfaced to a Raman spectrometer allows to obtain a map of the chemical and structural variations in these thin film systems and to follow the structural evolution on a local scale of the order of micrometers, with detailed description of the space variation and temperature evolution not possible by using the standard XRD equipment.

2. Sample synthesis

V_2O_5/TiO_2 sols were prepared by following the alkoxide route: titanium tetrabutoxide, $Ti(OBu)_4$, and vanadium oxo-triisopropoxide, $VO(OPr^i)_3$, in different molar ratio, were allowed to react with acetic acid in a 1/1 molar amount; after the weak exothermic reaction took place, 2-propanol was added as a solvent. Hydrolysis reaction was carried out after 2 hours using acidified water (HCl). Different pH values were tested for various samples. In the present work samples obtained after hydrolysis with pH=0 and pH=1 are analysed and compared.

After a couple of hours, layers of the thickness of about 120 nm were deposited on ITO-glass (20 $\Omega/\square$, Unaxis GmbH) substrates by means a dip-coating apparatus. To have more thick films, depositions of new layers, of about the same thickness, were repeated.
After the deposition, an exsiccation procedure was performed, at temperature not exceeding 120 °C, to expel most of the organic solvents residue from the films.

3. Experimental

DTA.-TG were performed on a Netzsch STA 409 instrument, gel samples were heated up to 900 °C with an heating rate of 10°C/min

A Raman microprobe ISA-Jobin-Yvon Labram was used equipped with a CCD detector and a He-Ne laser (632.8 nm emission). In all the experiments the power of the laser out of the objective, a 100x Mplan Olympus with NA of 0.90, was about 5 mW, and the focused laser spot had about 2-3 μm of apparent diameter. By assuming such values of spot diameter and laser power, we have for the unfiltered laser beam an irradiance of the order of 100 kW/cm^2 on the spot. When the 0.3 OD filter is used, such value is decreased of about 50%; for the filter 0.6 the effective power on the spot is about the 25% of the unfiltered power.

For the measurements on the mixed titania-vanadia films a special X-Y stage for automatic spectral mapping was used. The distances between the centre of any spot and the next one on the line were of 5μm, to avoid overlapping of the heating effect (the gaussian spatial distribution of the light intensity in the laser beam insures that points close to the diameter assumed above have a real light intensity quite lower than in the centre of the focus). Also the distance between lines was of the same order.

4. Results and discussion

4.1 Structural evolution

Because the main applicative interest of this research was to investigate the properties of the mixed vanadia-titania film as a function of the composition, and the samples having lower fraction of vanadia were preferred in principle for economic and environmental reasons, compositions ranging from pure titanium oxide up to a 1:1 atomic ratio between Ti:V were explored. The as-deposited mixed films obtained via the sol-gel route are structurally non-crystalline, as demonstrated by usual XRD analysis [4]. In a DTA study on the gels having different composition an exothermic peak can be always observed, due to the full crystallization of the gels into the standard anatase phase of TiO_2: it occurs at about 400 °C for pure titania films. For mixed titania- vanadia samples the structural evolution becomes more complex: the crystallization into the standard TiO_2 phases is inhibited, and separate crystallization of small nuclei of vanadium pentoxide can be expected, however, the main effect of the presence of vanadium oxide is the change of the thermal evolution of the mixture: the crystallization peak shifts toward increasing temperatures for increasing content of vanadia, as can be seen from Table 1, while some nucleation of a metastable monoclinic phase can be observed in XRD spectra of gels with atomic ratio Ti:V=1:1, fired at 400 °C [3]

Tab.1 Crystallization temperatures of the gels, determined by DTA, vs. the Ti/V atomic ratio.

Sample composition	T crystallization peak (°C)
pure TiO$_2$	**412**
Ti/V (25/1)	**430**
Ti/V (10/1)	**470**
Ti/V (1/1)	**565**

4.2 Spectroscopic analysis

A more detailed description of the existing structural configurations of the mixed films, on a local scale of the order of the micrometers, can be obtained by using the micro-Raman spectroscopy, coupled to a direct visual microscopic investigation.

Mixed films with various compositions are shown in the following: pure titanium oxide film, a low vanadium content film with atomic ratio Ti:V=25:1 and two films with high vanadia content, having atomic ratio Ti:V=1:1. In this latter case, the effects of pH during hydrolysis on the film structural configuration was also investigated, henceforth a film grown after pH=1 treatment was compared to a film obtained via a pH=0 hydrolysis.

The films dip-coated with gels of pure titanium oxide, as deposited, present to a visual investigation a homogeneous surface. A determination of the structural order by mean of the Raman spectroscopy is quite difficult, because the scattering intensity of the amorphous film is very poor, and the resulting spectra are undistinguishable from the spectrum of the ITO substrate. This is not due to the small amount of deposited material, because an appreciable Raman signal is found for the films having the same thickness, but crystallized after annealing at 400 °C: three well distinct peaks at 400 cm^{-1}, 515 cm^{-1} and 638 cm^{-1} are observed, in agreement with the reported spectra of crystalline TiO_2, anatase phase [9].

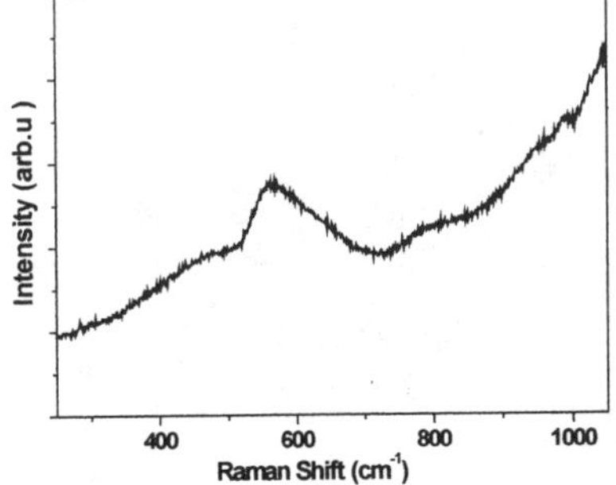

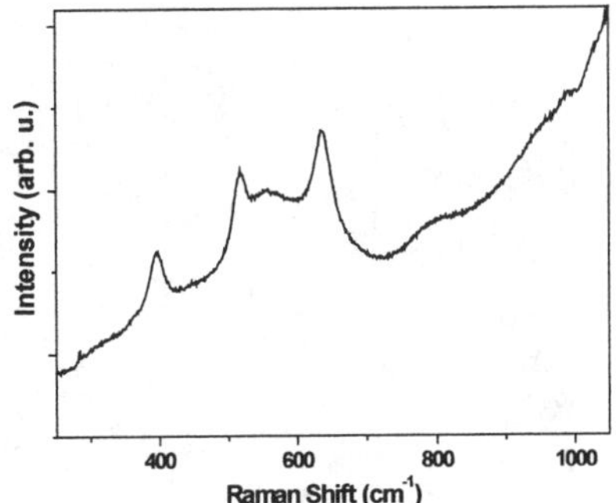

Fig.1a Raman spectrum of a pure TiO$_2$ film, grown by sol-gel route, as-deposited.

Fig. 1b Spectrum of an identical film, after annealing at 400 °C.

In fig. 1a and 1b are shown the spectra collected on a TiO$_2$ film, as-deposited, and from another identical film annealed at 400 °C, respectively. The spectra of annealed samples do not change in shape when the laser beam power is changed, thus demonstrating that the crystal phase is not due to laser induced crystallization. In conclusion, the spectrum of fig.1b is consistent with the DTA and XRD findings of almost complete crystallization of pure titanium oxide at 400°C, and the amorphous film of titanium oxide is characterized by a very small Raman cross-section.

A similar result can be found in a film deposited from a gel with atomic ratio Ti:V=25:1, as-deposited; no appreciable Raman signal assignable to the titanium oxide phases is observed in the homogeneous regions, beside the spectrum of glass and ITO coating. On the contrary, a low frequency contribution with maximum at about 210 cm^{-1} and a band centred at about 430 cm^{-1}, not assignable to the substrate, are observed along the lines separating the homogeneous regions, appearing to the direct microscopic observation as a lambda-shaped image. Thus, the micro-Raman spectral mapping of the surface shows a good correspondence with the microscopic images. In Fig. 2a and 2b are reported the micro-Raman mapping and the corresponding visual image, respectively. The former is collected from spots spaced each other of 5 μm, and represents on a false colour map the intensity ratio between the 180-250 cm^{-1} and the 250-320 cm^{-1} spectral regions, but the choice of other appropriate spectral slices in the ratio gives very similar results.

The spectra corresponding to various colours in the Raman map are reported in Fig 3. The more interesting is the one labelled as 1. The presence of broad bands, overlapping the substrate spectrum, suggests an amorphous aggregation and is consistent with similar findings [10] on pure sol-gel derived titania, treated at temperatures well below 400 °C, where the crystallization into anatase phase occurs. The overall spectral shape does not correspond to that of the common crystalline phases of TiO$_2$ (anatase or rutile), characterized by sharp, well defined peaks: three sharp peaks at 400, 515 and 638 cm^{-1} for the anatase crystal [8] in the frequency region here analysed, and three broader bands, peaked at 235, 447 and 612 cm^{-1} for the rutile structure [11]; no correspondence is observed also for less common brookite phase [12]. At higher frequency, only a very weak peak at about 950 cm^{-1} can be detected, constituting an indication for nuclei of a separate crystallization of the vanadium pentoxide in the regions where is appreciable the spectral contribution of the mixed film.

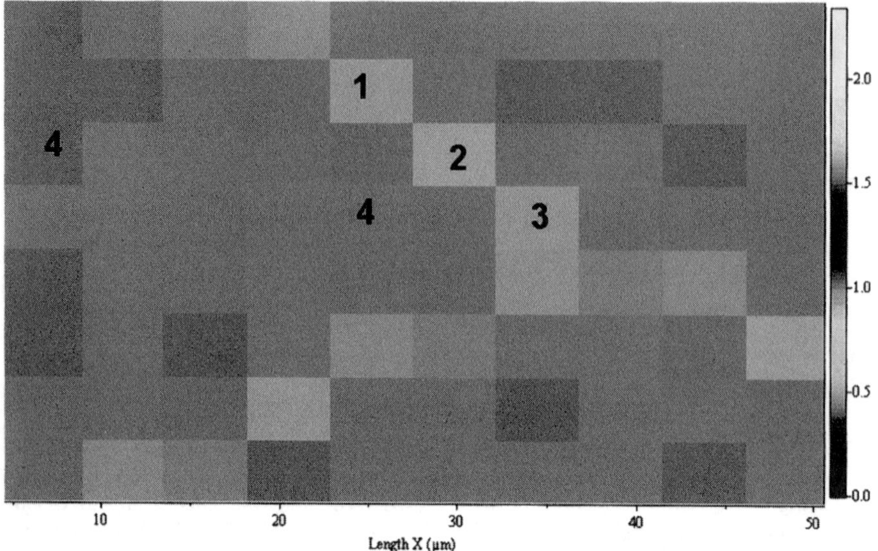

Fig.2 (a) Raman spectral map of a region of about 200x200 microns of a sol-gel dip-coated film of titania-vanadia mixed composition, having atomic ratio Ti/V=25/1. False colours scale represents the intensity ratio between the 180-250 cm^{-1} and the 250-320 cm^{-1} regions.

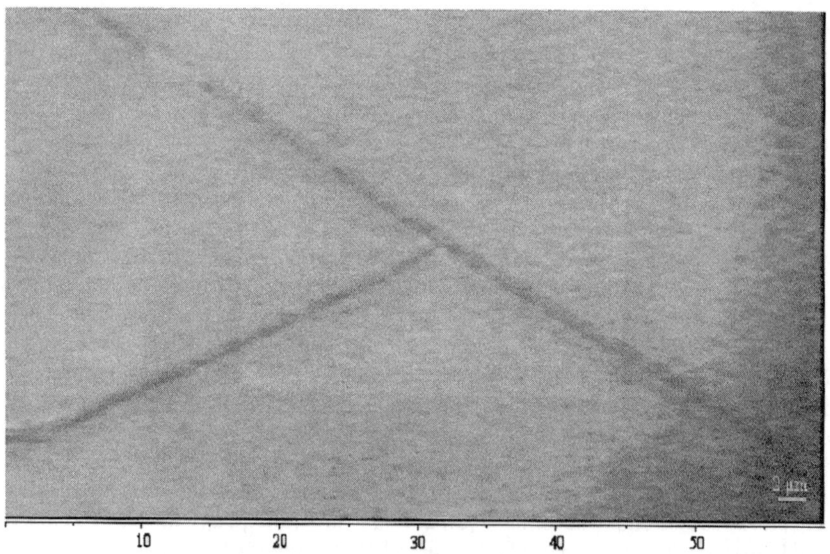

Fig.2 (b) Corresponding visual map of the same film region.

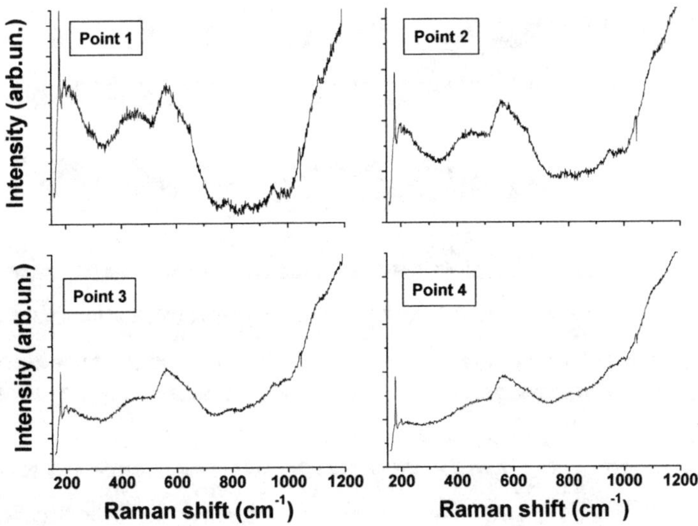

Fig.3 Different kinds of Raman spectra, labels indicate the correspondence to the points of the spectral map of Fig. 2a.

However, the spectral map is dominated by the kind of spectrum labelled as 4, corresponding mostly to the substrate contributions, while the zones labelled as 2 and 3 represent intermediate configurations, with a decreasing detectable Raman contributions of the film.

When the atomic ratio Ti:V=1:1 is reached, more complex findings are found, and the homogeneity of the film samples at microscopic level, as well as the structural conformation, appear to depend on the details of the sol gel process, in particular on the pH of the hydrolysing solution.

A micro Raman investigation has been performed on mixed films of Ti/V=1/1 composition, grown by a sol-gel route described above with a value of pH=1 during the hydrolysis step. The result seems to be a separate crystallization of the vanadium pentoxide in some spots of the film, even using low laser powers, and different spectral pattern appearing to the micro-Raman mapping. In fact, the micro-Raman study of an extended area shows the prevalence of amorphous configurations, with strong fluctuations between the relative spectral contributions of titania and vanadia and also regions with no appreciable Raman signal, indicating an almost pure amorphous titanium oxide deposition.

As it can be seen in Fig 4a, a large (200 μm X 200 μm) spectral map has been collected from a region of a mixed film by using unfiltered laser beam, with a separation between any focused spot of 5 micrometers.

To have a good mapping of the chemical and structural fluctuations, the intensity of the spectral slice 830-970 cm⁻¹ is represented, corresponding to a typical band of polyvanadate chains [13], and it has been normalized by the intensity of the 1000-1050 cm⁻¹ spectral slice, to discriminate from the spectra of the ITO substrate, having a monotonically increasing intensity in that spectral range. White spots correspond to a strong polyvanadate band, while dark regions exhibits spectra very close to that of the substrate (see spectrum A, Fig.5).

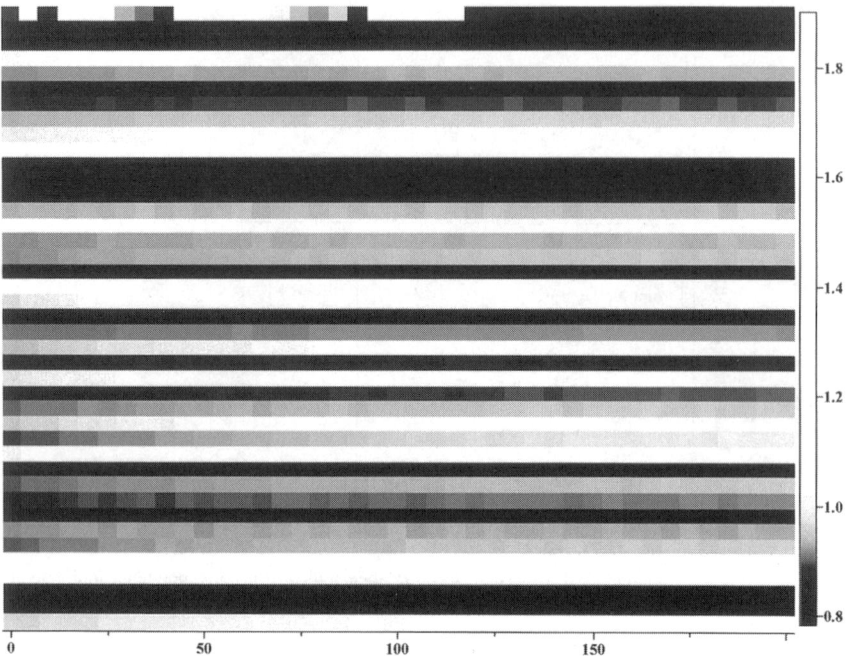

Fig.4 (a) Large spectral map, 200x200 microns, of the film sample with Ti/V atomic ratio 1/1, collected by using full laser power and a 30 s integration time for any spot. Any square of 5x5 micron corresponds to a Raman spectrum collected for 30 s. The colour scale represents the ratio between the 800-1000 cm^{-1} and the 1010-1050 spectral regions, in such a way to have white spots for strong and well defined Raman bands at 910 cm^{-1}.

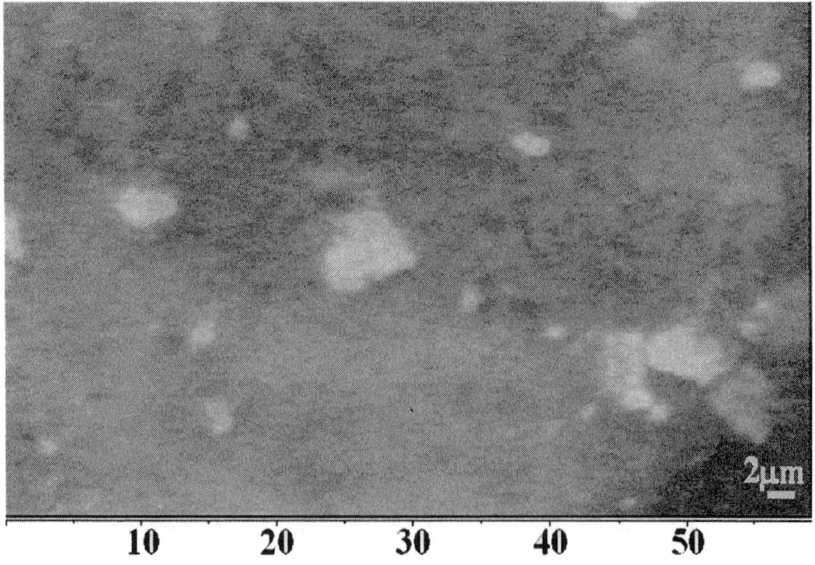

Fig.4 (b) Corresponding visual map, on a smaller region.

Spectrally homogeneous structures are observed along the X axis, corresponding to the direction of film extraction, while strong variations of the spectral shape, reflecting differences in the chemical or structural character, are observed along the Y axis. It is worth to remark that all structures are not observed in the direct microscopic image, reported in Fig. 4b. Some characteristic spectral patterns observed in the various region of the map can be observed in Fig. 5. Along the stripes showing strong Raman signal of the film four broad bands are observed, centred at about 235, 430, 670 and 920 cm^{-1} (see spectra 5D and 5E). The first three bands are consistent with those previously reported in gel solutions of pure titania after moderate annealing [10] and with the present results on the film with dominant titania composition, if the strong contribution of the substrate is accounted for. On the contrary, the broad band centred at about 910-920 cm^{-1}, stronger and stronger in spectra 5B, 5C and 5D, is related to the occurrence of polyvanadate chains in an amorphous configuration. [13]

The crystallized spots give the kind of spectrum depicted in Fig. 5F: here the narrow peaks of the polycrystalline V_2O_5 [14, 15] are superimposed to a background of broad bands typical of amorphous systems. In particular, the ratio between the Raman intensity of the highest frequency peak at about 990 cm $^{-1}$, usually assigned to the strongest V=O bond in crystalline vanadium pentoxide [16], and the polyvanadate band centered at 915 cm^{-1} is a good indicator of the extent of crystallization process. In Fig. 6 the evolution of the Raman spectrum due to laser-induced crystallization is shown, starting from low laser powers (filtered laser beam) up to the maximum available intensity (unfiltered laser). In the top spectrum the stable saturation result after 35 minutes of laser exposure is shown and the peaks of crystalline V_2O_5 reach the maximum intensity with respect to the broad background spectrum assignable to the amorphous mixture titania-vanadia. The existence of some spot sensible to the laser effect, while most of the surface do not change, even under excitation of the unfiltered laser, is an indication of spatial inhomogeneity of the film and structural instability. Similar effects have been found for sputtered film of WO_3, by using laser powers of the same order [17].

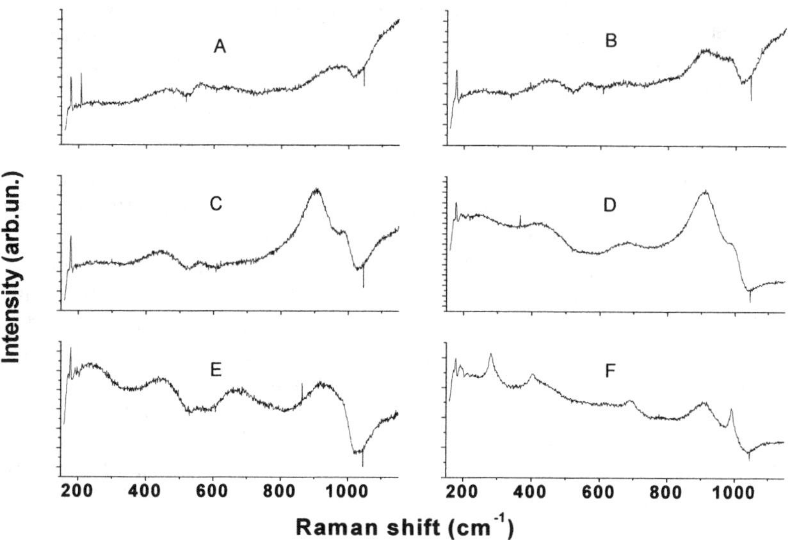

Fig.5 Typical Raman spectra from different spots on the map of Fig.6; the 6 classes here shown represent most of the spectra corresponding to the spots of the map (see text), and are mainly collected along the vertical line corresponding to X=110 μm.

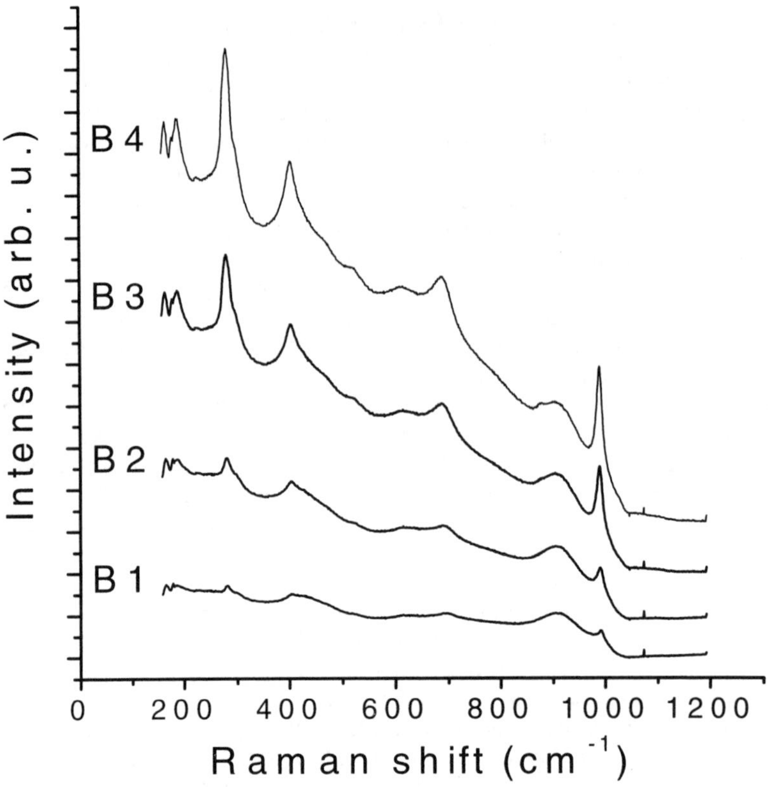

Fig.6 Series of Raman spectra collected at increasing laser powers, in a random spot, labelled B, of the same sol-gel grown film with atomic ratio Ti/V=1/1 appearing in the maps of Fig.4.the integration times are chosen to get a constant product with the laser intensity, for a better scaling of the spectra. B1, B2 and B3 correspond to spectra from spot B, performed for 25% (OD 0.6), 50% (OD 0.3) and full laser intensity, respectively. Spectrum B4 is collected with the same conditions of B3, but after 35 m of continuous exposure to full laser power.

The change of acidity of the hydrolysing solution during the sol-gel route, from the value pH=1 used for the previous samples to the value pH=0 strongly affects the structural conformation and the stability of the deposited films.

The micro Raman mapping of an "as deposited" film, with Ti:V=1:1 atomic ratio, but obtained from this modified sol-gel route, shows a quite different pattern. As it can be seen in Fig.7 no more stripes of different spectral shape are observed. A weak Raman signal from the film can be observed (see Fig. 8), in particular in the high frequency region (800-1000 cm^{-1}), assignable to the residual presence of polyvanadate chains and to a diffuse separate crystallization of vanadium pentoxide at nanoscopic level. However, no spots characterized by extended laser-induced crystallization of V_2O_5 are observed. In general, the observed variations of the intensity ratio between are small, due mainly to the fluctuation of the background.

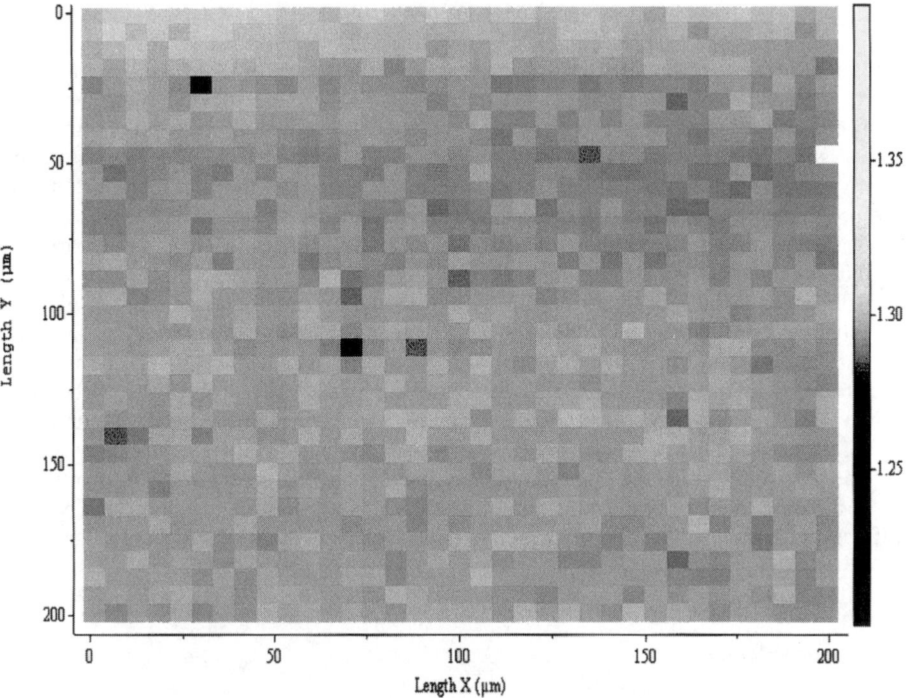

Fig.7 Micro-Raman mapping of a film as-deposited, atomic ratio Ti:V=1:1, from gel grown with pH=0. The false colour map represents the same intensity ratio as in Fig. 4(a).

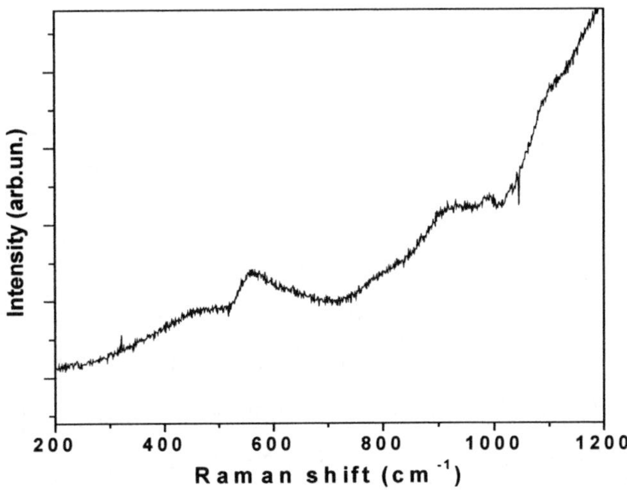

Fig.8 Typical spectrum collected in the Raman map of Fig. 7. The spectral shape does not change qualitatively through all the map.

5. Conclusions

The micro-Raman mapping has been applied to a series of oxide films of mixed compositions TiO_2-V_2O_5, grown by using a sol-gel route, and previously characterized [3,4] by various techniques. The increasing content of vanadium oxide, from nothing up to the atomic ratio 1:1, seem to increase the spatial inhomogeneity of the structural configurations and the structural instability in many regions of the film, demonstrated by the occurrence of laser-induced crystallization even at low powers of the exciting beam. However, a reasonable content of vanadia seems to be needed for some particular application, like rectifying layers in NLC cells.

Satisfactory results in terms of film homogeneity are thus obtained by changing the specific details of the sol-gel route; they appear very important to determine the structural conformation of the films: pH values of 1 in the hydrolysing solution induces anisotropic variations on microscopic scale of the structural conformation, not observed by the direct microscopic observation, but only via a micro Raman mapping. On the contrary the samples generated via a sol-gel process using a more acid hydrolysing solution (pH=0) appear more uniform to the micro-Raman mapping, showing a very weak Raman signal, as in the pure titanium oxide films, and a weak high frequency contribution, indicating a diffuse separate crystallization of vanadium pentoxide, but with very small crystal size.

6. Acknowledgments

This work, in particular for the travels of the authors, has been supported by the grant of Italian Ministero Istruzione Università Ricerca (MIUR), on the program COFIN 99.

The authors would like to thank also Dr. Danilo Bersani for the useful discussion and for kindly providing its Ph.D thesis.

References

[1]   M.S. Burdis, *Thin Solid Films* **311** (1997), 286

[2]   A. Surca, S. Bencic, B. Orel, B. Pihlar, *Electrochim. Acta*, **44**, (1999) 3075

[3]   R. Ceccato, G. Carturan, D. Anesi, F. Decker, F. Artuso, *"2° Italian Meeting on sol-gel"*, Parma, 21-22 september 2000, pp.117-124.

[4]   E. Cazzanelli, N. Scaramuzza, G. Strangi, C. Versace R. Ceccato, R. Carturan, *Molecular Crystal & Liquid Crystal*, to be published

[5]   G. Strangi, D. E. Lucchetta, E. Cazzanelli, N. Scaramuzza, C. C. Versace and R. Bartolino, *Applied Physics Letters* **74**, (1999) 534

[6]   E. Cazzanelli, N. Scaramuzza, G. Strangi, C. Versace, A. Pennisi, F. Simone, *Electrochimica Acta* **44**, n. 18, (1999) 3101

[7]   G. Strangi, N. Scaramuzza, C. Versace, E. Cazzanelli, F. Simone, A. Pennisi and R. Bartolino, *Ionics* **5**, (1999) 275

[8]   A. Antonaia , T. Polichetti, M.L. Addonizio, S. Aprea, C. Minarini, A. Rubino, *Thin Solid Films* **354** , (1999) 73

[9]   T. Ohsaka, F. Izumi and Y. Fujiki, *J. Raman Spectroscopy* **7**, (1978) 321

[10] D. Bersani, *Ph.D Thesis, University of Parma*,1997

[11] U. Balachandran and N.G. Eror ; *J. Solid State Chem.* **42**, (1982) 276

[12] G. A. Tompsett, G.A. Bowmaker, R. P. Cooney, J. B. Metson, K.A. Rodgers and J.M. Seakins, *J. Raman Spectrosc.* **26**, (1995) 57

[13] G .T. Went , S.T. Oyama and A.T.Bell, *J. Phys.Chem.* **94**, (1990) 4220

[14] E. Cazzanelli, G. Mariotto, S. Passerini and F. Decker, *Solid State Ionics* **70&71**, (1994) 412

[15] C. Julien, I. Ivanov and A. Gorenstein, *Mat. Science & Eng.* **B 33**, (1995) 168

[16] F. D. Hardcastle and I. E. Wachs; *J. Phys. Chem.* **95**, (1995) 5031
[17] E. Cazzanelli, L. Papalino, A. Pennisi and F. Simone, *Electrochim. Acta* **46**, (2001) 1937

GNSR 2001
G. Messina and S. Santangelo (Eds.)
IOS Press, 2002

Near-field Raman spectroscopy:
an experimental set-up

S. Patanè

*Dipartimento di Fisica della Materia e Tecnologie Fisiche Avanzate e
INFM, Università di Messina, S.ta Sperone 31, I-98166 Messina, Italy.*

P.G. Gucciardi, S. Trusso, C. Vasi

*Istituto di Tecniche Spettroscopiche CNR,
Via La Farina 237, I-98123 Messina, Italy.*

M. Allegrini

*Dipartimento di Fisica e INFM, Università di Pisa, Piazza Torricelli 2,
I-56126 Pisa, Italy.*

Abstract. An experimental apparatus capable to perform Raman spectroscopy with a
sub-diffraction spatial resolution is presented. Based on a home made near-field scanning
optical microscope two different experimental configurations have been exploited in
order to maximize the Raman signal. Test measurements on organic crystal material, were
performed in both the experimental configurations in order to assess the lateral resolution.

1. Introduction

Raman spectroscopy is one of the most powerful tools among the experimental techniques
used for materials characterization, due to its ability to give structural and chemical
information as well as to its non-invasive and non-destructive nature. The advent of
micro-Raman spectroscopy, in which a Raman spectrometer is used in conjunction with an
optical microscope, permitted to collect information on a length scale of the order of $\lambda/2$,
where λ is the wavelength of the exciting light, which is the ultimate resolution limit
imposed by the diffraction optics [1]. Nevertheless science and technology continuously
ask for sub-micron scale investigation techniques for studying biological systems, single
molecules or quantum structures and devices, only to cite a few applications. In the last
years a new optical microscopy technique, the near-field scanning optical microscopy
(NSOM or SNOM) was developed [2-4]. It's ultimate spatial resolution is well beyond the
diffraction limit and artifact-free images have been obtained with a spatial resolution of
the order of ~20 nm [5]. Moreover this technique offers the exciting combination of a high
spatial resolution together with an amount of spectroscopy contrast mechanisms coming
out from the use of light as a probe, offering the possibility of a wealth of novel
experimental configurations. However, the sub-wavelength optical aperture, required to
achieve a so high resolution, causes the main limitation to the technique: a very low light
throughput. Hence innovative experimental setups are needed to perform fluorescence or

Raman spectroscopy, with particular care for the aspect of light collection in order to gain information on the local characteristics of the sample. Several experimental set-up have been presented able to perform Raman scattering experiments using a SNOM apparatus [6-8]. In this picture we present a homemade setup based on commercial grade aperture probe, able to perform at the same time both spectroscopy and topography mapping with nanoscale resolution. The instrument allows for the recording of the Raman signal coming out from the sample in both illumination and collection mode, with transmission and reflection configurations leading to an extremely versatile setup. An intensified and thermo-cooled CCD is used to collect all spectra at one time. Besides, a photomultiplier is employed to obtain 2D Raman maps, at a chosen Raman shift, with much shorter acquisition times. As tests we present the Raman spectra and maps performed on a thin film of an organic crystal.

2. Instrument description

The experimental set-up is based on a homemade scanning near field microscope, details of which can be found in refs. [9,10]. Here we give a brief description of the modifications made to the SNOM apparatus to perform Raman scattering experiments. A block diagram of the instrument is reported in Fig. 1. The SNOM probes we use are commercially available single mode optical fibers Cr/Al coated, in such a way to leave an apical aperture with a nominal diameter of about 100 nm, purchased from Nanonics Ltd. The fiber is glued on one arm of a quartz tuning fork for non-optical shear-force detection purposes [11].

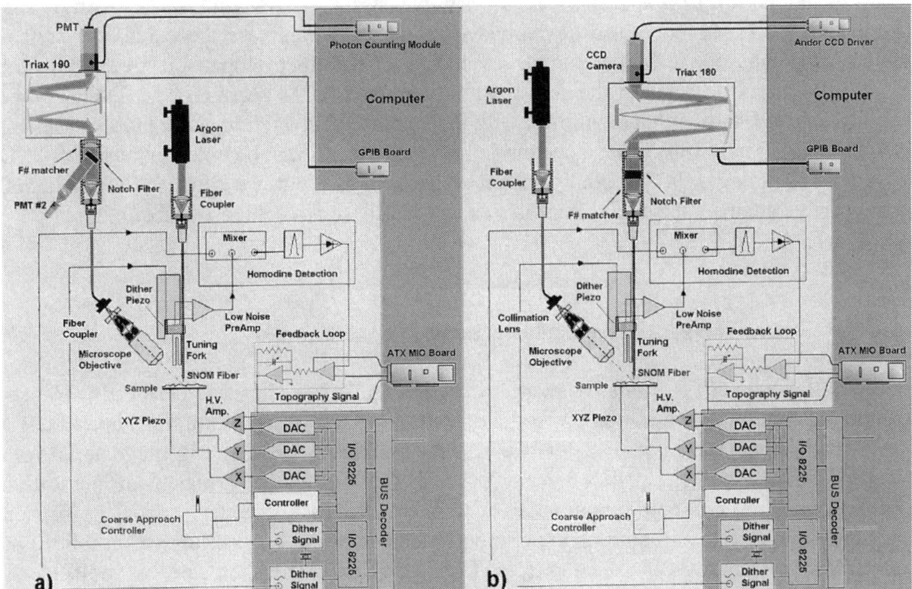

Fig.1: Schematic view of the experimental setup. (a) Illumination mode apparatus: the optical fiber is used to shine the sample surface while the Raman signal is gathered by the microscpoe objective. (b) Collection mode apparatus : the excitation light is focused onto the sample surface by means of a microscope objective, while the scattered radiation is collected through the optical fiber.

A piezo slab is used to dither the tuning fork at resonance ($\sim$ 32 kHz). The oscillation amplitude of the system is thus monitored by measuring the voltage present at the electrodes of the tuning fork. The signal is amplified by a low-noise preamplifier positioned close to the tuning-fork and, finally, lock-in techniques are exploited to detect the vibration of the probe-fork system. This signal is fed to a feedback loop acting on the vertical electrode of the scanning piezo in order to keep the probe-sample assembly within the near-field region all along the scan. As a by-product the feedback signal furnishes details of the sample surface topography acquired simultaneously with the spectroscopic information.

As reported above, the instrument can perform Raman scattering experiments both in collection and illumination mode. In illumination mode [Fig. 1(a)] the excitation light coming from an Ar^+ laser (λ=514.5 nm) is coupled with the SNOM fiber probe. Only few mW of the laser power can be coupled with the fiber, in order to avoid thermal damages. As a consequence excitation powers as low as few hundreds of nW, depending on the actual optical fiber aperture and on the effective taper-cone angle, can be only be achieved. The scattered light is collected by a long working distance microscope objective (Olympus 50X, WD=10.6 mm, NA=0.50), and then driven, through a multimode optical fiber, to the entrance slits of a 190 mm focal length monochromator (Jobin-Yvon Triax 190) equipped with a 1200 lines/mm ruled grating blazed at 500 nm. The angle between the objective and the probe axis is set to 45°, yielding the best collection geometry. Rejection of the elastically scattered radiation is accomplished by a notch filter (Kaiser Optics). The detector is a photomultiplier (Hamamatsu H6240-01) that operates in photon counting mode. In this configuration the Raman spectrum at a given point on the surface of the investigated sample can be recorded by scanning the monochromator between two wavelengths. While a bi-dimensional map of the Raman efficiency of the sample at a given wavelength can be constructed performing a scan over the sample surface. A second setup operating in collection mode has been also developed [Fig. 1(b)]. In this case the excitation light is focused on the surface of the sample by means of the microscope objective, while the scattered light is collected in the near-field through the SNOM aperture. The SNOM probe is thus coupled, through a notch filter, with the entrance slits of an imaging monochromator (Jobin-Yvon Triax 180). The detector, in this case, is a thermo-cooled intensified CCD (Andor Technology). In this configuration a spectrum in the range between 250 and 3500 cm^{-1} can be obtained in a single measurement with a spectral resolution of about 20 cm^{-1}.

3. Nano-Raman imaging of molecular crystals.

The efficiency of the Raman scattering in solids can be very low and performing Raman scattering experiments using an optical fiber with a 100 nm or less of nominal aperture is a very challenging task. In order to test the instrument, then, we used an organic material with a good Raman efficiency in order to estimate the overall performance of the set-up employing both the configurations and to estimate the maximum spatial resolution achievable in a Raman scattering experiment. Measurements were performed on thin films of 7,7',8,8'-tetracyanoquinodimethane, or TCNQ for short. The choice of this molecule as test sample is related on its good Raman efficiency as well as to the possibility to grow both as a single crystal, by precipitation from a solution, and as a thin film by means of vacuum evaporation. Furthermore it can form charge-transfer metal complexes, e.g. reacting with copper (CuTCNQ), in which some of the vibrational frequencies are shifted providing a potential source of chemical contrast. The most intense Raman lines of the TCNQ are

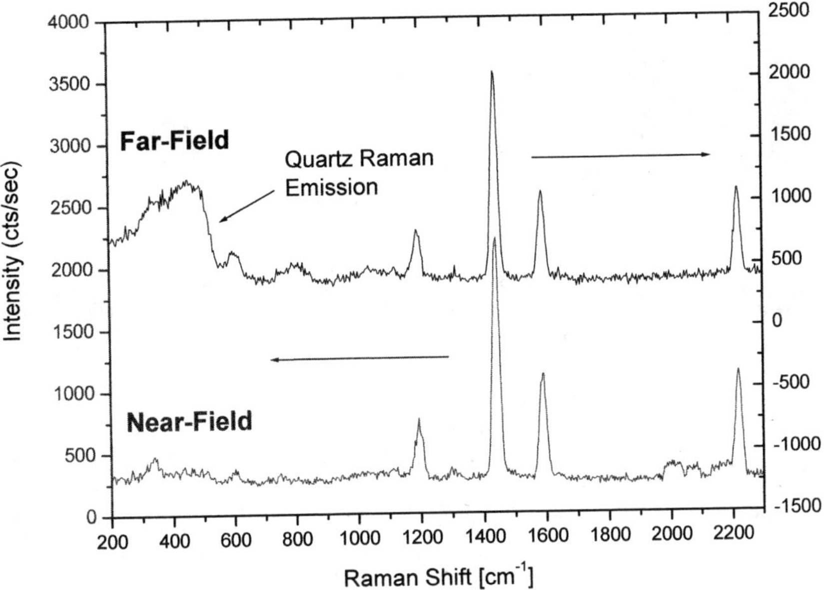

Fig.2 Raman spectra of a TCNQ film acquired in illumination mode with the optical fiber tip held far and near the sample surface. Far-field and near-field spectra intensities are reported on the right and left ordinate axes respectively.

related to stretching modes of the C-N, C-CN and C=C bonds positioned at 2220, 1453 and 1618 cm^{-1} respectively. In Fig.2 we report the spectra of a TCNQ crystal carried out in the illumination configuration with the fiber tip held far (few μm) and near (~15 nm) the sample surface. As can be seen the two spectra appear similar with the exception of the contribution, at low frequency, of the Raman signal arising from the optical fiber [12]. Four peaks are clearly observable, related to the above cited stretching. These spectra were acquired with an integration time of 100 ms and provide an estimated spectral resolution of about 25 cm^{-1}.

In Fig. 3 are reported the topography (a) and the Raman map (b) of a 2.4×1.1 μm^2 area. The Raman signal was acquired on a grid of 128×58 points with an integration time of 100 ms per point. The Raman map was acquired at the wavelength corresponding to the Raman shift of 1453 cm^{-1}, i.e. looking at the C=C stretching vibration. The topography is characterized by a rather flat area (dark zone located on the right), while bump like topographic feature is evident on the left hand side (white area), about 350 nm high. The corresponding Raman map shows a different behaviour: the intensity of the Raman signal is quite uniform across the sampled area, with the notably exception of the region in correspondence of the edge of the topography feature, where a clear decrease of the Raman intensity can be observed. It's worth noting that the Raman signal never goes to zero in that region but just a decrease of the intensity is observed. In order to get more insight about the Raman signal fluctuations, we have carried out several spectra [Fig. 4(b)] at fixed points (indicated by the arrows) along a line profile showing such modulations which is reported in Fig. 4(a). The profile length is 11 μm and is part of a Raman map (not reported) carried out at 1450 cm^{-1}. Each point spectrum was collected in the 1400-2300 cm^{-1} range. We can see that the intensity modulations measured from the line profile, between 5000 and 4200 count/sec with a typical background of ~600 counts/sec, are recovered when looking at the

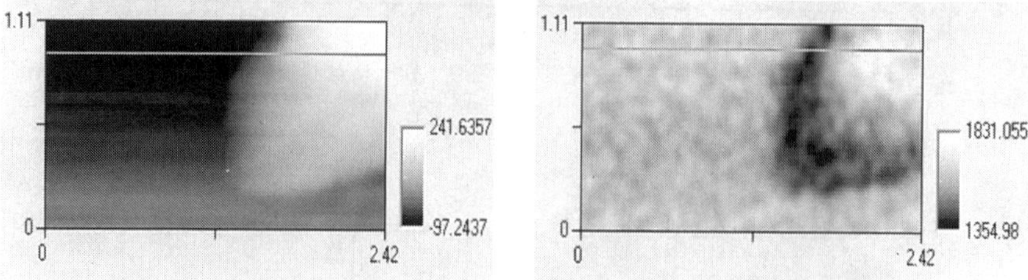

Fig.3 (a) Topography map of a 2.42×1.11 μm² area of a defect arising on a TCNQ crystal surface. The corresponding Raman intensity map (b) collected at 1453 cm⁻¹. The straight lines indicate the position of the profiles reported in fig. 6.

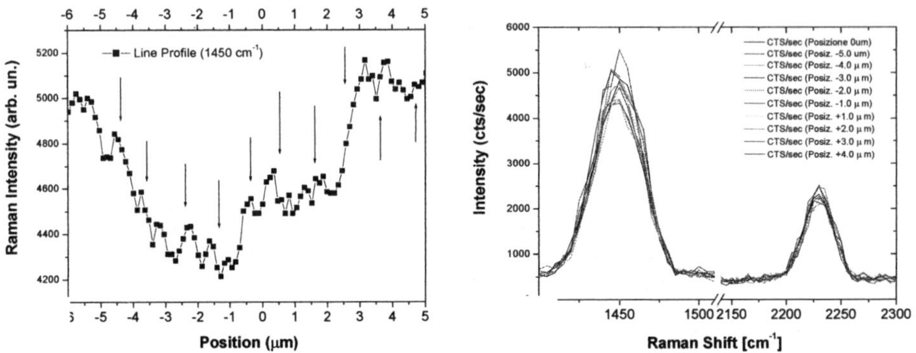

Fig.4 (a) Intensity variation of the Raman signal measured at 1450 cm⁻¹ collected along a line profile 11 μm on the sample surface. The Raman spectra (b) were collected at the points indicated by the arrows in (a) and show only a fluctuation of th epeaks intensity without any appreciable energy shift.

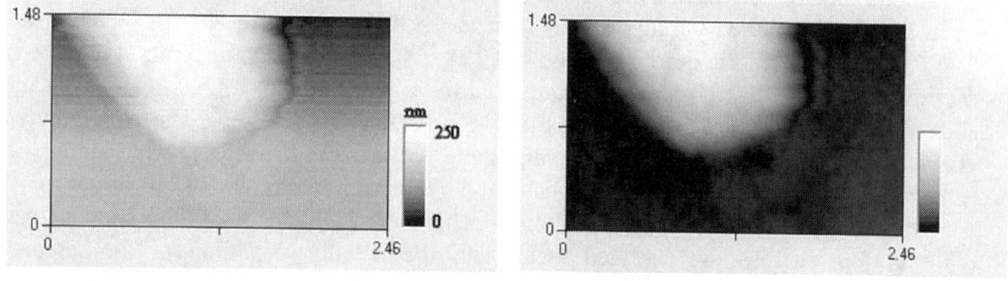

Fig.5 Topography (a) and reflection (b) images of an area displaying a topographic feature similar to the one shown in Fig. 4. There is no evidence of a systematic decrease of the reflection signal at the boundaries.

fluctuations of the 1450 cm⁻¹ peak intensity. No Raman shift along the spatial points can be appreciated within our spectral resolution. We can note that the background intensity, due both to the dark noise of the detector and to some unfiltered stray light, is constant from point to point and thus does not contribute to the recorded modulations of the Raman signal in the maps. The origin of the decrease of the Raman intensity occurring at the boundaries of the topography features is not clear up to now. A main hypothesis deals with the presence of an amorphous phase of TCNQ characterized with a lower Raman efficiency.

We would like point out that geometrical effects, related to the topography, could also influence the intensity of the collected Raman signal. In order to exclude the presence of such artifacts we carried out a measurement of the elastically scattered light in a region presenting similar topographic features as the one shown in Fig. 3. The results are shown in Fig. 5. As can be seen there is no evidence for a decrease of the intensity of the elastically scattered light [Fig. 5 (b)] at the boundaries of the bump [Fig. 5 (a)]. These findings can exclude, with good confidence, any fictitious origin of the observed modulations of the Raman signal.

The above cited Raman measurements can be used to draw conclusions concerning the spatial resolution achievable with our instrument when performing Raman scattering imaging experiments. In Fig.6 are reported the topography (a) and the Raman intensity (b) profiles collected along the line shown in Fig. 3. The profiles were determined collecting the signal at points about 20 nm apart. The lateral spatial resolution can be determined in correspondence of the signal decrease and can be evaluated with the help of the conventional 10-90% method or by measuring of the full width at half maximum (FWHM). The first method gives a value of 170 nm while the second one furnishes a value of 250 nm. These values agree well with the estimated aperture of the optical fiber used to illuminate the sample surface, i.e. about 200 nm.

To conclude, in Fig. 7 we show a spectrum acquired with the collection mode experimental setup on the same sample. The integration time was set to 10 s. As can be seen the signal to noise ratio is worse than the one of the illumination mode apparatus even thought that the integration time is ten times longer. Nevertheless the evident advantage is that a complete spectrum is obtained at one shot. In this respect the two experimental apparata presented in this paper can be viewed as complementary tools: while the one operating in illumination mode is able to perform Raman imaging with reasonable acquisition times, the one operating in collection mode can be used to obtain a more detailed spectral information in spite of longer measurement times.

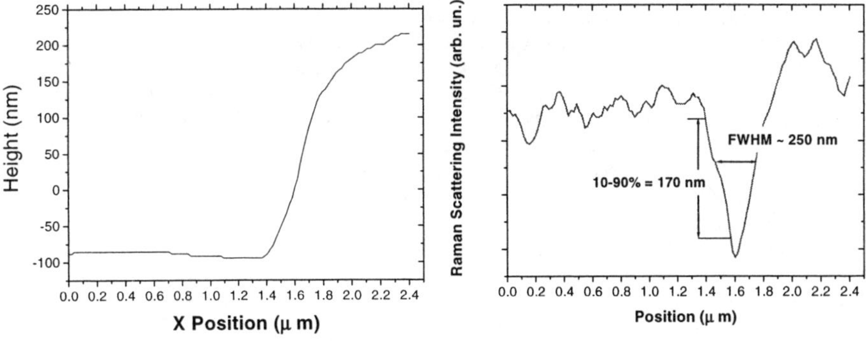

Fig.6 Topography (left) and corresponding Raman intensity profile (right) collected along a line. The estimated lateral spatial resolution by the 10-90 and the FWHM methods are also shown.

4. Conclusions

An experimental apparatus based on a SNOM microscope capable to perform Raman scattering experiment has been developed. It can operate both in illumination and collection mode using two different detection systems namely a photomultiplier or an intensified CCD. Raman spectra have been obtained on a sample with good Raman efficiency. Imaging capabilities with very short acquisition times, of about 1 hour for a picture of 128×128 points, have been also demonstrated. The spatial resolution has been estimated to be 200 nm i.e. of the same order of the optical fiber aperture. The use of chemically etched fibers and improvements in the signal detection system are planned in order to enhance the system performances.

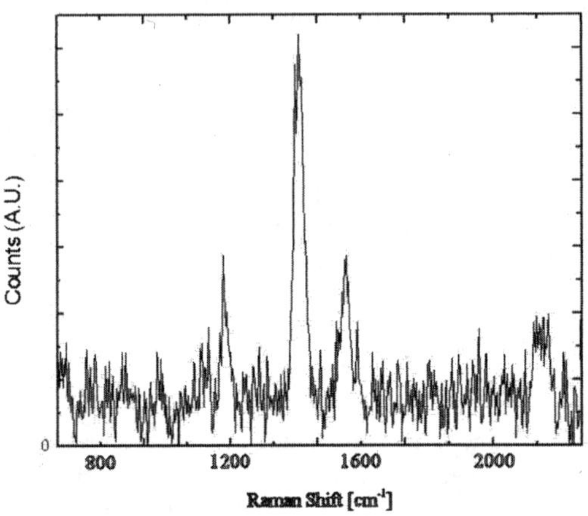

Fig.7 Raman spectrum of a TCNQ film obtained with the collection mode apparatus. The integration time is 10s.

References

[1]   E. Abbe, Ark. Microscop. Anat. , 9 (1873) 413.

[2]   D. W. Pohl, W. Denk and M. Lanz, Appl. Phys. Lett., **44** (1984) 651.

[3]   A. Lewis, M. Isaacson, A. Harootunian and A. Murray, Ultramicroscopy **13** (1984) 227.

[4]   E. Betzig, J. K. Trautman, T. D. Harris, J. S Weiner and R. L. Kostelak, Science **251** (1991) 1468.

[5]   M. Labardi, P. G. Gucciardi, M. Allegrini, C. Pelosi, Appl. Phys. A, **66** (1998) S397.

[6]   C. L. Jhancke, M.A. Paesler and H. H. Allen, Appl. Phys. Lett., **67** (1995) 2483.

[7]   S. Webster, D. N. Batchelder and D. A Smith, Appl. Phys. Lett., **72** (1998) 1478.

[8]   J. Wang, H. Yan, Y. Deng, H. Li, Y. Zhang, F. Zhang, Z. Xia, Q. Gao, W. Du, H. Zhou and Y. Zou, Solid State Commun., **115** (2000) 173.

[9]   P. G. Gucciardi, M. Labardi, S. Gennai, F. Lazzeri and M. Allegrini, Rev. Sci. Instrum. **68** (1997) 3088.

[10] M. Labardi, P. G. Gucciardi and M. Allegrini, La Rivista del Nuovo Cimento **24**, Vol. 1 (2000) 1.

[11] K. Karrai and R. D. Grober, Appl. Phyts. Lett., **66** (1995) 1842.

[12] J. Grausem, B. Humbert, A. Burneau, J. Oswalt, Appl. Phys. Lett. **70** (1997) 1671.

GNSR 2001
G. Messina and S. Santangelo (Eds.)
IOS Press, 2002

Optical and electronic characterization of UV detectors based on synthetic diamond

Andrea Pini[*], Emanuele Pace

Dip. Astronomia e Scienza dello Spazio. Università di Firenze.
Largo E. Fermi, 5. Firenze 50125 Italia.
[*]*E-mail address: apini@arcetri.astro.it, Tel +39(055)27.52.208*

Abstract. Recently, several researches about the use of synthetic diamond films have been performed, being the diamond a sensible material in the revelation of UV radiations. Indeed, diamonds offer good characteristics of rejection to the possible visible spectral components and a low presence of dark current, showing a gap of $5,48eV$. Unfortunately, a complete utilization of this material is actually limited by the slowness of devices and the limited collection length of photogenerated charges. As a matter of fact, a high presence of defects and impurities persist in the synthetic diamond, because of its polycrystalline nature and consequently to grain-borders; defects and impurities limit its use as a detector.
The homoepitaxial single crystal diamond can be considered a possible candidate for the solution of these problems. The absence, in this material, of grain-borders defects can result into higher collection lengths and, consequently, into better performances of devices.
Therefore, the present project has carried out a comparative study of the structural properties of a single crystal Ib diamond sample, produced by HPHT technique, and of a polycrystalline one, grown up through CVD technique. In particular, it has pursued measures of photoluminescence, Shift Raman, and absorption in the infrared through FTIR technique. The results of these measures have been successively analysed in order to evaluate the quality of crystals and to relate it with the final performances of the devices. Thus, the photocurrent answers in the 40-220nm range have been measured, and the quantum efficiencies and temporal answers of the over mentioned samples calculated.

1. Introduction

The not common physical properties of synthetic materials and the progressive decrease in their costs of production, caused by the development of extremely efficient growth techniques, make diamonds very promising materials for the revelation of *UV* radiations. They prove to be excellent especially in adverse environments, thanks to the large energetic gap, that results into a high rejection of the visible light (no filter is required) and into an extremely low dark current (no cooling system). Furthermore, diamonds are the ideal candidates for the realization of detectors acting in "aggressive" environments, thanks to their extreme resistance to highly energetic radiations and to their chemical inertia. Unfortunately, the state of art of material synthesis and devices realization does not allow the direct application of the diamond in those fields where, in theory, it proves to have exceptional characteristics. Indeed, as far as material synthesis, the most used and efficient techniques (CVD) produce polycrystalline films with high densities of defects, which limit the final performances of detectors. Other techniques (HPHT, single crystal CVD) are able to obtain single crystal diamonds of high quality and could give interesting results. Nevertheless, they are not very used yet, because of the costs and the reduced dimensions of diamonds.

In this study, we have realized a performances comparison of those devices created by the use of both polycrystalline materials and commercial single crystal Ib diamonds. For this reason, thanks to the optimisation of the growth technique MWPE-CVD, a film of diamond of high quality has been produced, and its structural characteristics have been studied through techniques of micro and macro Raman spectroscopy, and photoluminescence, through measures of optical absorption and infrared absorption with FTIR.

The same study was carried out for commercial samples of Ib diamond, together with a comparison of the optical and electronic properties of the two mentioned films. Afterwards, photo-detectors have been realized through photolithographic techniques of deposition of electric contacts, and several measurements have been pursued, such as IV in both dark and UV light regimes characteristics, quantum efficiency measures in the gap between 50 and 250nm and temporal measures. Thanks to a comparative analysis of the given data, it was possible to create a model of operation for both typologies of material, in order to have important information concerning the quantities of mobility of the photo-generated charges and of their medium lifetime, and to link them with the structural characteristics previously shown.

2. Diamond film and devices

A diamond film was grown up through MWPE-CVD technique, over a silicon substrate of p type, using the common method of abrasion, in order to increase the crystallites nucleation. For the synthesis, a mixture of 1% of CH_4 in H_2 was used to obtain a rhythm of growth of 0.8 μm/h, circa. The growth time was fixed around 50 hours and the substrate temperature was maintained at 800°C. After the deposition, a thermal treatment in air at 500°C was realized, in order to remove hydrogen terminations on the surface, from the growth side. The medium largeness of micro crystals was evaluated to be 20 μm, circa. As for the single crystal sample, it was bought by Sumitomo Inc. and is an Ib diamond, that is to say a diamond with a high level of nitrogen, which is normally commercialised by this company.

Gold inter-digitated contacts were placed on both materials, on a 2 mm^2 surface for the polycrystalline diamond and 9 mm^2 for the single crystal. The inter-electrode gap is 20 μm; each electrode is 15 μm large and 40 μm thick.

3. Experimental set-up

As for the experimental set-up used for the Raman spectroscopy measures, a double S. A. Ramanor U1000 monochromator was used, equipped with a BX40 Olympus lens for microscopes. An argon laser with 5145.5 nm emission was the source and the spot used in the macro Raman configuration was 100 μm circa, while the over mentioned lens brought to 1 μm the spot diameter in the micro-Raman configuration.

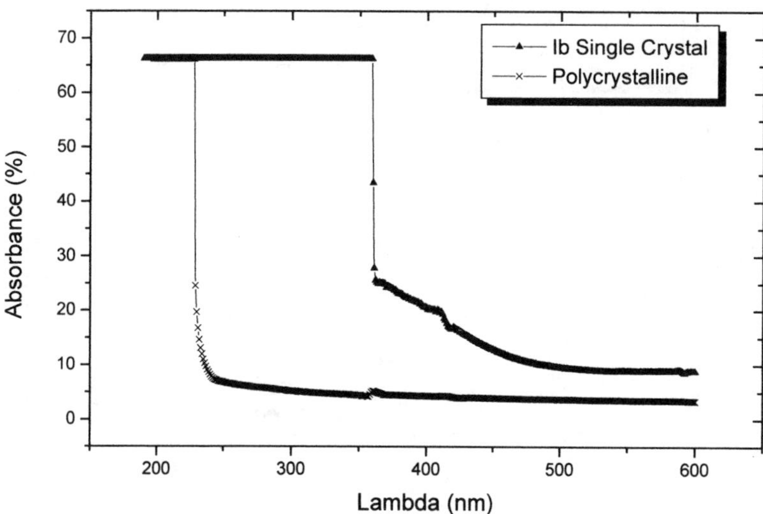

Fig.1 Optical absorption spectra

Thanks to a co-focal 100 μm opening, it was possible to obtain a 2 μm resolution under the surface. The spot laser power, that reaches the sample surface, is 10 mW.

For the photoluminescence measures, a two-stadium laser - the first one was an Ar^+ and the second one was a self-mode-locked titanium-sapphire (Spectra-Physics, Tzunami) – was used, together with a part of the over illustrated set-up. Afterwards, the beam was frequency duplicated several times in order to have a 210 nm radiation. The set-up used for photocurrent measures is composed of a Mc Pherson UHV monochromator, a deuterium lamp with a window in MgF2 and a Keythley 6517a electrometer. A hollow cathode lamp, alimented with He, Ar and N, was used to explore the region of wavelengths included between 50 nm and 140 nm.

4. Structural characterisation measures

An absorption spectrum of the two utilised samples – the CVD polycrystalline and the single HPHT crystal – is given in fig. 1. The single crystal HPHT sample presents an extremely low absorption until 470 *nm*, with a slight increase for minor wavelengths, a weak absorption peak at 420 *nm* and a rush step at 360 *nm* circa: this structure is a sign of the presence of nitrogen aggregates. Furthermore, this structure confers the diamond a yellowish colour. For the polycrystalline sample, it is possible to observe its good properties of transparency in the region from 230 *nm* to 900 *nm*, that is to say from the energies of band transition in the near *UV* to the near infra-red.

Now, if we observe fig. 2 - where the absorption spectrums in the infrared region, realized through the FTIR technique, are reported -, it is worth noticing that the single crystal sample shows an absorption spectrum which, for the two and three phonons area, appears

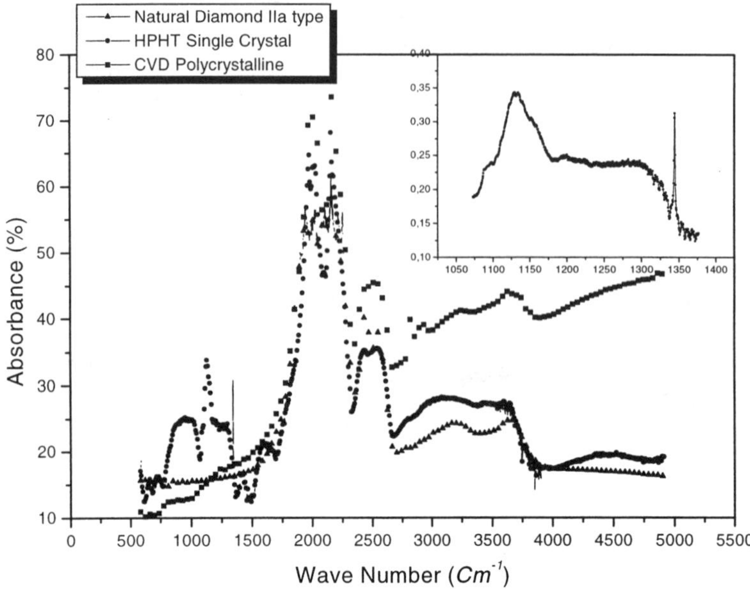

Fig.2 FTIR infrared absorption spectra measure. In the insert the typical diamond Ib structure is reported.

to be similar to that of natural diamonds. In the one phonon zone, there is the classical absorption structure with a high quantity of nitrogen, also present in Ib diamonds. By analysing the spectrum movement of the polycrystalline sample, it is possible to observe that it appears to be analogous to the natural IIa diamond, in one and two phonons areas. In the two phonons area, it is noticeable an increase of absorption, due to the presence of structural chaos of the crystal, correlated with its polycrystalline nature. Furthermore, at 2800cm^{-1} circa, there is a structure caused by the presence of hydrogen, probably a superficial residuum of the growth.

In fig. 3, it is possible to notice photoluminescence spectrums for the single crystal HPHT sample (a) and for the polycrystalline CVD one (b). If we focus our attention on fig. 3 (a), we can see that the single crystal diamond is characterized by many structures: the A band, centred at 2,5 eV circa, related to the presence of a states distribution that is caused by grain-border defects; a peak at 2,53 eV circa, due to the residual presence of nickel, used during the growth process as a catalyst; a prominent peak at 2,46 eV and its two phononic replicas. The last mentioned structure is denominated, in literature, H3 centre, and it is related to the presence of nitrogen-vacancy-nitrogen aggregates in the crystal. A slight peak is also present at 2,3 eV, related to the occurrence of the H3 centre, but generally visible only in cathode-luminescence and observed because the excitation energy of the laser was higher than that of the band direct transition, reproducing the excitation dynamics proper to that technique.

If we pay attention to fig. 3 (b), it is noticeable that the polycrystalline sample is

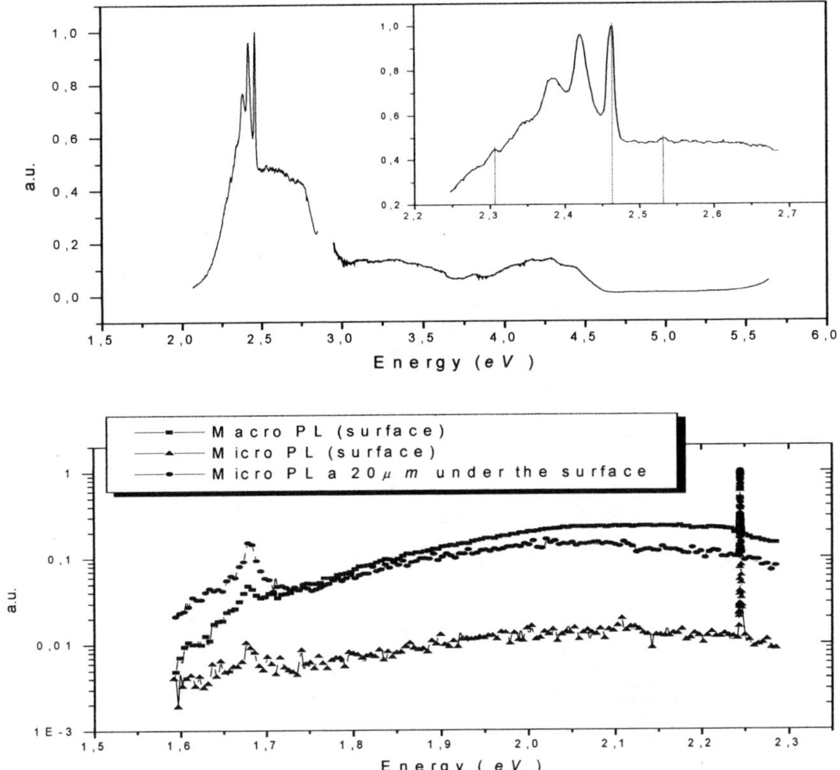

Fig.3 Photoluminescence spectra. a) single crystal, b) polycrystalline

characterized by single crystallites of extremely high quality and this fact is witnessed by an almost total absence of luminescence bands in the spectrum related to a superficial micro photoluminescence. Even the peak at 1,68 eV, linked with the presence of silicon, diffused in the crystal during the growth process, is nearly absent. The surface macro photoluminescence spectra and the 20µm deep micro photoluminescence show many differences from those previously analysed. In such measures, a large photoluminescence band is present and it is related to the presence of an almost continuous distribution of trap intra-gap states; the silicon peak is stronger and it increases its presence moving towards the substrate side of the sample.

In fig. 4, the Raman shift measures of the polycrystalline sample (a) and of the single crystal specimen (b) are shown. These figures clearly illustrate the high quality of both diamonds. Furthermore, for the single crystal we have noticed a high structural homogeneity, witnessed by the similarity of the three measures related to three different zones on the sample surface. Moreover, the structural stress for this sample is almost absent and the Raman peak is well centred at 1331,5 cm^{-1} and very narrow (2,3 cm^{-1}). For the polycrystalline diamond, we can confirm the extremely high quality of the micro-crystallites from the analysis of the micro-Raman peak positioned at 1332,7 cm^{-1} and with a FWHM of 2,4 cm^{-1}. Getting 20µm deeper under the sample growth surface, the Raman spectra shows a significant decreasing of the diamond quality and it moves to 1331,6 cm^{-1} with a FWHM of 3,3 cm^{-1}, increasing both directional and non-directional structural stress.

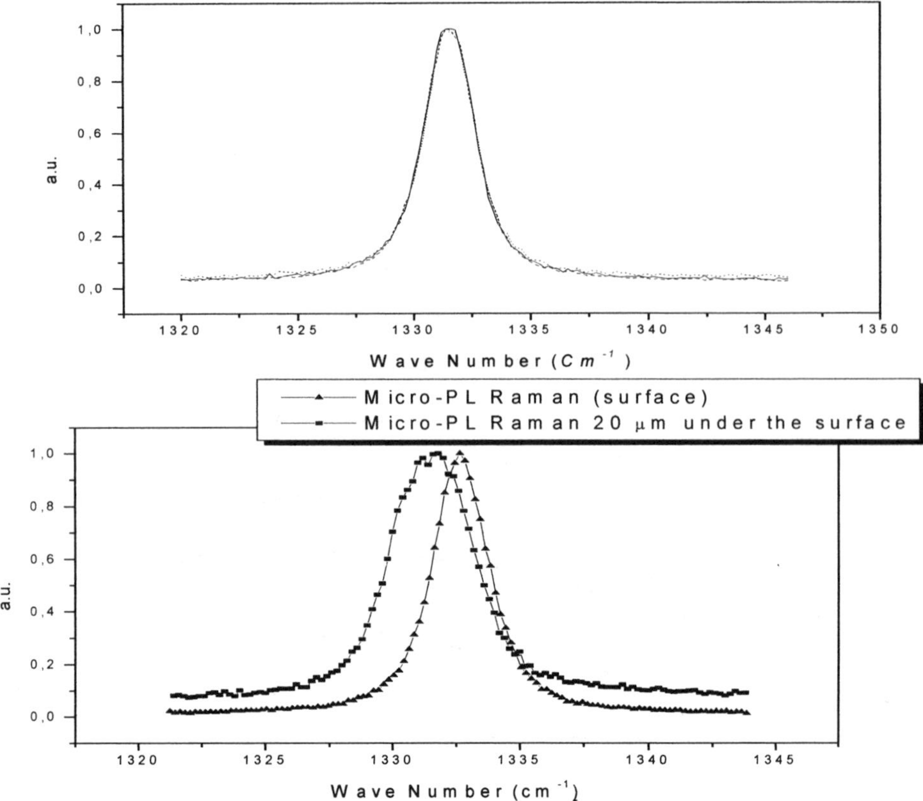

Fig.4 Raman Spectra. a) Three measures related to the single crystal diamond. b) Two measures related to the polycristalline sample: a superficial micro-Raman and a 20 µm deep micro-Raman.

5. Electrical measures

The analysis of the dark IV characteristics of the device has shown a very high resistance of the material, confirming the very low conductivity of the diamond surface. The devices have reported a resistance of 10^{13} and 10^{14} respectively for the homoepitaxial HPHT single crystal and for the CVD polycrystalline.

In fig.5, three IV curves have been reported: they are related to the illumination with three different wavelengths of the HPHT diamond-based device. The used wavelengths are 120nm, 160nm and 200nm. We know that it is possible to express the photocurrent by:

$$I_f = qN_p X_{abs} Y_{coll}$$

where q is the elementary charge, N_p is the number of incident photons, X_{abs} is the fraction of absorbed photon by the material and Y_{Coll} is the collection efficiency of the device. Writing the collection efficiency of the device as

$$Y_{coll} = \frac{\mu E}{\left(1 + \dfrac{\mu E}{v_{sat}}\right)} \cdot \frac{\tau \cdot (1 + r)}{d_{electrodes}}$$

where μ, E, v_{sat}, τ, r and $d_{electrodes}$ are respectively the charge mobility, the applied electric field, the saturation velocity, the ratio between the electrons mobility and the hole mobility (which can be calculated from literature and is equal to 0,8), the charge mean lifetime and the inter-electrodes distance. Many fitting sessions of the experimental data have been carried out using the previously reported equation, and the parameters have been calculated fixing the saturation velocity at 10^7 cm s ec^{-1}. The calculation led to a value of mobility of about 150 cm^2 V^{-1} sec^{-1} for the single crystal sample and about 100 cm^2 V^{-1} sec^{-1} for the polycrystalline diamond, while the mean lifetime of the charges is 0,1 µsec for the single crystal device and 100 psec for the polycrystalline specimen. According to the literature, the reported charge lifetime values for the single crystal sample are related to high quality diamond samples while the value calculated for the polycrystalline is reported for medium quality materials. In fig. 6, the curves related to the external quantum efficiency of the devices are shown. The maximum of the quantum efficiency is revealed by the single crystal device, with a value of about 700 e$^-$/photons, while the polycrystalline CVD has shown a maximum of about 1 e$^-$/photons. Such values are in good agreement with those calculated by the previously mentioned fitting sessions. Furthermore, a good visible rejection is clear from the curves and both devices loose three order of magnitude of efficiency passing from 200nm to 250nm, confirming the visible blindness of the material.

It is now important to analyse the devices behaviour under subsequent UV light pulses. In fig. 7, it is possible to see the curves related to the single crystal and polycrystalline devices, after 160nm pulsed light excitation. It is worth mentioning that the pulses length was different from each other. Indeed, first of all the photocurrent of the single crystal detector shows a slow decrease. Secondly, after the pulse end and the beginning of a new pulse, the current rapidly recovers the previous value and the decrease starts again with the same characteristics. The final stable value of the photocurrent is about the 17% of the starting one and the constant time of the exponential decrease is about 600sec. The polycrystalline-based photo-detector (fig 7 b) is characterized by a decrease of the rise time and a very slow increase of the measured photocurrent. The previously mentioned behaviours can be explained as consequences of the following transition models. In the single crystal diamond, the high presence of nitrogen is the main characteristic impurity, noticeable by photoluminescence measures. The photocurrent decrease phenomenon has to be imputed to the subtraction of a part of the available carriers by the trapping process, related to meta-stable nitrogen intra-gap states. The effect of those traps, with a very long mean lifetime, may influence the photo-response after some seconds from the beginning of the exposition to the UV light. The band, introduced by nitrogen, is sufficiently deep and has a very low nitrogen-related states-valence band transition probability. This phenomena leads to the accumulation of charges in the traps and to the manifestation of the observed photocurrent decrease. Now, if we analyse the curve related to the polycrystalline material, we can explain the observed behaviour with the progressive decrease of the effect of the charge subtraction phenomenon, due to the almost-continuous distribution of intra-gap states. The presence of such distribution has been noticed in the photoluminescence measures and correlated with the effect of the grain-border defects.

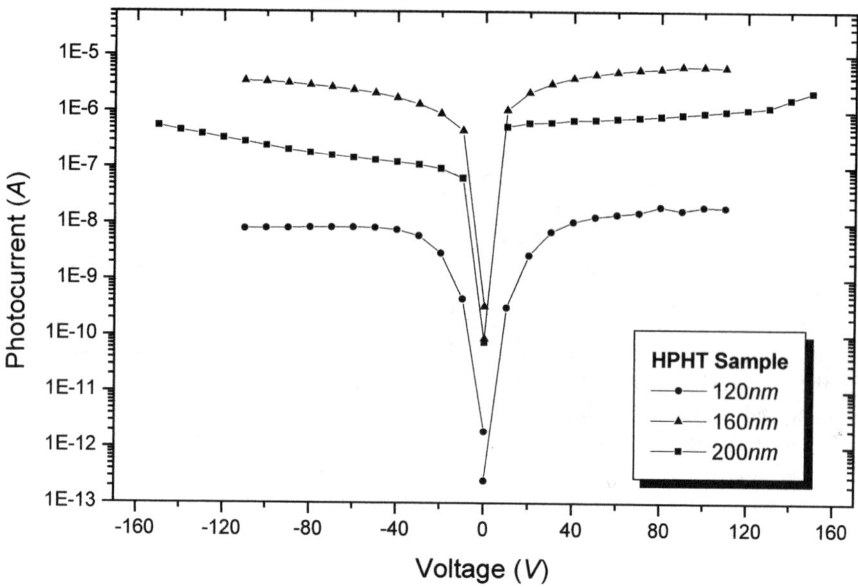

Fig.5 IV Characteristics under different UV wavelength for the single crystal diamond

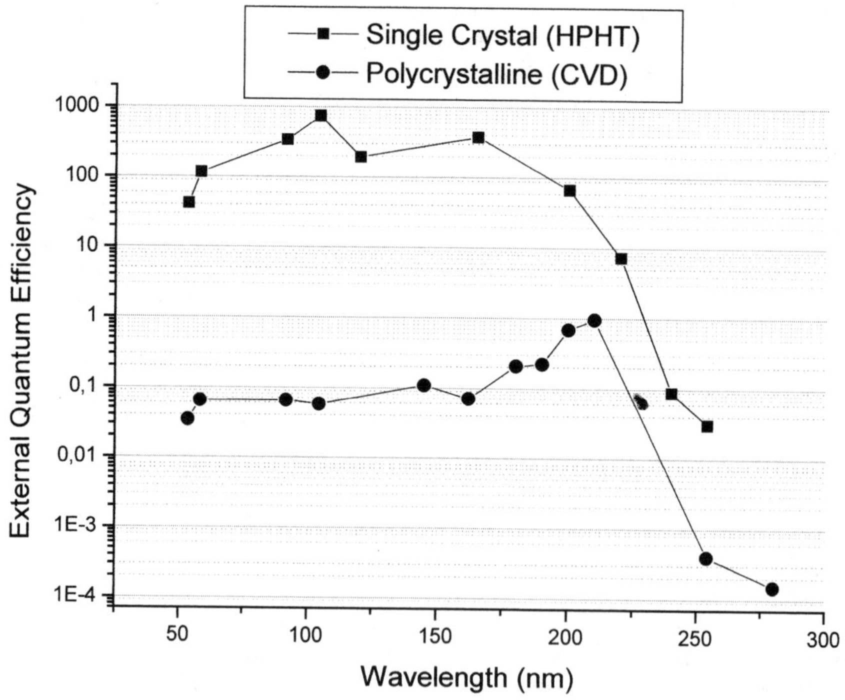

Fig.6 External quantum efficiency measures

6. Conclusion

In this paper, the performances related to two different devices based on two different diamond growth techniques were analysed. The first device was based on catalysed HPHT (single crystal Ib) and the second one on MWPE-CVD (polycrystalline diamond sample). A full session of optical measurements have been performed on both specimens, in order to identify the peculiarity of each growth product and to relate those characteristics to the final devices performances. Both diamond samples have resulted in a very high crystal quality, but with a strong presence of nitrogen in the single crystal sample and a presence of an almost-continuous distribution of trap states, due to the existence of grain border defects, in the polycrystalline specimen. The electrical measures have confirmed the extremely high material resistance in absence of UV radiation. The measured photocurrent behaviour has allowed the calculation of the carrier mobility and their lifetime, through a fit session with some analytic functions. The obtained values have shown better performances for the single crystal device and are in good agreement with the direct measures of the devices quantum efficiency. As far as the time related measures, it has been noticed that the photocurrent behaviour is extremely different between the devices, and these differences are related to the presence of nitrogen in the single crystal, and of grain boundaries in the polycrystalline material. In particular, the presence of a nitrogen related band, in the Ib single crystal diamond, leads to a strong decrease of the photocurrent during long steady state exposions, while the polycrystalline nature of the CVD synthesized material has a decrease of the rise time, due to the progressive filling of the almost-continuous distribution of traps.

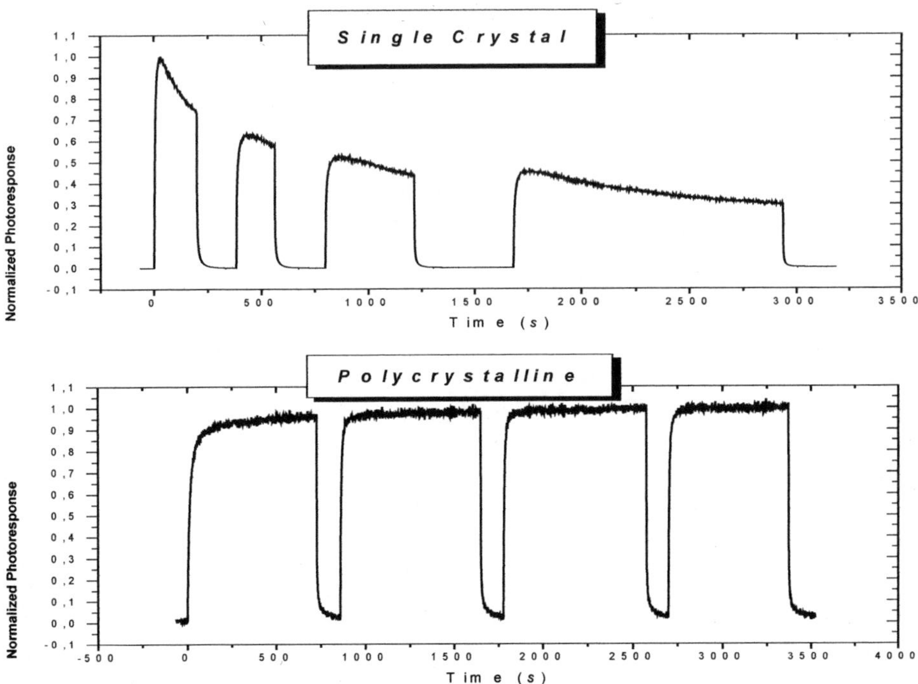

Fig.7 Time response measures. a) single crystal sample. b) CVD polycrystalline sample

References

[1] E. Pace, A. Vinattieri, A. Pini, F. Bogani, M. Santoro, Y. Sato, *Structural and functional characterization of HPHT diamond crystals used in photoconductive devices*, Physica Status Solidi A, **Vol. 181 n.1** (2000).

[2] G. Faggio, M. Marinelli, G. Messina, E. Milani, A. Paoletti, S. Santangelo, A. Tucciarone, G. Verona Rinati, *Comparative study of band-A cathodoluminescence and Raman spectroscopy in CVD diamond films*, Diam. Relat. Mater., **8** (1998) 640 – 643.

[3] L. Bergman, M. T. McClure, T. J. Glass, R. J. Nemanich, J. Appl. Phys., **76** (1994) 3020.

[4] L. Bergman, B. R. Stoner, K. F. Turner, T. J. Glass, R. J. Nemanich, J. Appl. Phys., **73** (1993) 3951.

[5] R. D. McKean, R. B. Jackman, Diam. Relat. Mater., **7** (1998) 513.

[6] P. Gonon, S. Prawer, Y. Boiko, D. N. Jamieson, *Electrical conduction in polycrystalline diamond and the effect of UV irradiation*, Diam. Relat. Mater., **6** (1996) 860-864.

[7] S. C. Binari, M. Marchywka, D. A. Koolbeck, H. B. Dietrich, D. Moses, Diam. Relat. Mater., **2** (1993) 1020.

[8] D. Tromson, P. Bergonzo, A. Brambilla, C. Mer, F. Foulon, *Influence of electrical defects on diamond detection properties,* Diam. Relat. Mater., **9** (2000) 1091-1095

[9] S. Salvatori, M. C. Rossi, F. Galluzzi, D. Riedel, M. C. Castex, *Transient photoresponse of CVD diamond-based detectors in the time domain 10^{-9} s-10^3 s.,* Diam. Relat. Mater. **8** (1999) 871-876

Optical Spectroscopy Studies of Single Layers and Superstructures of Porous Silicon

Giorgio Mattei

Istituto di Metodologie Inorganiche e dei Plasmi, CNR-IMIP,
(IMAI) Area della Ricerca di Roma 1, Montelibretti,
CP 10, 00016 Monterotondo Sc. (Roma), Italy

Abstract. In this paper, after a brief introduction describing the main properties of porous silicon (PS) that have attracted a great interest in view of possible applications in several scientific and technological fields, some recent studies by optical spectroscopies (Raman, IR, SHG) on PS systems, carried on by our group and collaborators, are reviewed.

In the first part, dedicated to the study of single layers of PS, the spectra obtained by micro-Raman spectroscopy from some thin mesoporous samples at different laser power are reported. The thermal effects induced by the laser heating are discussed and it is shown how it is possible to evaluate, among other things, the thermal conductivity of the measured PS material. By FTIR reflection spectroscopy the oxidation processes of PS in humid air with and without pyridine vapor have been studied. The main results are the description of the time evolution of the oxidation processes and the discovery of the catalytic role played by the organic base that accelerates the oxidation owing mainly to the water present in the air.

In the second part, concerning the works on PS superstructures (stacks of PS layers of different thickness and dielectric function), the study by FTIR spectroscopy of the dispersion of cavity-polaritons, due to the interaction of the Si-H absorption bands with the cavity mode in a PS Fabry-Perot is presented. Finally the enhancement of some order of magnitude of the signal (IR, Raman and SHG) in PS Fabry-Perot systems, due to the electric filed confinement in the cavity and to the resonance with the incident and/or the emitted light, is reported and discussed.

1. Introduction

Porous silicon (PS) is obtained by partial anodical dissolution of a Si wafer in a solution containing HF in a suitable current density regime below the electropolishing threshold [1,2]. Its "sponge like" structure with dimension in the range from microns to nanometers was early recognized [3,4]. The more recent discovery of a strong photoluminescence (PL) and electroluminescence (EL) in the visible of porous silicon at room temperature [5-7] attracted a great interest. This fact triggered an intense research activity devoted to the preparation, characterization and study of this material, also in view of possible applications in different technological fields. Since then, some thousands papers and several books [8-10] and review papers have been published.

The morphology and the dimensions of pores and of the Si skeleton depend on the Si properties and on the preparation conditions. PS materials can be divided in three groups depending on their pore size distribution, namely: microporous <2nm, mesoporous 2-50 nm and macroporous >50 nm. Microporous materials are usually characterized by strong visible luminescence that can be low or absent in the other types of PS. Without touching the debate on the origin of this PL and EL (quantum size confinement [5] or surface species [11,12]) it is important to stress here that this is the main property of PS since it opens potentially the door to the dream of realizing an all-silicon Si based optoelectronics. Beside the PL, other PS properties are also very important. In fact the PS can present a surface area as large as 600 m^2/g and because its surface reactivity can be an ideal material in the sensors field. Moreover, since for mesoporous and microporous materials the average dimension of pores and of the skeleton are much lower than the wavelength of light in the visible and in the infrared, those PS materials can be regarded as continuous and homogeneous composite media with a dielectric function intermediate between that of the air filling the pores and the one of the crystalline Si of the skeleton. This allows one to realize PS layers with well-defined optical properties that depend on the material porosity (defined as the ratio between the volume of empty space and the total volume) and hence on the preparation conditions. Since by modulating the current during the electrochemical process it is possible to fabricate PS stacks made of layers of different optical thickness and different index of refraction, any kind of optical filters and superstructures can be made with PS material [13,14]. Such PS systems can find application in optics, spectroscopy [15], laser technology [14], energy conversion (anti reflection coatings for solar cells) [16], environmental monitoring (e.g. gas sensing) [17] and optoelectronics [18-21].

The optical spectroscopies (luminescence, UV-NIR absorption, infrared, Raman, etc.) have been extensively used to study and characterize PS, being investigation techniques that provide fundamental information (band gap, impurity level, morphology, porosity surface composition, optical constant, layers thickness, etc) on these materials. In this paper the results of recent studies on PS single layers and superstructures produced in our group performed by different optical spectroscopies (Raman, FTIR and SHG) also in collaboration with other groups, are reviewed and presented together with some unpublished data. Emphasis is given to the novelty of the results and to their direct or potential implications for a better knowledge of PS properties and for the development of PS practical applications.

The rest of this paper is divided in two parts, although not indicated by headings, followed by the conclusive section. The first part consists of two sections and is devoted to the study on single Ps layers of the thermal properties by Raman spectroscopy and of the oxidative process in humid air (with and without pyridine vapor) by FTIR. The second part consists of three sections and concerns the study by FTIR, Raman and SHG spectroscopy of PS superstructure (Fabry-Perot). Details on the preparation procedure are not reported. The interested reader can find the relevant information in the cited references.

2. Study by Micro-Raman Spectroscopy of the Thermal Properties of PS layers

During the last few years intensive studies have been devoted to the characterization of porous silicon (PS) materials. Whereas electronic and optical properties of this material have been deeply investigated [10], thermal properties, in spite of their relevance for the performance of PS devices, have attracted much lower attention. By measuring the electrically induced thermal waves, thermal conductivity data strongly dependent on the morphology of the studied PS layers have been obtained [22,23]. Modulated photothermal reflectance [24] as well as photoacustic spectroscopy [25-27] have been also used to measure thermal properties of PS layers. Finally, some data on the thermal behavior of a PS layer on Si substrate, obtained by a Raman investigation, have been reported [28].

With the aim of obtaining information about the thermal stability and the thermal conductivity of single layers of different porosity and mesoporous morphology (average pore size 2 –50 nm), that are the main components of PS and optical filters, we carried out a micro-Raman study of PS membranes at different laser power density [29].

Mesoporous PS layers of thickness~30 μm were obtained by anodic dissolution of heavily doped p^+ type (100) Si plates in an electrochemical cell containing an ethanoic HF solution. By varying the current density, samples of different porosity were obtained. At the end of the required dissolution time, the current density was suddenly increased to break the PS/Si interface and hence to allow the easy removing of the PS layers from the Si substrates. After drying and storing for one month in dry air, the free standing films were used in the spectroscopic measurements. By comparing the infrared spectra with the ones calculated according to a Bruggeman model [30] the index of refraction, the absorption coefficient as well as the thickness and the porosity of the different samples were obtained. Confocal micro Raman measurements were performed by focusing the 514.5 nm radiation of an Ar^+ laser into a 2μm (in diameter) spot onto the PS membranes.

To measure the temperature of the sampled volume of the membranes the ratio of the integrated intensity of the anti-Stokes and the Stokes Si bands was employed [31,32]. To correct the data for the different overall instrumental sensitivity in the different spectral regions, anti-Stokes and Stokes Raman measurements of toluene, used as reference sample at room temperature, were performed. The uncertainty in the PS temperature measurements was estimated [33] to be 1% in the range 300 – 900 K.

As expected, PS material is very sensitive to the laser power irradiation that induces a local increase of the temperature with significant changes in the Raman spectra, as shown in Fig. 1(a). In this figure the Stokes spectra of a 43% porosity PS membrane for three different laser powers on the sample are presented. The main effects are a red-shift and a slightly asymmetrical broadening of the Si Raman band as the laser power increases.

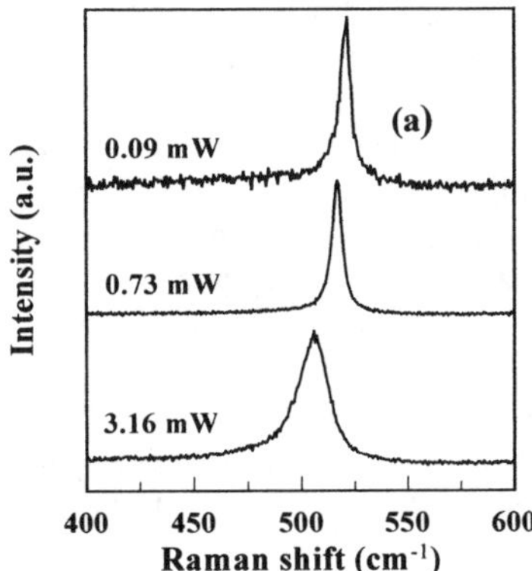

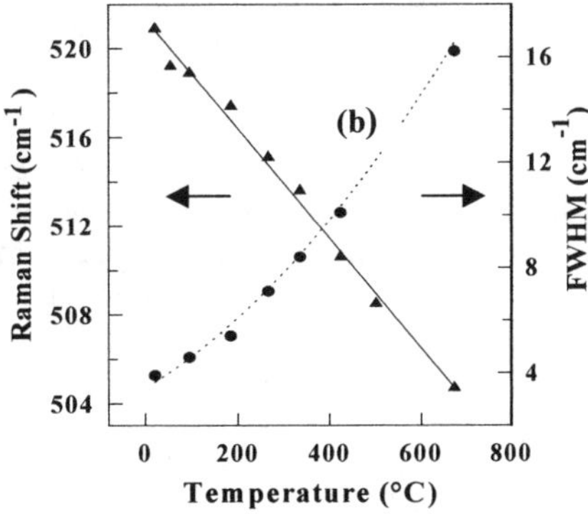

Fig. 1 For the sample of 43% porosity: (a) Micro-Raman spectra for three values of the exciting laser power; (b) Raman shift (•) and FWHM (•) of the Si band versus the measured temperature.

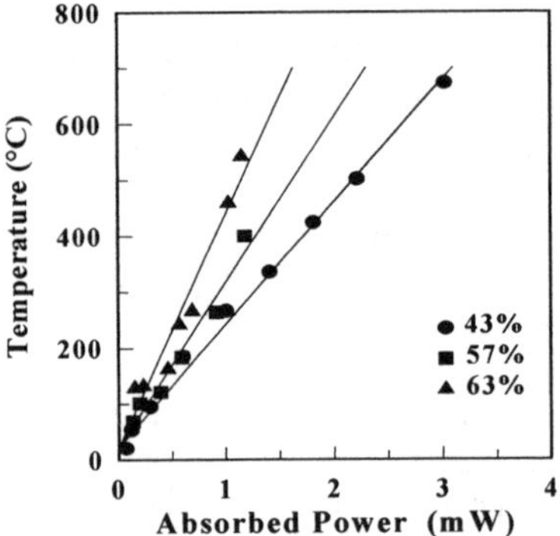

Fig. 2 Temperature versus adsorbed laser power for three PS membranes of different porosity, as indicated.

These modifications are reversible with the laser power density at least up to a maximal value that, for this sample, is about 10^5 W/cm^2. This power density value, defining a thermal stability range, is lower for higher porosity materials; for example it is $\sim 3 \cdot 10^4$ and $\sim 5 \cdot 10^4$ W/cm^2 for 57% and 63% porosity membranes, respectively.

These results should drive the attention to the need of using always an adequately low laser power in the Raman measurements to avoid the heating effect that can affect and modify strongly the spectrum of the PS material.

For the same sample of Fig. 1(a), the dependencies of the Raman shift and the Raman line-width on the measured temperature are presented in Fig. 1(b). The straight line in Fig.1(b) has an angular coefficient equal to 0.024 cm^{-1}/°C. This value is the same in the errors for all the measured membranes of different porosity and is very close to that for crystalline Si [34], in agreement with previous results [18]. Besides, it should be noted that in the limit of very low laser power the straight line points to a frequency very close to that of crystalline Si (520.5 cm^{-1}). This findings suggest that the vibrational properties of our mesoporous PS material are mainly determined by the crystalline Si properties and are not strongly affected by the morphology.

Fig. 2 shows, in the thermal stability range, the measured temperature as a function of the laser power absorbed by the sample, for three membranes of different porosity. As it is clear, for the same laser power the temperature increases as the porosity increases. Moreover, the data are fitted quite well by straight lines that point, as it is expected, to room temperature for zero laser power. While it is easily understandable that higher porosity should imply lower thermal conductivity and hence higher temperature for the same laser power irradiation, the linear dependencies deserve some additional comments.

Let us consider the sampled volume in the micro-Raman measurement as an hemisphere of radius R in thermal equilibrium at temperature T, under laser irradiation of power P, as indicated in Fig. 3

Neglecting the back irradiated heat and assuming that the incoming power is dissipated outside the hemisphere mostly via the PS thermal conductivity λ_{ps} (considered temperature independent) and that the other side of the membrane (of thickness d) is at room temperature, T_0, it results:

$$(T - T_0) = P/(2\pi\lambda_{ps} R)$$

if R<<d, as it is in this case (R/d $\cong$ 1/30).

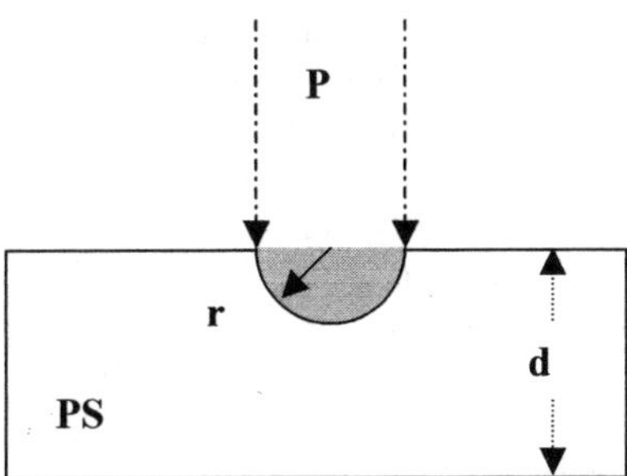

Fig. 3 Scheme of the micro-Raman experiment. The gray area represents the illuminated volume, considered here an hemisphere of radius r at temperature T; d is the thickness of the PS membrane; P is the power of the laser beam.

According to this simple model the angular coefficient of the straight lines in Fig. 2 depends not only on λ_{ps} but also on the radius R. This quantity, actually the "thickness" of the sampled volume in the material by micro-Raman, depends on the laser wavelength and on the optical constant of the material [35]. For our samples, by changing the porosity and hence by modifying also the refraction and absorption indexes, this thickness, mainly governed by the optical absorption, assumes different values, being ~1.0, 1.6 and 2.1μm for the samples of porosity 43%, 57% and 63%, respectively.

On this base, the values of the effective thermal conductivity λ_{ps}, obtained from the T vs P dependence, were: 0.7, 0.3 and 0.2 $Wm^{-1}K^{-1}$, for the three samples of increasing porosity. These values are several orders of magnitude lower than that of crystalline Si (150 $Wm^{-1}K^{-1}$ at 273 K) and are even lower than that of silicon oxide (1.4 $Wm^{-1}K^{-1}$ at 273 K). On the other hand, they are of the same order, although slightly lower, of those reported for fresh p^{+} PS as measured by dynamic 3ω technique at room temperature (e.g. $\approx$ 0.7 $Wm^{-1}K^{-1}$ for 64% porosity) [36]. It should be noted, however, that in our case the samples, measured after storing them for 30 days, were characterized by the presence of SiO_2 in the material, as evidenced by the infrared spectra.

Considering all of these results, we suggest that the thermal behavior of our PS membranes can be interpreted considering their skeleton as made by small Si crystallites connected by thin bridges, where the SiO_2, even though present in very small average amount in the whole material, becomes the main component and acts as heat barrier. Then, the overall thermal conductivity is strongly influenced not only by the microstructure, but also by the thermal properties of silicon oxide. Besides, the opposite behavior of the Si and the SiO_2 thermal conductivity as a function of the temperature could compensate each other and justify the negligible influence of the temperature on the effective thermal conductivity of the studied PS membranes. Independently, the same method has been used by other authors to study the thermal conductivity of fresh PS layers on Si wafer [37].

3. Study by FTIR of the Oxidative Process of PS Layers in Humid Air Alone or with Saturated Vapor of Various Organic Bases

In this section the main topic is the study of the oxidation by FTIR spectroscopy of PS layers. The oxidation of porous silicon is an important issue to be considered for a better understanding of the physical and chemical properties of this material. Intentional oxidation is used to stabilize the PS properties through the formation of passivating surface oxide layers. Moreover, there is a great interest in Si/SiO_2 systems in view of new applications in the field of optoelectronics [38]. Oxidation of PS has been studied in various conditions: in oxygen at middle and high temperatures [39], in liquid water at room temperature (RT) [40], in UHV chamber exposed to water vapor at various temperatures [41], in humid and dry air at different temperatures including RT [42-44], in HNO_3 [45] and in water solution of KNO_3 and alcohol [46].

Recently, we have discovered that the presence of pyridine vapor in air at RT accelerate the oxidation of PS layers [47]. Here the main results of a comparative FTIR study on the oxidation of PS in humid air alone or saturated with vapor of pyridine [48] or of other organic bases are briefly presented and discussed.

In Fig.4 the IR reflection spectra of a sample exposed to air saturated with water vapor at RT for three different exposure time t_e are presented. For the sample as prepared ($t_e = 0$) the main absorption bands are due to the Si-H vibrations of SiH_x (Si-H stretching in SiH at 2080 cm^{-1} and in SiH_2 at 2106 cm^{-1}), as expected for the hydrogen terminated PS surface. With increasing t_e the reflection minimum due to the interference moves to higher frequency owing to the reduction of the effective refractive index of the PS layer due to the substitution of Si with silicon oxide. Besides, new bands appear and the relative intensities of all the bands change. Strong silicon oxide absorption bands appear in the region 950 – 1250 cm^{-1} (that includes the bulk as well the surface Si-O-Si stretching vibrations [39]) and their intensity increases with the exposure time. In the region 2100 – 2300 cm^{-1}, bands attributable to the Si-H stretching for silicon back-bonded to oxygen atoms, O_ySiH_x (Si-H stretching in $OSiH_3$ at 2136 cm^{-1}, in O_2SiH_2 at 2200 cm^{-1} and in O_3SiH at 2249 cm^{-1} [42]), grow up. In the 3000 – 3700 cm^{-1} region weak and broad bands due to the O-H vibration also appear. While the intensities of all the above mentioned new bands tend to increase

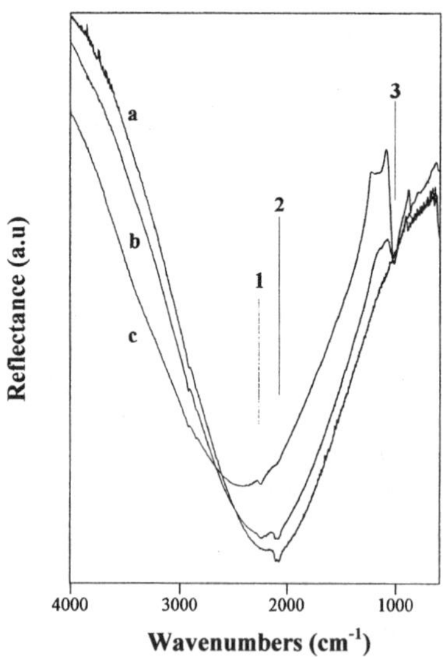

Fig. 4 FTIR reflection spectra of a PS sample exposed at RT to air saturated with water vapor for 0 min (a), 3880 min (b) and 12372 min (c). The vertical lines indicate the positions of the main bands: O_ySi-H_x (1), Si-H_x (2) and Si-O-Si (3).

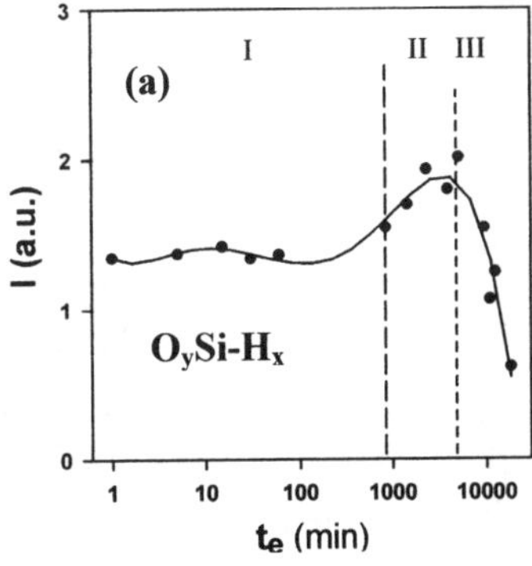
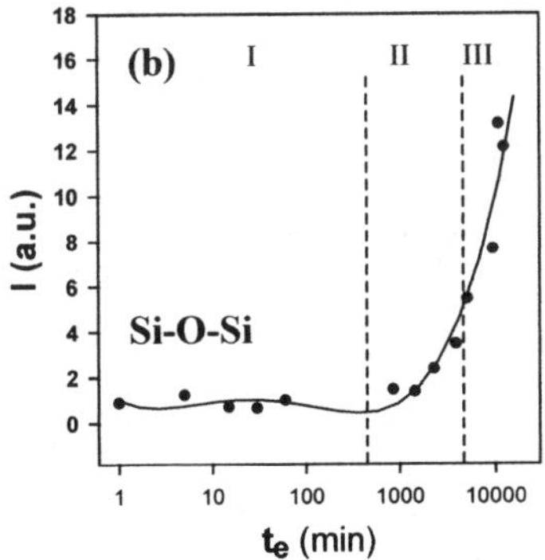

Fig. 5 Integrated absorption intensity of O_ySi-H_x (a) and Si-O-Si (b) infrared bands as a function of the time of exposure to air saturated with water vapor at RT. The solid lines are guides to the eye.

with t_e, the Si-H stretching bands due to the original H terminated PS surface have opposite behavior.

The time dependencies for the O_ySiH_x and Si-O-Si bands are shown in Fig.5 where the total integrated absorption intensities, obtained by spectral treatments and by adding the contributions of all the bands present in the specific spectral regions mentioned above, are reported as a function of t_e.

All the process can be divided in three stages. Stage I, the first 1000 min, corresponds to the formation of Si-OH species. The O-H is the only signal (not shown here) increasing during this time. It is due to the formation of OH species on the surface through the dissociative interaction of water molecules [49] that is slowed down by the H terminations of the PS surface. A similar preliminary slow oxidation step has been observed in the oxidation of PS in pure liquid water [40]. As it is known from studies on the reaction of water with crystalline silicon [50-51], the Si-OH is a special effective site for the attack of water to silicon back-bonds to form Si-O-Si. This process, stage II, gives rise to an increase

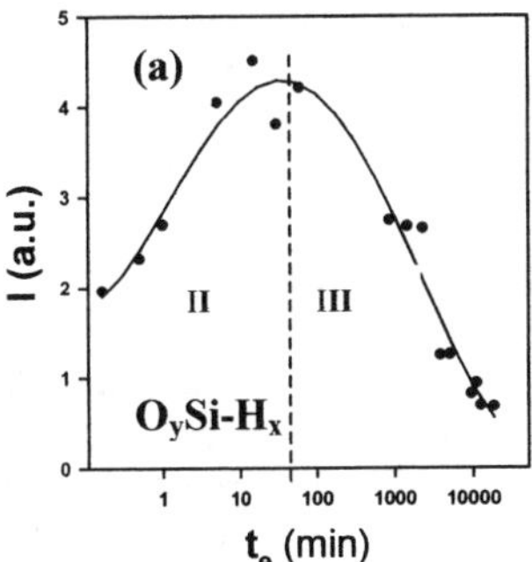
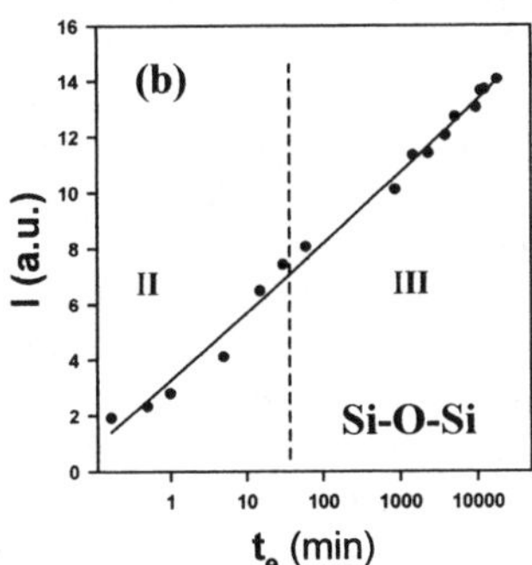

Fig. 6 Integrated absorption intensity of O_ySi-H_x (a) and Si-O-Si (b) infrared bands as a function of the time of exposure to humid air saturated with pyridine vapor at RT. The solid lines are guides to the eye.

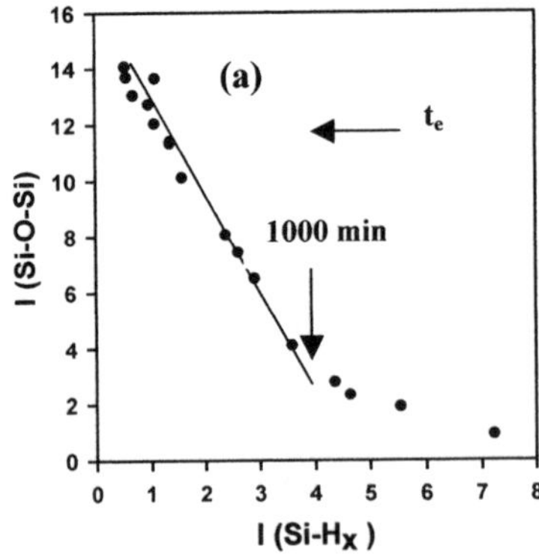
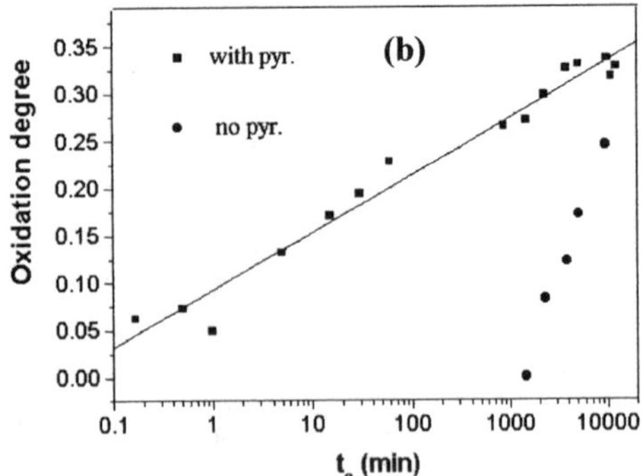

Fig.7 (a) Integrated absorption intensity of Si-O-Si bands versus integrated absorption intensity of Si-H$_x$ bands. The horizontal arrow indicates the direction of increasing exposure time in humid air alone.
(b) Oxidation degree for two parts of the same PS sample treated in air saturated with water vapor (•) and in humid air saturated with pyridine vapor (•) at RT as a function of the exposure time.

in intensity of the absorption bands of the silicon oxide and O$_y$SiH$_x$ species and to a decrease of the SiH$_x$ signal (not shown here). After about 5000 min (stage III), the amount of H atoms bonded on the PS surface is significantly reduced because of the formation of Si-OH and surface Si-O-Si bonds. The oxidation process continues: the silicon oxide signal is still growing up while that of the O$_y$SiH$_x$ species decreases.

From the comparison of these results with those of other authors of studies on the PS oxidation in dry and humid air [42,44] and in liquid water [40,50] we infer that the water is the main oxidant agent of PS in humid air at RT.

If PS is exposed to air (25% humidity) saturated with pyridine vapor at RT the oxidation process is similar to that discussed above but is accelerated. The infrared spectra show the same absorption bands, in addition some very weak new bands attributable to the adsorbed pyridine are present. As shown in Fig. 6, the step II of the process is already started after 10 seconds and after 20 minutes there is the transition to the finale stage III.

To interpret this fast oxidation of PS in air in presence of pyridine, we first note that, as indicated in Fig. 7(a), there is a clear correlation between the decreasing of the Si-H signal with the exposure time and the increasing of the silicon oxide signal that becomes a linear relation after stage I. This correlation that holds in both cases (with and without pyridine vapor), although with different time scale, proves that the H terminated sites on the PS surface are the places where the oxidation of the material initiates and develops. Nevertheless the S-H bond is quite resistant and in normal condition, in air at RT, the oxidation process is quite slow. In presence of pyridine the oxidation conditions can chance significantly. The pyridine is a well known aromatic heterocyclic base that can establish an attractive interaction through its nitrogen lone pair with the hydrogen atoms bonded to the silicon and present in large extent on the PS inner surface. This interaction, proved by the slight change of the frequency of some pyridine absorption bands, weaken the Si-H bond promoting the interaction with the oxidant agent (mainly the water molecules) and in such way developping and sustaining the observed fast PS oxidation. Finally, we note that, in spite of the different rates of the processes, the final oxidation degree (the percentage of silicon converted in silicon oxide) is the same in both cases for the same PS material, as shown in Fig. 7(b), being likely related to the original amount of Si-H species in the PS layer.

Recently, we have started an infrared study of some pieces of the same mesoporous PS sample exposed to humid air saturated with the vapor of different organic bases (e.g. lutidine, picoline, mono-, di- and triethylamine, etc.) at RT [52]. The preliminary results indicate that the main parameter affecting the acceleration of the PS oxidation is the concentration in vapor phase of the used organic molecules.

4. Investigation by FTIR of the Dispersion of Cavity-Polaritons in a PS Fabry-Perot

This and the following sections concern the study by optical spectroscopies of PS superstructures and in particular of PS Fabry-Perot micro-cavities.

As indicated in the introduction, by controlling the current and the anodization time, layers of well defined porosity and thickness can be produced. By governing the porosity it is possible to select a desired value for the effective index of refraction of the PSi layer in a wide continuous range from 1.1 up to 2.6 or more [15]. Refractive index, porosity and growth rate ca be obtained as functions of the current passed through the electrochemical cell. These results ca be than used as calibration for the preparation of PSi superstructures whose optimal characteristics are designed by a computer simulation. By this procedures several PSi samples were prepared and characterized: single layers, high reflection Bragg mirrors and optical microcavity systems.

Distributed Bragg reflectors designed by alternating high and low refractive index layers of $\lambda/4$ optical thickness were obtained. As an example in Fig.8 the infrared spectrum of a PS Bragg reflector, centered around 2000 cm^{-1}, with reflectivity higher then 95% is compared with the calculated one. The two curves overlap almost completely except in the central region of the high reflectivity plateau where several weak absorption bands due to the Si-H vibrations appear in the measured spectrum while were not considered in the calculation.

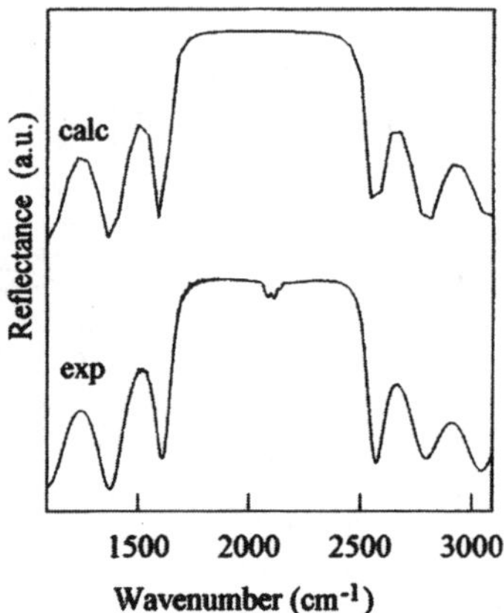

Fig.8 Comparison between measured and calculated infrared TM reflection spectra for a diffuse Bragg reflector made by four repetitions of alternate PSi layers of low and high index of refraction. Reflectivity at the top plateau is 95%.

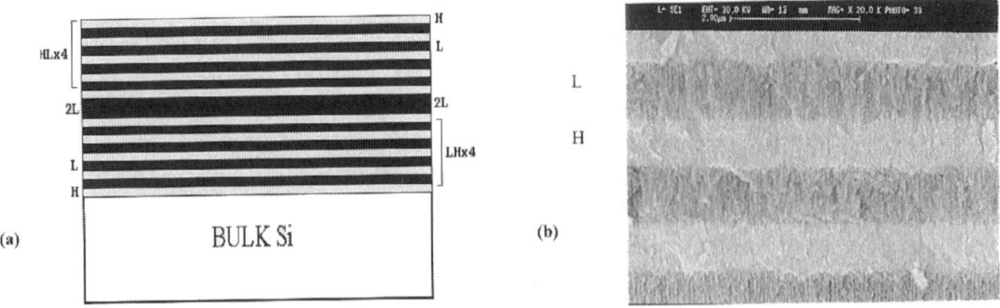

Fig. 9. (a) Scheme of a FPMC system: H and L indicate the high and low refractive index of refraction layers, respectively; (b) SEM picture of part of the cross-section of a FPMC system showing the different PSi layers.

PS Fabry–Perot microcavities, hereafter indicated as FPMC, with two distributed Bragg reflectors used as mirrors have been made. The scheme of a typical FPMC structure is shown in Fig. 9(a). The $\lambda/2$ thick region (2L) between the to mirrors is the cavity layer. In Fig. 9(b) a SEM picture of the cross section of such FPMC is shown. In spite of the defects and damaging introduced by the cutting, the regularity of the PSi different layers can be appreciated.

The TM infrared spectra of a PS FPMC centered at normal incidence around 2000 cm^{-1} measured at four angles of incidence are presented in Fig. 10(a) [15]. The bands around 2100 cm^{-1} due to the Si-H vibrations are clearly detectable. As shown by the figure, by tilting the sample, namely by increasing the angle of incidence, the filter band moves to higher frequency and the cavity mode, represented by the sharp deep on the high reflectivity

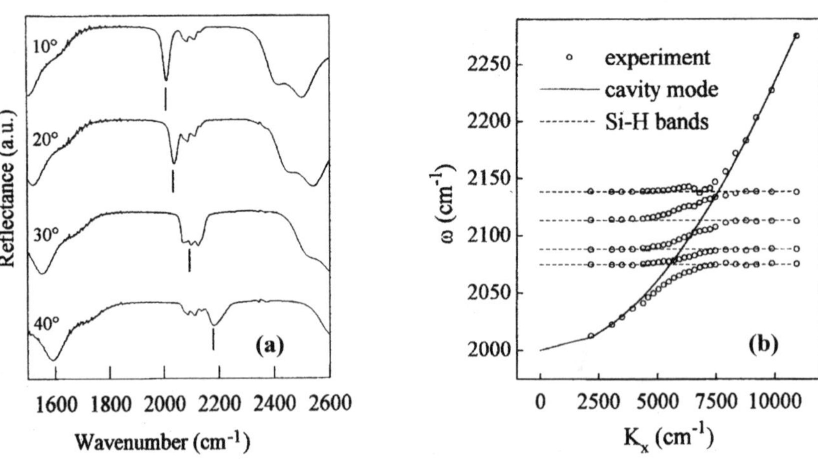

Fig.10.   (a) TM infrared spectra at four angles of incidence for the PSi FPMC centred around 2000 cm^{-1} at normal incidence. The vertical bars indicate the frequency position for the unperturbed cavity mode at each angle. (b) Dispersion curves for the coupled modes (cavity mode/Si-H vibrations) for the FPMC discussed in the text. The circles are the experimental values, the solid line is the calculated dispersion for the unperturbed cavity mode while the broken lines are the dispersion curves for the unperturbed Si-H vibrations.

plateau, crosses the Si-H absorption bands giving rise to coupled modes. These coupled modes present very interesting features in their dispersion as shown in Fig.10(b), where the frequencies of these modes are plotted as a function of the wavevector, K_x, parallel to the interfaces (that is proportional to the sinus of the angle of incidence). Cavity mode shifts with the angle of incidence while Si-H band positions do not depend on this angle until cavity mode crosses them. At the crossing, the modes couple and gaps in the dispersion curves appear as shown in Fig. 3. These curves resemble the dispersion behavior of polariton modes in a film but, being coupled modes in an optical cavity, they have a smaller radiative width (i.e. longer lifetime) as a result of the high reflectivity at the film interfaces (the Bragg mirrors) and hence of the low radiative leakage. Because of these characteristic features we named these modes as "cavity polaritons".

We have observed similar coupled modes also in a subsequent study of the physisorption of dodecane molecules, $C_{12}H_{26}$, in a PS FPMC [53], as shown in Fig. 11(a). In this case three gaps in the dispersion curves are clearly seen (corresponding to the three C-H vibrations at 2855, 2924 and 2960 cm^{-1}) For weak coupling (small oscillator strength due to the low adsorbate concentration in the PS pores) no gap in the dispersion curve appears and the curves intersect with a slight distortion. For intermediate concentrations, only some gaps disappear starting from weaker bands. The interaction of microcavity mode and dodecane's bands is present even when the concentration becomes very small (0.7%).

Another important feature of the intersection is the enhancement of absorption by vibrational bands as a result of photon confinement in microcavity acting like multipass cell. Fig. 11(b) demonstrates this effect for one of the modes. Maximum enhancement appears at the intersection region. This effect, due to the electric field enhancement caused by the multireflection in the cavity at the optical resonance condition, i.e. around the cavity mode frequency, is also responsible for the enhancement of the PL and EL emission in PS microcavities [19,54] and the enhancement in the Raman intensity in suitable semiconductor Fabry-Perot systems [55]. This infrared absorption enhancement (up to orders of magnitude) due to the interaction with cavity mode, once tuned in the frequency range of interest, has allowed us to detect very weak absorption features down to the dodecane concentration 0.7%.

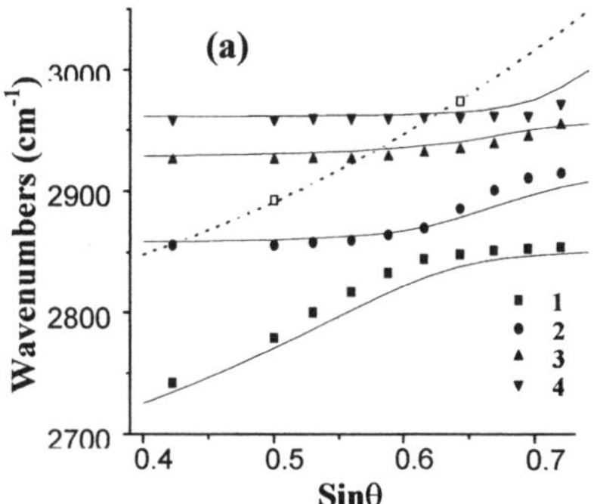
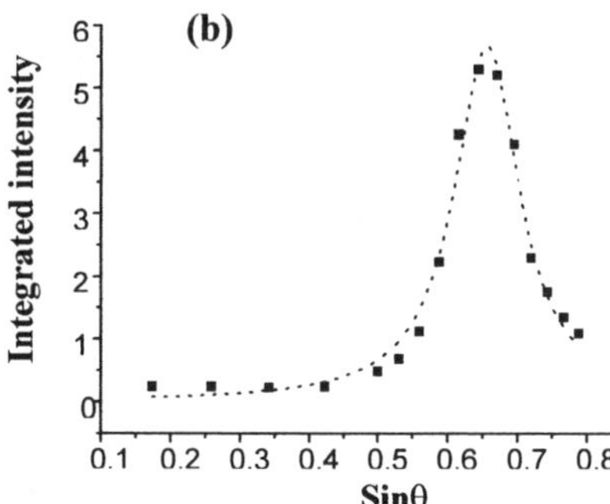

Fig.11 (a) Calculated (solid lines) and experimental (filled symbols) dispersion curves for the cavity polaritons (cavity mode/C-H vibrations) at high dodecane concentration. The dashed line and open squares are the calculated and the experimental values, respectively, for microcavity before dodecane introduction. The symbols are numbered to indicate the different branches of experimental dispersion curves starting from that at lower frequency. (b) Integrated intensity of the absorption band of the C-H at 2855 cm^{-1} (unperturbed frequency) as a function of the sine of the angle of incidence θ. The black square are the experimental values while the dotted line represents a fitting curve.

5. Enhancement of the Raman Signal in PS Fabry-Perot Structures

As it is well known Raman spectroscopy is characterized by a quite low cross section (e.g usually several order of magnitude lower than for IR absorption). For this reason is quite difficult to measured the Raman signal of very thin films and of adsorbates without some enhancement (e.g. SERS , Resonant Raman). In this section we shall show how it is possible to take advantage from the e.m. field enhancement in the cavity in a FPMC to improve significantly the sensitivity of Raman spectroscopy. We shall consider first the case of an all-PS FPMC (subsection 5.1) and then case of an hybrid PS/Ag FPMC (subsection 5.2).

Unlike reflection spectroscopy, where only one angle (the angle of incidence equal to the angle of reflection) plays the main role in the enhancement of the observed signal from a FPMC, in Raman experiment the angle of incidence of the exciting radiation and the angle of collection of the scattered light are independent and double resonance conditions can be achieved. These resonance conditions can be better understood by considering Fig. 12 where are sketched the ideal reflection spectra of a FPMC at three angles of observation. Using a FPMC with the cavity mode centered at frequency ν_0 at normal incidence, $\theta = 0$, it is possible by increasing the angle of incidence to an optimal value θ_L, to blue-shift the cavity mode frequency to match that, ν_{laser}, of the excitation laser. Being the stokes radiation at lower frequency ν_s then, at a suitable angle θ_s lower then θ_L, the scattered light can be also collected in resonance condition with the cavity mode. Owing to this double resonance mechanism, enhancement of the Raman signal up to four orders of magnitude has been reported for a high quality (GaAs/AlAs) FPMC [55].

Some general remark should be made: i) only a relatively small frequency range of the stokes radiation is efficiently enhanced so that, to optimize the Raman feature of interest, the FPMC must be specially designed; ii) in case of wide range Raman spectrum at a defined θ_s this frequency dependent enhancement can introduce significant modifications in the intensity distribution of the bands so that care should be taken in the spectrum interpretation.

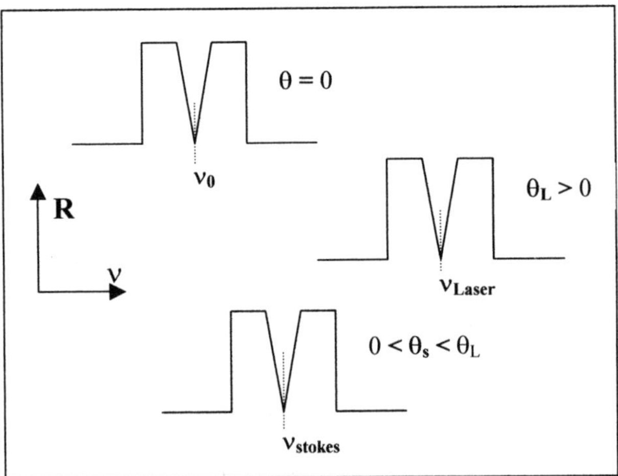

Fig. 12. Sketches of the reflection spectra (R vs. ν) of an ideal FPMC at normal incidence $\theta = 0$, with the laser radiation of frequency ν_{Laser} incident in resonance condition at the angle θ_L and with the scattered stokes light of frequency $\nu_{stokesr}$ collected in resonance condition at the angle θ_s.

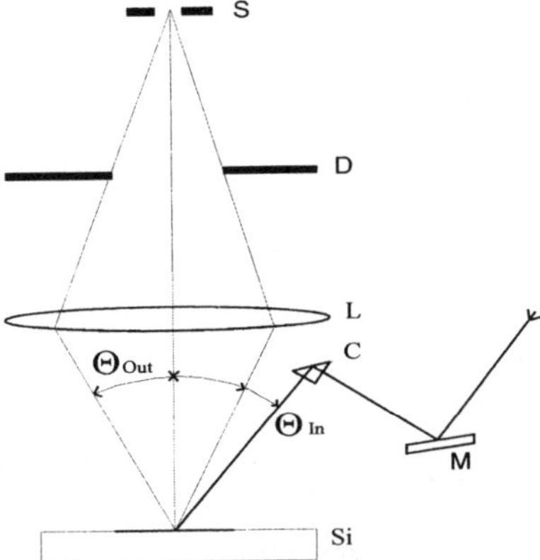

Fig. 13. Optical scheme of the Raman experiment: the laser beam strikes on the sample (Si or PS) at the angle Θ_{in} and the scattered light is collected by the lens L and sent to the entrance slits S of the spectrometer after passing through the diaphragm of a diameter D which determines the angle Θ_{out}. M is a mirror and C is a prism.

5.1 Case of all-PS Fabry-Perot Micro-Cavities

As indicated above to realize the optimal double resonance Raman conditions it is necessary to design the PS FPMC by taking into account the frequency of the excitation laser and of the scattered light as well. The scattering geometry must be settled consequently.

In a first work [56] we have designed and fabricated a PS FPMC for studying the enhancement of Raman signal of the silicon phonon as a function of the incidence and scattering angles excited by the green light (514.5 nm) of an argon ion laser. To reduce the luminescence background a mesoporous PS material was used. The optical scheme of the Raman experiment is presented in Fig. 13 (for more details see ref. [56]).

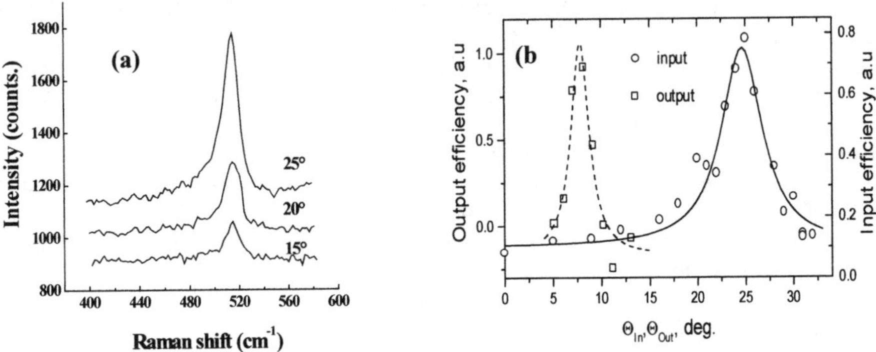

Fig. 14   (a) Raman spectra of the PS structure excited at different angles of incidence Θ_{In} of the laser beam. (b) Raman output dependence on Θ_{In} – open circles and on the angle of scattering Θ_{Out} – open squares. The lines show Lorentz fits of these dependencies.

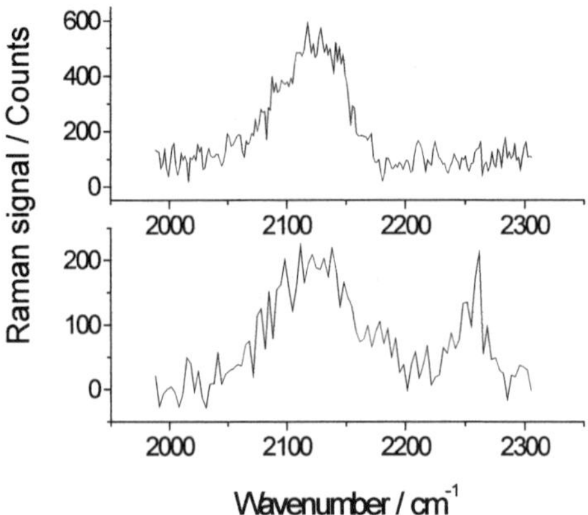

Fig. 15. Raman spectra of stretching Si-H vibrations in the fresh (top) and oxidized (bottom) PS FPMC in double resonance conditions with the cavity mode.

Some Raman spectra collected along the normal to the surface at different angles of incidenceare shown in Fig. 14(a).

From the obtained spectra we have calculated the integrated intensity of the silicon band at 520 cm^{-1}. To demonstrate experimentally an enhancement of the Raman output due to the microcavity, the dependencies of the integrated band intensity of the PS structure on both the incidence and the scattering angles were measured. These dependencies, normalized with respect to that of crystalline silicon (measured in the same conditions), to take into account the geometry of the experiment, are presented in Fig. 14(b). It is clear from these results that the best angles for double resonance condition are near 25° for the incident laser photons and 8° for the Stokes scattered photons. The first angle corresponds to the minimum in the measured reflectivity of our PS structure for the laser light. We have estimated for the enhancement of the Raman signal, arising from the double resonance with the cavity mode, a value of 30 or higher.

Once demonstrated the enhancement of Raman intensity in PS structure, we have applied the method to the study of others species in PS material. Raman spectra of the Si-H stretching vibration were measured for an as-prepared FPMC [57]. The main features shown on Fig. 15, bands in the region 2100 cm^{-1}-2150 cm^{-1} (Si-H$_x$), are the same as in the IR spectra of the same sample. For oxidized sample these bands are weaker and a new band at 2250 cm^{-1} (OSi-H$_x$) appears. It is important to note that the observed Raman signal is quite weak and without the enhancement due to the coupling with the cavity mode, namely out of the resonance conditions, is undetectable in this FPMC.

5.2 Case of a hybrid Fabry -Perot (PS/C$_{60}$/Ag) micro-cavity

Sometimes it is difficult to introduce the molecules of interest deep inside porous silicon structures but it is possible to adsorb them on the surface and in the pores of the top layer. In such case a hybrid metal-porous silicon structure can be used to enhance Raman signal [57].

We have designed and fabricated a hybrid FPMC by preparing first a PS structure made with at the bottom a Bragg reflector and with a λ/2 layer of high porosity on the top. Then

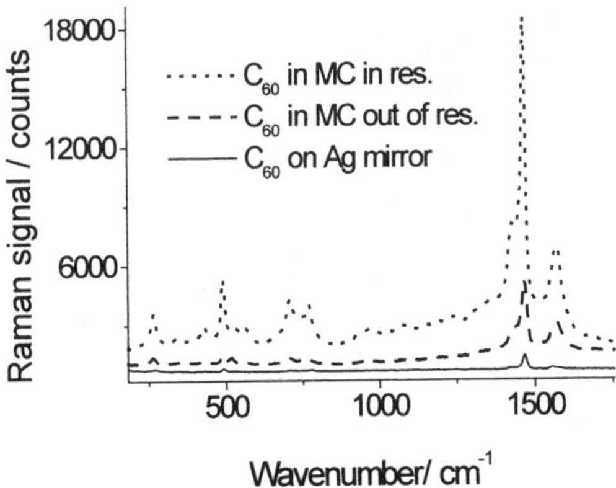

Fig . 16　　Raman spectra of fullerene, C_{60}, on the silver mirror and in the hybrid PS/Ag FPMC.

we have deposited a 10 nm thick fullerene film and a semitransparent 30 nm thick continuous silver film as a top reflector.

In Fig. 16 the Raman spectra of fullerene in the FPMC measured in resonance and out of resonance conditions are compared with the spectrum of the same fullerene film on the same Ag mirror. As shown in the figure, this hybrid PS/Ag structure also gives a considerable enhancement of Raman signal in the resonance conditions.

In conclusion, our results demonstrate the possibility of double resonance enhancement of the Raman signal in PS structures and hence provide a way to increase the sensitivity of the Raman spectroscopy for studying the species inside porous silicon which can considerably influence the properties of this material and hence of the PS based devices.

6. Enhancement of the SHG Signal in a PS Fabry-Perot Superstructure

A recent work has been published that was devoted to the investigation of the non linear optical response of a photonic crystal fabricated from a centrosymmetric materials, namely a porous silicon Fabry-Perot microcavity [58]. The intensity spectrum of the second harmonic (SH) generated in a PS FPMC was reported. The used FPMC was designed to work with the pump wavelength in the range 730 – 1100 nm.

As an example of the obtained results, the dependence of the SH intensity $I_{2\omega}$ on the pump wavelength λ_{ω} measured at the angle of incidence $\theta = 45°$ in the s-p geometry is presented in Fig. 17(a). For comparison the reflectivity R_s of the s-polarized pumping radiation is shown in Fig. 17(b). As indicated by the figure, the enhancement of the SH response was observed near the cavity mode at 780 nm (200 enhancement factor) and at the edge, at 910 nm, of the high reflectivity plateau (the "photonic band gap").

In Fig. 17 the experimental data are compared with the calculated curves obtained using a suitable model for the FPMC optical response. This model demonstrates that the enhancement of SH response at the cavity mode frequency differs in nature from the enhancement at the edges of the photonic band gap. The SH resonance with the cavity mode

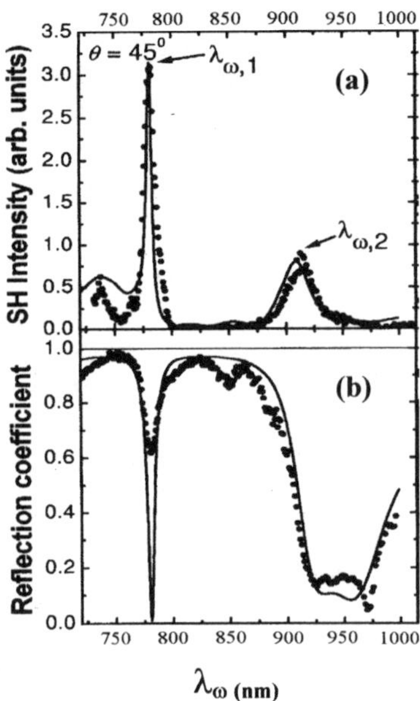

Fig.17 (a) SH intensity, $I_{2\omega}$, of the p-polarized SH radiation as a function of the wavelength λ_ω of the s-polarized pump radiation measured in the PS FPMC at angle of incidence $\theta = 45°$. (b) Reflection coefficient at 45° of the s-polarized pump radiation. Solid line are the results of model calculation.

is caused by the localization (amplification) of the standing pump wave in the vicinity of the cavity layer, whereas the SH resonance at the edge of the photonic band gap is due to the uniform amplification of the pump field in the distributed Brag mirrors of the FPMC.

7. Conclusions

In this paper several applications of the optical spectroscopies to the study of PS systems have been presented.

It has been shown how it is possible to evaluate the thermal conductivity of PS single layers from their spectra obtained by micro-Raman spectroscopy. The thermal effects induced by the heating due to the laser exciting radiation has been also discussed.

The importance of the oxidation process in PS material has been stressed and the results of a FTIR study on the effects of the exposure of PS layers to humid air alone or in presence of vapor of some organic base (e.g. pyridine) have been shown. A complete and detailed description of the evolution with the exposure time of the IR absorption bands of the main species involved in the oxidation processes (namely: $Si\text{-}H_x$, $O_ySi\text{-}H_x$, $SiO\text{-}H$ and $Si\text{-}O\text{-}Si$) has been also presented. Besides, it has been demonstrated that the organic molecules act as catalyst by accelerating the oxidation process provoked by the water present in the air.

The second part of the paper was devoted to the study of PS superstructures (Fabry-Perot microcavity).

The dispersion curves of the cavity-polaritons due to the interaction of the IR absorption bands (due to $Si-H_x$ or C-H of adsorbed dodecane molecules) with the cavity mode in PS Fabry-Perot systems were obtained by FTIR spectroscopy. It has been also shown that in such PS micro-cavities the IR signal, when in resonance with the cavity mode, can be enhanced by some order of magnitude.

An analogous enhancement effect has been also reported for the Raman signal in PS micro-cavities. It as been proved that the maximum enhancement is achieved in double resonance conditions, i.e. when both the incident exciting radiation and the scattered light are at suitable angles in resonance with the cavity mode. The dependencies on the incidence and scattering angles of the Raman signal of the Si vibration near 520 cm^{-1} have been presented. This cavity enhanced Raman effect has been used to study the oxidation effect in the Raman spectrum of a PS Fabry-Perot. Besides, the enhanced Raman spectrum of fullerene molecules in a hybrid ($PS/C_{60}/Ag$) Fabry-Perot micro-cavity has been shown.

Finally, the enhancement (by a factor $2x10^2$) of the SHG signal in a PS Fabry-Perot micro-cavity has been presented. The causes of this effect have been briefly discussed.

Acknowledgments

The author wish to acknowledge all the colleagues that where co-authors of the works discussed in this paper. In particular: A. Marucci, M. Pagannone and V. Valentini of the "Istituto di Metodologie Avanzate Inorganiche, CNR-IMAI, Montelibretti (Roma), Italy", E; Alieva, L. A. Kuzik, J.E. Petrov and V.A. Yakovlev of the "Institute of Spectroscopy, Russian Academy of Sciences, Troitsk, (Moscow region) Russia", T.V Dolgova, A.I.Maidikoskii, M.G. Martem'ianov, A.A. Fedyanin and O.A. Aktsipetrov of the "Moscow State University, Vorob'ery gory, (Moskow) Russia" and finally G. Marowsky and D. Schumacher of the "Laser-Laboratorium Gottingen, (Gottingen) Germany". Finally, special thanks to V. Valentini for her precious assistance in the preparation of the manuscript.

References

[1]   A. Uhlir, *Bell Syst. Tech. J.* **35** (1956) 333.
[2]   D. R. Turner, *J. Electrochem. Soc.* **105** (1958) 402.
[3]   G. Bomchil, A. Alimaoui and R. Hérino, *Microelectr. Eng.* **8** (1988) 293.
[4]   G. Bomchil, A. Alimaoui and R. Hérino, *Appl. Surf. Sci.* **41/42** (1989) 604.
[5]   L. T. Canham, *Appl. Phys. Lett.* 57, **1990**, 1046.
[6]   A. Halimaui, C. Oules, G. Bomchil, B. Bsiesy, F. Gespard, R. Hérino, M Ligeon, and F. Muller, *Appl. Phys. Lett.* **59** (1991) 304.
[7]   L.T. Canham, W.Y. Leong, M.I.J. Beale, T.I. Cox and L. Taylor, *Appl. Phys.Lett.* **61** (1992) 2563.
[8]   Z.C. Feng and R. Tsu (Editors), Porous Silicon, World Scientific, Singapore, 1994.
[9]   J-C Vial and J. Derrien (editors), Porous Silicon Science and Technology, Springer-Verlag, Berlin Heidelberg and Les Editions de Paris, Les Ulis, 1995.
[10]   L. Canham (ed.), Properties of Porous Silicon, EMIS 18, INSPEC, London 1997.
[11]   S.M. Prokes, O. J. Glembock, V.M. Bermudez, R. Kaplan, L.E. Friedersdorf and P.C. Searson, *Phys Rev. B* **45** (1992) 13788.
[12]   M. Stutzmann, M.S. Brandt, M. Rosenbauer, J. Weber and H.D. Fuchs, *Phys Rev. B* **47** (1993) 4806.
[13]   M.G. Berger, C. Dieker, M. Thönissen, L. Vescan, H. Luth, H. Munder, W. Theiß, M. Wernke and P. Grosse, *J. Phys. D – Appl. Phys.* **27** (1994) 1333.
[14]   G. Mattei, A. Marucci and V. A. Yakovlev and M. Pagannone, *Laser Phys.* **8** (1998) 755.
[15]   G. Mattei, A. Marucci and V. A. Yakovlev, *Mat. Scienc. & Engineering B* **51**, 158 (1998).
[16]   P. Menna, G. Di Francia and V. La Ferrara, *Sol. Energ. Mat. and Solar Cells*, **37** (1995) 13.
[17]   M.P. Stewart and J.M. Buriak, *Adv. Mater.* **12** (2000) 859.
[18]   S. Frohnhoff and M. G. Berger, *Adv. Mater.* **6** (1994) 963
[19]   V. Pellegrini, A. Tredicucci, C. Mazzoleni and L. Pavesi, *Phys. Rev. B* **52** (1995) 14328.
[20]   L. Pavesi, R. Guardini, C. Mazzoleni, *Solid. State Comm.* **97** (1996) 1051.

[21]   M. Araki, H. Koyama and N. Koshida, *J. Appl. Phys.* **80**, (1996) 4841.

[22]   A. Drost, P. Steiner, H. Moser, W. Lang, *Sens. Mater.* **7** (1995) 111.

[23]   W. Lang, A. Drost, P. Steiner and H. Sandmayer, *Mater. Res. Soc. Symp. Proc.* **358** (1995) 561.

[24]   G. Amato, L. Boarino, G. Benedetto and R. Spagnolo, *Thin Solid Films* **255** (1995) 111.

[25]   A. N. Abraztsov, V. Yu. Timoshenko, H Okushi and H. Watanabe, *Semiconductors* **31** (1997) 534.

[26]   G. Benedetto, L. Boarino and R. Spagnolo, *Appl. Phys. A* **64** (1997) 155.⁹I. V. Blonskij, M. S. Brodyn,

[27]   V. A. Tkhoryk, A. G. Filin and Ju. P. Piryatiaskij, *Semicond. Sci. Technol.* **12** (1997) 11.

[28]   E. D. Obraztsova, L. P. Avakyants and G. B. Demidovich, *J Electr. Spec. and Relat Phenomena* **64/65** (1993) 587.

[29]   G. Mattei, A. Marucci, M. Pagannone and V.A. Yakovlev. In: A.M. Heyns (ed.), Proceedings of the XVI International Conference on Raman Spectroscopy. John Wiley & Sons, Chichester-New York-Weinheim-Brisbane-Singapore-Toronto, 1998, pp.640 –641.

[30]   T W. Theiss, *Surf. Sci. Reports* **29** (1997) 91.

[31]   B. M. Malyj and J. E. Griffiths, *Appl. Spectr.* **37** (1983) 315.

[32]   S B. J. Kip and R. J. Meyer, *Appl. Spectr.* **44** (1990) 707.

[33]   F. LaPlant, G. Laurence and D. Ben-Amotz, *Appl. Spectrosc.* **50** (1996) 1034.

[34]   R. Tsu and J. G. Hernandez, *Appl. Phys. Lett.* **41** (1982) 1016.

[35]   G. Turrell and P. Dhamelincourt. In: J.J. Laserna (ed.), Modern Techniques in Raman Spectroscopy. John Wiley & Sons Chichester-New York-Weinheim-Brisbane-Singapore-Toronto, 1996, Chapt.4.

[36]   G. Gesele, J. Linsmier, V. Drach, J. Fricke and R. Arens-Fisher, *J. Phys. D* **30** (1997) 2911..

[37]   S. Perichon, V. Lysenko, B. Remaki, D. Barbier and B. Champagnon, *J. Appl. Phys.* **86** (1999) 4700.

[38]   L. Pavesi, L. Dal Negro, C. Mazzoleni, G. Franzò and F. Priolo, *Nature*, **408** (2000) 440.

[39]   D. B. Mawhinney, J. A. Glass, Jr., and J. T. Yates,Jr., *J. Phys. Chem. B* **101** (1997) 1202.

[40]   J. E. Bateman, R. D. Eagling, B. R. Horrocks, A. Houlton and D. R. Worrall, *Chem. Commun.,* **23** (1997) 2275.

[41]   P. Gupta, A. C. Dillon, A. S. Braker and S. M. George, *Surf. Sci.* **245** (1991) 360 .

[42]   Y. Ogata, H. Niki, T.Sakka and M. Iwasaki, *J. Electrochem. Soc.* **142** (1995) 1595.

[43]   J. Salonen, V.-P. Lehto and E. Laine, *Appl. Phys. Lett.* **70** (1997) 637.

[44]   Y. H. Ogata, T. Tsuboi, T. Sakka and S. Naito, *J. of Porous Mat.* **7** (2000) 63.

[45]   F. Leisenberger, R. Duschek, R.Czaputa, R.Netzer, F.P. Beamsom and J.A.D. Matthew, *Appl. Surf. Sci.* **108** (1997) 273.

[46]   Salonen, V.P. Lehto and E. Laine, *Appl. Surf. Sci.* **120** (1997) 191.

[47]   G. Mattei, E. V. Alieva, J. E. Petrov, and V. A. Yakovlev, *Phys. Stat. Sol. (a)* **182** (2000) 139.

[48]   G. Mattei, V. Valentini and V.A. Yakovlev, accepted for publication in *Surf. Sci.* (2002).

[49]   R. L. Smith and S. D. Collins, *J. Appl. Phys.* **71** (1992) R1.

[50]   D. Gräf, M. Grundner and R. Schulz, *J. Vac. Sci. Technol. A* **7** (1989) 808.

[51]   E. P. Boonekamp, J. J. Kelly, J. Van de Ven and A. H. M. Sondag, *J. Appl. Phys.* **75** (1994) 8121.

[52]   V. Valentini and G. Mattei, to be published.

[53]   G. Mattei, E. V. Alieva, J. E. Petrov, and V. A. Yakovlev, *Surf. Sci.* **427-428** (1999) 235.

[54]   L. Pavesi, C. Mazzoleni, A. Tredicucci and V. Pellegrini, *Appl. Phys. Lett.* **67** (1995) 3280.

[55]   A. Fainstein, B. Jusserand and V. Thierry-Mieg, *Phys. Rev. Lett.* **75** (1995) 3764.

[56]   L.A. Kuzik, V.A. Yakovlev abd G. Mattei, *Appl. Phys. Lett.* **75** (1999) 1380.

[57]   G. Mattei, L.A. Kuzik and V.A. Yakovlev. In , Shu-Lin Zhang and Bang-fen Zhu Editors, Proceedings of the XVIIth International Conference on Raman Spectroscopy. John Wiley & Sons, Chichester-New York-Weinheim-Brisbane-Singapore-Toronto, 2000, pp. 498-499.

[58]   T.V. Dolgova, A.I. Maidikovskii, M.G. Martem'yanoy, G. Marovsky, G. Mattei, D. Schumacher, V.A. Yakovlev, A.A. Fedyanin and O.A. Aktsipetrov, *JEPT Letters* **73** (2001) 6-9.

GNSR 2001
G. Messina and S. Santangelo (Eds.)
IOS Press, 2002

Pulsed Laser Deposition
of superlattices and diamond-like
carbon films

S.Lavanga[a], P.G.Medaglia[a], G.Messina[b], S.Santangelo[b], A.Tebano[a]

[a] *INFM, Dipartimento di Scienze e Tecnologie Fisiche ed Energetiche,*
Università di Roma Tor Vergata - 00133 Roma
[b] *INFM, Dipartimento di Meccanica e Materiali,*
Università di Reggio Calabria, loc. Feo di Vito – 89060 Reggio Calabria

Abstract In this paper it will be described the deposition by Pulsed Laser Deposition and Laser Molecular Beam Epitaxy techniques of high temperature superconducting cuprates superlattices on their structural and superconducting properties, and on the deposition and characterisation of Diamond-like Carbon (DLC) films.

1. Introduction

The Pulsed Laser Deposition (PLD) technique, and its evolution the Laser Molecular Beam Epitaxy (Laser-MBE) possesses many unique properties that have, led up to now, to fabrication of excellent quality films of a variety of materials.

In the following we will present first a brief overview on what the PLD and Laser-MBE techniques mean (Section 1). Afterwards, in Sec. 2, will be presented the growth of the superlattice films. In Sec. 3 the deposition and the characterisation of hydrogenated and hydrogen-free amorphous carbon films will be presented. We will show also our study on the effect of excimer laser irradiation on films, to study conditions for device patterning. Finally, Sec. 4 will be devoted to concluding remarks.

2. Pulsed Laser Deposition and Laser MBE Techniques

Both conceptually and experimentally PLD is extremely simple, probably one of the simplest among all thin film growth techniques [1]. It is based on the interaction between a laser beam, with wavelength ranging from 193 nm to 1064 nm, and the target material. In particular, in the case of thin film growth, the useful wavelength interval lies between 200 and 400 nm. Over this range many materials used in deposition exhibit strong absorption, since their absorption coefficients tend to increase as one moves towards the shortest wavelength. This fact causes a reduction of the penetration depth into the target material and the possibility to more easily ablate thinner layers of surface. The stronger absorption at the shortest wavelength also results in a decrease of the

                    S. Lavanga et al. / Pulsed Laser Deposition

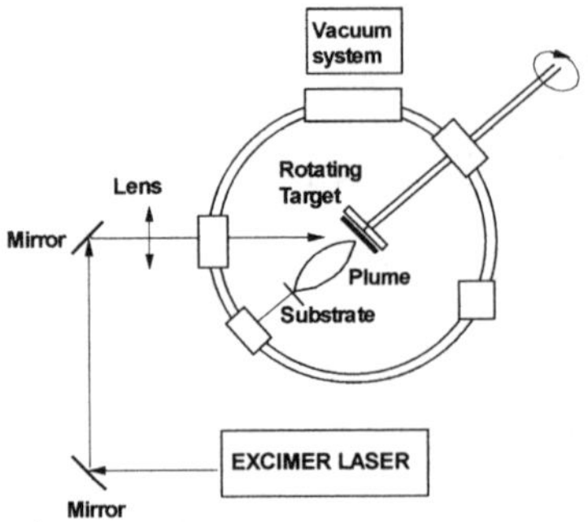

Fig.1 A schematic diagram of a typical PLDexperimental set-up

ablation fluence threshold (λ=200 nm is the lower limit imposed by absorption band of molecular oxygen).

From the reasons just mentioned, it has turned out that excimer lasers represent a favourite choice for PLD work. They emit radiation directly in the UV range and deliver high outputs, typically 200 mJ/pulse. They can also achieve high repetition rate (up to several hundred hertz).

In Fig. 1 is shown a schematic diagram of a typical experimental set-up. It consists of: a target holder, which can rotate about its main axis during the irradiation, and a substrate holder, both housed in a vacuum chamber; an external laser used as power source to vaporise material and deposit thin films; and finally, a set of optical components to focus the laser beam over the target surface.

The decoupling of the vacuum hardware from the evaporation power source makes this technique so flexible that it is easily adaptable to different operational modes, without the constraints imposed by the use of internally powered evaporation sources.
Film growth can be carried out in an environment containing any kind of gas, with or without plasma excitation. It can also be conducted in conjunction with other types of evaporation sources in a hybrid approach. In contrast to the hardware simplicity, the laser target interaction is quite a complex physical phenomenon. Theoretical descriptions are multidisciplinary and combine equilibrium and non-equilibrium processes.
The mechanism that leads to material ablation depends on laser characteristics (wavelength, fluence, spot size, etc.) as well as on optical, topological and thermodynamical properties of the target. In the ablation process one can distinguish three main steps:

a) Interaction of the laser beam with the target material, which causes "evaporation" of the surface layers.

b) Interaction of the emitted material with incident radiation and subsequent formation of a plasma.

c) Anisotropic plasma expansion followed by deposition of ablated materials.
The PLD method offers many advantages. Among them we can count simplicity and flexibility, due essentially to the fact that laser beams are easier to transport and manipulate, and allow decoupling of the vacuum chamber from the power source. Compared with other sputtering processes, it is rather clean, thanks to the confinement of the laser-target interaction region. All these properties also give PLD a good degree of

reproducibility. In any case its key feature is the possibility to reproduce the stoichiometric structure of the target material, under appropriate conditions.

PLD offers the opportunity to grow multilayer and artificial structures with a very high degree of complexity by positioning different targets under the laser beam. This can be accomplished by mounting a multitarget rotating holder inside the chamber.

In 1981, the first observation [2] of oscillations in the Reflection High Energy Electron Diffraction (RHEED) intensity during the epitaxial growth of GaAs, offered a new tool to control thin film growth with atomic layer precision. In the last few years several research groups have been able to use this powerful diagnostic technique, *in situ*, in combination with PLD, i.e. the Laser MBE technique, to obtain the two-dimensional (2D) growth of artificial materials otherwise difficult or impossible to synthesise. The advantage of the Laser MBE technique is due to the possibility to monitor the RHEED intensity oscillations during the growth process, enabling very precise control of the layer-by-layer epitaxial growth (phase locked growth) thus reducing as much

as possible occasional fluctuations in the thickness of each constituent layer.

Presently the Pulsed Laser Deposition and the Laser-MBE techniques are used to grow a wide variety of thin oxide films, most of them with a complex structure: high T_c superconductors [3, 4], ferroelectrics [5, 6], piezoelectrics [7], colossal magnetoresistance oxides [8, 9], electro-optics materials [10], amorphous diamond-like carbon. With the possibilities offered by the utilisation, in situ, of the RHEED diagnostic, it is possible to predict the realisation of functional materials of very good quality, with very interesting applications in several fields [11].

3. Super lattices

All existing high temperature superconducting (HTS) cuprates exhibit a layered structure and can be represented by the general formula $A_mB_2D_{n-1}Cu_nO_{2n+2+\delta}$ with a stacking sequence $[BO-A_mO_\delta-BO]-[(CuO_2)_nD_{n-1}]$ where A, B, D represent appropriate cations [12]. The structure of the HTS cuprates is usually considered as made of two distinct structural subunits having different functions: the $[BO-A_mO_\delta -BO]$ layer, called the charge reservoir (CR) block and the $[(CuO_2)_nD_{n-1}]$ layer, called the infinite layer (IL) block. The CR block can easily include in the structure excess oxygen ions or substitutional cations which uncompensate its electrical charge. Some of the excess charge (holes) can be transferred to the IL block giving rise to superconductivity. Despite of many efforts in the last years, the search for new HTS compounds is still based on few simple structural considerations: good superconducting properties (high Tc, high critical fields and current densities) are generally thought to be connected with: a) IL blocks with a low content of structural defects, b) CR blocks thin and highly metallic [13].

Recently the use of high pressure (several GPa) synthesis techniques has made possible to discover a large number of new metastable HTS cuprates which cannot be synthesized by conventional solid state reaction. Among them particular interest have those belonging to the Ba-Ca-Cu-O family [14, 15, 16]. These compounds (general formula $CuBa_2Ca_{n-1}Cu_nO_y$) have been reported to have low anisotropy, high T_c (up to 118 K), high critical current densities and high irreversibility fields [17, 18]. Such characteristics were attributed to the short spacing between adjacent IL layers, even shorter than in the case of $HgBa_2Ca_2Cu_3O_y$ and $TlBa_2Ca_3Cu_4O_y$. The Ba-Ca-Cu-O system forms a homologous series with n (number of CuO_2 planes in the IL block) ranging from 3 to 6. The superconducting transition temperature reaches a maximum value (for n=3\div 4) of 118 K. The chemistry of the CR\ block for these compounds is quite complicated and not yet fully understood.

Considering also the absence of volatile toxic elements, these compounds offer, in perspective, very interesting possibilities for applications in the field of thin film devices. However, presently, these materials, due to their high instability, can be obtained in the form of polycrystals or small single crystals only by high pressure synthesis.

On the other hand, advances in the layer by layer growth of epitaxial thin films, have opened new perspectives for the atomic engineering of artificially layered cuprate superconducting structures [19]. A crucial step in this direction has been in the past the growth of epitaxial films of the "infinite layers" (IL) $Ca_{1-x}Sr_xCuO_y$ compounds. The deposition of simple IL thin films has been achieved now by several researchers in the case of the $(Ca_{1-x}Sr_x)CuO_2$ compound (x ranging from 0 to 1) using suitable monocrystalline substrates to stabilize the otherwise metastable IL phase. Much more difficult is the growth of the $BaCuO_2$ IL structure. Such IL structure is by far less stable than those based on Ca. The IL structure, which consists of square CuO_2 layers alternately stacked with alkaline earth cations (Ca, Sr and, to a lower extent, Ba), is the simplest structure which contains the CuO_2 planes considered essential for high T_c superconductivity. This structure, however, is not superconductive and its resistivity shows a steep increase at low temperatures ("semiconducting" behavior). Successively the two compounds $SrCuO_2$ and $CaCuO_2$ [20, 21] of this family were used as building modules for the growth of $SrCuO_2/CaCuO_2$ artificial superlattices [22, 23]. However the $SrCuO_2/CaCuO_2$ superlattices did not show any trace of superconductivity. Such result is expected because in these superlattices both the $(CaCuO_2)_m$ and the $(SrCuO_2)_n$ layers are well charge compensated and neither of them is expected to behave as a charge reservoir block. Much more interesting is the case of superlattices containing $BaCuO_2$ IL layers. The Ba based IL structure is very unstable. It can be grown as a thin film because of the stabilizing effect of the $SrTiO_3$ substrate only for thicknesses smaller than 400Å [24]. Indeed this IL structure can easily include excess oxygen in the Ba planes ($BaCuO_{2+\delta}$) which, in turn, can act as a charge reservoir layer for the second constituent IL layer $((CaCuO_2)_m$ or $(SrCuO_2)_m)$. The artificially layered $BaCuO_2/(CaSr)CuO_2$ [25] thin films appear to be closely related to the above mentioned compounds obtained by high pressure synthesis. Therefore the optimisation of these artificial structures can be of major interest and could allow obtaining thin epitaxial films with T_c higher than 110 K and high critical current densities.

The first observation of RHEED intensity oscillations on oxide superlattices was carried out by A.Gupta et al. [26] on $CaCuO_2/SrCuO_2$ using during the deposition additional atomic oxygen source. Recently has been reported the observation of RHEED intensity oscillations on $(BaCuO_2)_2/(SrCuO_2)_2$ not superconducting superlattices [27]

An important requirement for superconductivity is that there must be no defects in or near the CuO_2 layers. In Ref. 28 it was pointed out that structural disorder strongly affects the superconducting properties of these superlattices. Defects in the planes that separate adjacent CuO_2 layers in the IL block have been shown to lower T_c and even destroy superconductivity [29, 30]. Under this respect the laser MBE technique could allow a better control of the interface disorder.

The depositions were performed using an excimer laser charged with KrF, generating 248 nm wavelength pulses of 25 ns width with 1Hz repetition rate. The laser beam, with an energy of 60 mJ per pulse, was focused in a high vacuum chamber on to the computer controlled multi-target rotating system. Substrates used for deposition were (100) $SrTiO_3$

single crystal, placed at a distance of about 70 mm from the target on a heated holder. Before growth, substrates were chemically etched in a buffered solution of NH_4F-HF (pH=4.6) for 8 minutes. This etching treatment leaves a terminating substrate layer of TiO_2 and decreases the surface roughness [31].

Before starting the superlattice deposition, a few monolayers of $SrTiO_3$ were deposited, to have an optimal 2D starting deposition surface. The incident RHEED electron beam was parallel to the [100] substrate direction. To grow the $BaCuO_x$/$CaCuO_2$ superlattices, $BaCuO_x$ and $CaCuO_2$ targets, prepared by solid-state reactions were used. Targets were prepared according the following procedure: stoichiometric mixtures of high purity $CaCO_3$, $SrCO_3$, $BaCO_3$ and CuO powders were calcined at 860°C in air for 24 h, pressed to form a disk and finally heated at 900°C for 12 h. Pellets obtained according this procedure were not single phase. The deposition temperature was decreased to 500°C and the molecular oxygen pressure increased to 10^{-4} mbar. The homoepitaxial $SrTiO_3$ deposition was performed under a molecular oxygen pressure of 10^{-5} mbar and with the substrate temperature at 640°C. Four monolayers of $SrTiO_3$ were deposited monitoring four RHEED intensity oscillations of the specular spot. The $SrTiO_3$ deposition was stopped when the specular spot intensity reached the fourth maximum, after about 200 laser pulses (inset Fig.1).

The deposition was started with the $BaCuO_x$ layer. In these conditions, the growth rates of $BaCuO_x$ and $CaCuO_2$ were calibrated, monitoring the RHEED intensity oscillations of the specular spot. The layer by layer deposition of $BaCuO_x$/$CaCuO_2$ superlattices was carried out stacking in sequence m $BaCuO_x$ unit layers and n $CaCuO_2$ unit layers, (the so called mxn superstructure, $[(BaCuO_x)_m/(CaCuO_2)_n]_S$, where S represents the total number of deposition cycles). The bi-dimensionality of the growth process is proved by the RHEED pattern which, at the end of the growth, shows typical 2D features (Fig. 2).

Monitoring the RHEED intensity oscillations, the laser pulses on the two targets were adjusted to obtain the $(BaCuO_x)_2/(CaCuO_2)_2$ superlattice. In Fig. 1 the RHEED intensity oscillations of the specular streak are shown during four of the overall 15 cycles of the $(BaCuO_x)_2/(CaCuO_2)_2$ superlattice deposition. The time evolution of RHEED intensity in a single cycle remains unchanged throughout the superlattice growth. Its major feature is the sizeable variation of intensity when the growth is switched from the $BaCuO_x$ oxide to

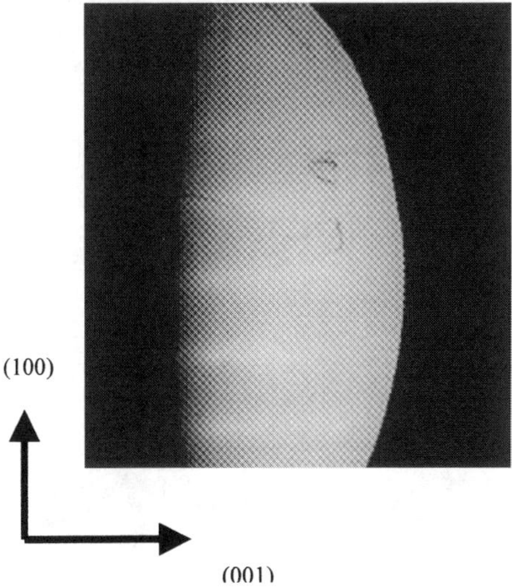

Fig.2 RHEED pattern at the end of the deposition process of a $[(BaCuO_x)_2/(CaCuO_2)_2]_{15}$ artificial superlattice.

the $CaCuO_2$ oxide. This effect can be possibly ascribed to two causes. The first cause is the electron scattering factors of atoms in the two layers, which differ in their absolute magnitude, in the phase shift upon scattering and phase shift due to the height difference. The second cause is the difference in the average surface quality between the $BaCuO_x$ and $CaCuO_2$ layer. In addition to the layer oscillations caused by the composition of the up-most layer, weaker oscillations can be clearly recognised, in the time evolution pattern of the RHEED intensity. During the $CaCuO_2$ deposition it is possible to note two weak RHEED intensity oscillations corresponding to the growth of a 2D layer with a thickness of two unit cells. During the $BaCuO_x$ deposition there was a first faint RHEED intensity oscillation followed by a more evident one indicating, also in this case, the growth of a 2D layer with a thickness equal to two $BaCuO_x$ cells. The clear asymmetry between the first and the second oscillation during the deposition of the $BaCuO_x$ layer gives qualitative support to the results of a recent EXAFS study on these superlattices. In this study it was shown that the structural unit of the $BaCuO_x$ layer consists of four atomic planes ($CuO_2/BaO/CuO_x/BaO$) rather than two identical IL blocks ($CuO_2/Ba/CuO_2/Ba$) [32]. Such RHEED intensity oscillations were observed during all of the 15 cycles. From the period of the RHEED intensity oscillations, a growth rate of 20 laser pulses per unit $BaCuO_x$ layer and 50 laser pulses per unit $CaCuO_2$ layer, was estimated.

In Fig. 3 the θ-2θ x-ray diffraction (XRD) spectrum of the same superlattice is shown. Peaks have been indexed using the superlattices convention: SL_0 indicates the average structure peak, while $SL_{1,2}$ indicate, respectively, the first and the second order satellite peaks. From the angular distance between the first order satellite peaks (SL_{-1} and SL_{+1}) it is possible to obtain the period Λ of the superlattice, $\Lambda=\lambda/(sin\theta_{+1} - sin\theta_{-1})$, where θ_{+1} and θ_{-1} represent the angular positions of the first order satellite peaks and λ represents the x-ray wavelength. The average lattice parameter, $\bar{c}$, can be estimated from the angular positions of the zeroth order SL_0 (001) peak, θ_0, $\bar{c}=\lambda l/2sin\theta_0$.

The total number of layers composing the superlattice, $N=m+n$, can then be calculated, $N=\Lambda/\bar{c}$. A large difference in the intensity of the average structure peak can be noticed between spectra (a) and (b): while the intensity of the SL_0 peak in the case of the $(SrCuO_2)_2/(BaCuO_2)_2$ superlattice is about one order of magnitude larger than those of the satellite peaks, in the case of the $(CaCuO_2)_2/(BaCuO_2)_2$ superlattice, all peaks have comparable intensity. The angular position of the average structure peaks depends on the average lattice parameter $c=(d_{(Ca,Sr)CuO2}+d_{BaCuO2})/2$, while its intensity depends both on the thicknesses and the structure factors of the two constituents layers $(CaCuO_2)_2$ and

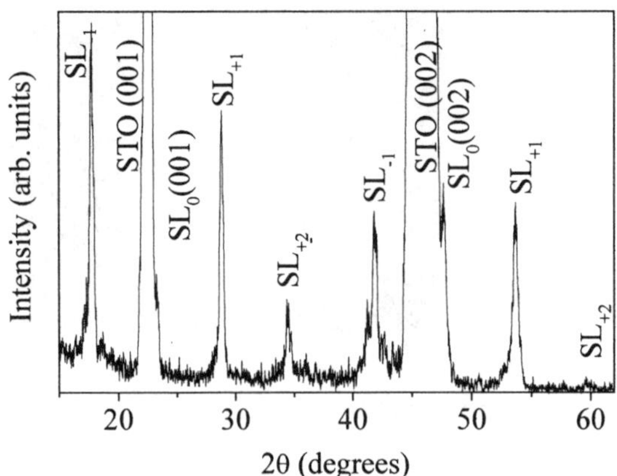

Fig. 3 XRD experimental spectrum of the $BaCuO_x/CaCuO_2$ superlattice.

$(BaCuO_2)_2$. The intensity of the average structure peak can be comparable with those of the satellite peaks if the differences in thickness and structure factor between the component layers are large (which is the actual case). On the other hand, if the thickness and the structure factor of the two constituent layers are comparable, the intensity of this peak can be much larger relative to the satellite peaks. From this simple analysis of the spectrum, it was calculated that $\Lambda=16.25\ \AA$ and $N=4.25$, only slightly greater than the value aimed for (N=4).

The full width at half maximum (FWHM) of the rocking curve for the $SL_{-1}(001)$ peak of the grown film is about 0.07°, approximately the same as the (002) $SrTiO_3$ FWHM peak.

A more accurate analysis of the x-ray spectrum was performed using the procedure described in Ref. [33]. In this reference, numerical simulations of x-ray spectra were carried out following a kinematical approach using a simplified model structure. In this approach a two dimensional layer by layer growth is considered. In such a case, if m and n remain perfectly constant during the growth of the superlattice, but not corresponding to an integer number of unit layers, the composition of the mixed layer, formed at the interface, is always well defined and varies in a controlled fashion throughout the film thickness (controlled disorder) [34]. The model also assumes that mixed composition layers can be corrugated to adjust the internal stresses due to the large mismatch between the constituent oxides. An additional random disorder is added to take into account the experimental dispersion in the amount of deposited material in each iteration. Namely, in each deposition cycle, m and n take random values that follow a Gaussian distribution with dispersion σ around the mean values $<n_{BaCuOx}>$ and $<n_{CaCuO2}>$.

The simulated spectrum (full line) is shown in Fig. 4 together with the experimental data (empty dots). In order to simulate the experiment the Ca-Ca distance was fixed at 3.20Å, in accordance with the c lattice parameter of the IL $CaCuO_2$ phase [21]. On the other hand, due to the mobility of oxygen ions in the $BaCuO_x$ block, Ba-Ba distances can vary depending on growth conditions (O_2 pressure and growth temperature). From the fit it has been concluded that the Ba-Ba distance in this case was about 4.56Å (slightly greater

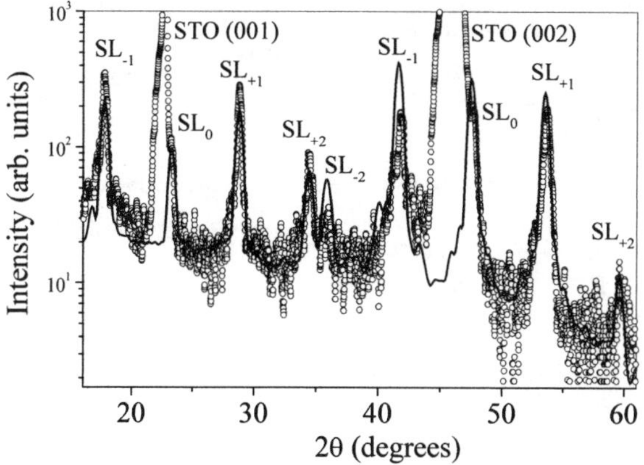

Fig.4 XRD experimental spectrum of the $(BaCuO_x)_2/(CaCuO_2)_2$ superlattice (open circles) compared with the simulated spectra (continuous line).

than the value found in Ref. 35). The composition of the superlattice was shown to be $[(BaCuO_x)_{1.95}/(CaCuO_2)_{2.3}]$.

A further important result is that the random disorder parameter σ is more than one order of magnitude smaller, relative to the one used to simulate similar structures grown without in-situ RHEED diagnostic. In the present case, the peak broadening arises mainly from the small film thickness and the controlled disorder, and in practice σ can be considered negligible.

The resistivity versus temperature behavior of the $BaCuO_x/CaCuO_2$ superlattices was measured by the standard four probe DC technique. It was found that the resistivity values increase at low temperature following roughly a Variable Range Hopping mechanism (Fig. 5a). Such a result was expected due to the low oxygen growth pressure ($\approx 10^{-4}$ mbar). For comparison in Fig. 5c) is reported the $\rho(T)$ curve for a superconducting $(BaCuO_x)_2/(CaCuO_2)_2$ superlattice grown without in situ RHEED diagnostic at higher oxygen pressure ($9x10^{-1}$ mbar) and for the pressures equal or lower than $5x10^{-2}$ mbar it was not possible to observe any superconducting transition up to 20 K (Fig 5 b). At low temperatures the $(CaCuO_2)_2/(BaCuO_2)_2$ superlattice, grown at higher oxygen pressure, shows a full transition to superconductivity. Higher pressures (up to 0.4 mbar) resulted in higher transition temperatures. For the $(BaCuO_x)_2/(CaCuO_2)_2$ superlattice grown without in situ RHEED diagnostic at higher oxygen pressure ($9x10^{-1}$ mbar) preliminary measurements have been also carried out to investigate the dependence of the transition temperature on the number n of CuO_2 layers in the IL block. T_c seems to stay about constant for n=3,4 while it decreases for higher or lower n values.

Clearly in order to take advantage from the very good structural quality of the superlattices grown with the in situ RHEED diagnostic at lower oxygen pressure, the degree of oxidation during the growth, has to be still strongly increased.

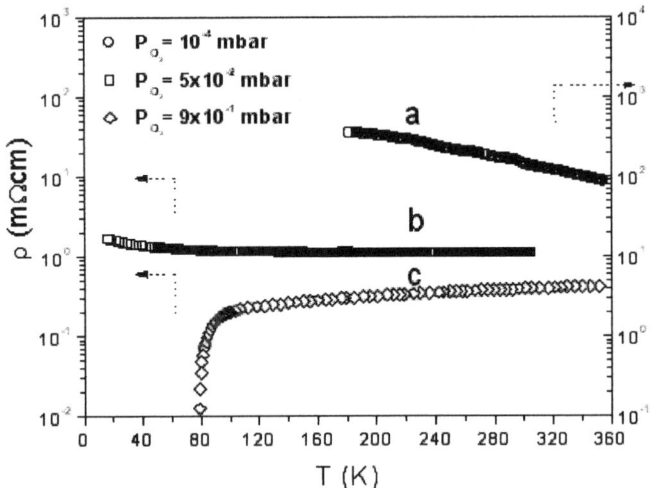

Fig.5 Resitivity vs. temperature measurements for the $BaCuO_x/CaCuO_2$ superlattice grown in different conditions: (a) 10^{-4} mbar of oxygen pressure (Laser MBE), (b) $5x10^{-2}$ mbar of oxygen pressure (PLD), (c) $9x10^{-1}$ mbar of oxygen pressure (PLD).

4. Amorphous carbon films

In the last years, a growing attention has been focused on amorphous diamond-like carbon. The amorphous carbon is an aggregation state of carbon with very interesting properties varying between those of graphite and those of diamond [36]. The tribological, electrical and optical properties of the amorphous carbon thin films, such as chemical resistance, thermal conductivity, mechanical hardness, electrical insulation, optical transparency, low friction coefficient, good adherence to the substrate, and low roughness, make this material very unusual and attractive for technological applications [37, 38].

The amorphous carbon films are one of the more interesting candidates for the realisation of cold cathodes, based on the Electron Field Emission (EFE). The EFE technology, presently, is used by several industries for applications that need a cold cathode as an electron source. These devices can find important applications in different fields: microwave devices, liquid crystal displays and flat panel displays. The cold cathode application fields represent a very important and rapidly increasing market. The actual electron emitter devices are essentially based on the microtips production. This kind of production has several disadvantages, i.e. necessity of microtips uniformity and high costs. For these reasons there is a big effort to develop new technologies that can offer the advantages of the microtip technologies with lower costs and more easy production. Alternative to this kind of technologies is the utilisation of thin films that spontaneously emit electrons with very low applied electric fields. These are the materials with low or negative electron affinity. The hydrogenated diamond films possess negative electron affinity [39] and in principle they can emit electrons. The problem of diamond for some applications, in the microcathods realisation, can be represented from the high deposition temperature of the films. Recent studies suggest that the electron affinity is fundamentally connected to the chemical bond type and not to the cristallinity of the material, in this way is possible that also the amorphous carbon films have a low electron affinity. The amorphous carbon film microcathode is much bigger than the emitting microtips and there is no need of lithographic efforts. The amorphous carbon films has also the advantage of the deposition temperature close to room temperature making the deposition possible on materials as plastics or glasses.

The interest for these films is strongly related to the very low applied electric field (1-20 V/μm) necessary for the field emission and to the possibility to have electron emission from flat surfaces [40, 41, 42, 43]. The emission mechanism is not completely known and this makes difficult the optimisation of the film quality (emission uniformity and density of sites from which the emission comes). Several emission models were proposed up to now [40, 44, 45], but no-one of them is demonstrated definitely correct, it is therefore possible that several mechanism are simultaneously involved.

The techniques used for the deposition of amorphous carbon film are different; one of these is the PLD. PLD has attracted much effort in the synthesis of amorphous carbon films for optical and electronic applications; the hardness and the wear resistance plus the qualities of high adhesion and low substrate temperature during film growth make PLD of the amorphous carbon films a superior technology for tribological applications.

The doping effect in the films due to the nitrogen incorporation changes the electronic properties as dc conductivity, activation energy [46], and also reduce the internal compressive stress of the amorphous carbon films [47, 48].

We performed depositions using an excimer laser charged with two different gas mixtures XeCl, and KrF generating respectively 308 nm and 248 nm wavelengths. The laser beam was focused on 99.99% purity graphite to achieve ablation power density per pulse of $3*10^8$ W cm^{-2}. The substrate was Si (100) (1 cm^2 surface area.). After etching in hydrofluoric acid solution and cleaning in acetone and alcohol the substrate was heated, in the deposition chamber, at 500°C for 30 min, after that, the temperature was set at the desired deposition temperature. We deposited films at different temperatures in between R.T. and 600°C. The base pressure in the chamber was 10^{-7} mbar. We studied the effect on the film characteristics of different atmospheres: vacuum, 2 mbar partial pressure of hydrogen and partial pressure of nitrogen, the role of the substrate temperature, of the radiation laser wavelength and the effect of the laser irradiation on the films characteristics.

The surface homogeneity and the thickness, of the films, were estimated by Scanning Electron Microscopy (LEICA Cambridge 260). A technological application requires homogeneous films with low roughness. The film roughness is connected with the growth process and with the structural film characteristics; recent studies on the amorphous carbon film growth process are based on morphology studies [49]. The AFM microscopy is a very powerful technique to investigate the morphology characteristics of the amorphous carbon films and to obtain information on the growth process [50, 51, 52]. The AFM apparatus has been described elsewhere [53]. Images were obtained in the repulsive mode with a force of about 1-4 nN from zero cantilever deflection. Gold-coated silicon nitride tips with a spring constant of 0.16 N/m from Park Scientific Instruments, were used. All the images shown are untreated apart from rigid plane subtraction. A grey map has been superimposed that varies from black for the deeper zones to white for the highest. The surface vertical roughness, $(R=[\Sigma(Z_i-Z_{ave})^2/N]^{1/2}$ were Z_{ave} is the average Z height value within a given area, Z_i is the current Z value, and N is the number of points within the given area) was measured and averaged by the system.

The Raman spectroscopy is one of the more common techniques used for the evaluation of the carbon films quality [54, 55]. The Raman spectrum of amorphous carbon films consists of a broad peak centred at about 1550 cm^{-1} (G peak). An asymmetric broadening of the Raman peak and the appearance of a shoulder in the 1350 cm^{-1} region (D peak) is due to an increase of the sp^2 bonds. The intensity ratio between the intensity of the D and the G peak, I_D/I_G, is correlated with the degree of graphitisation of the sample [56]. The Raman characterisation was carried out with Jobin Yvon Ramanor U-1000 monochromator. An Ar ion laser with a wavelength of 514 nm was used. The Raman spectra were fitted by Lorentzian curve for the D peak, Breit-Wigner-Fano curve for the G peak and linear background, with the software ORIGIN.

The mechanical hardness for the amorphous carbon sample was determined by Vickers method. Taking into account the very low thickness of the amorphous carbon films, we tried to obtain information as much as possible quantitative on the hardness characteristics of the films and to compare the results obtained on the different films.

One of the simplest and more effective technologies available for patterning and machining of the amorphous carbon films is the excimer laser ablation. For the excimer lasers, each UV laser photon possesses sufficient energy to break the bonds responsible for holding together the material. UV excimer lasers offer an alternative route by providing a resist-free, simple and precise patterning process that can be utilised either in direct-write mode or by projecting the laser light via a lens or a mask. In order to implement this technique under optimised conditions, we studied the effect of laser irradiation in air on amorphous diamond. The excimer laser exposure experiments were carried out with the same XeCl laser used for the film deposition.

5. Results and discussion

The Raman spectra of films deposited in hydrogen atmosphere are strongly dependent on the substrate deposition temperature, fig. 6.
In the Table 1 we report the ratio I_D/I_G and the D peak FWHM values obtained from the fits on the hydrogenated samples. For the films deposited at a higher temperature the D peak at 1350 cm^{-1} is more evident and is connected to a higher content of graphitic bonds. The spectra of the samples deposited at 25, 180, 300 and 450 °C are characterised by an increasing value of I_D/I_G, in the spectrum of the film deposited at 580 °C is present a distinct and separated D peak. Also for the films deposited in vacuum between R.T. and 150°C there is a gradual worsening of the Raman characteristics with the increase of the deposition temperature, fig. 7. From the Raman analysis it is possible to conclude that increasing the deposition temperature, the spectra show evident modifications, due to the increase of the graphitic content.

To have more information on the diamond-like qualities of the amorphous carbon films we characterised the deposited films by hardness measurement. The hydrogen-free films showed a much higher hardness than the hydrogenated ones for all the deposition temperatures. The obtained hardness values for the vacuum deposited films are reported in fig. 8. The higher hardness value was obtained for a deposition temperature of 60°C.The AFM study revealed that the lower roughness on the amorphous carbon films was obtained for the films deposited at lower temperature in fig. 9 (a) is showed the AFM image of the film surface deposited in vacuum at 60°C using laser radiation of 308 nm and fluence of 200 mJ/shot, in this case the roughness was about 5 nm. For the films deposited at 300°C and at 500°C the roughness calculated was about 30 nm and 65 nm respectively. The amorphous carbon films with a high content of sp^3 bonds are associated with a surface atomically smooth; this can be interpreted as an internal film growth process [57, 58, 59]. In fig. 9 (b) is showed the AFM image of an amorphous carbon film deposited in vacuum at 60°C using laser radiation at 248 nm at energy density of 200 mJ/pulse. The roughness is about 1 nm.

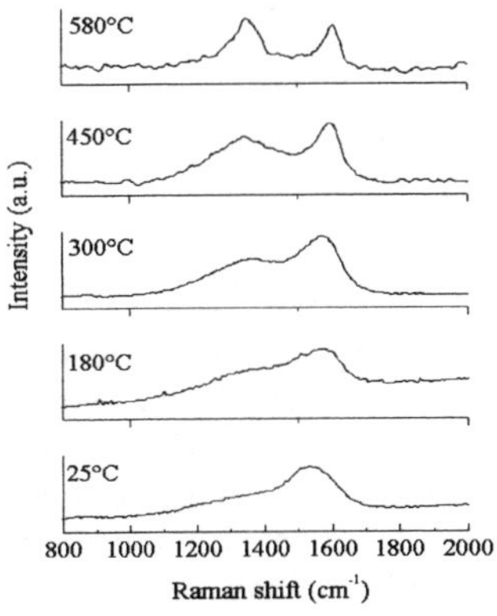

Fig.6 Raman spectra of the films deposited in hydrogen atmosphere at different temperatures

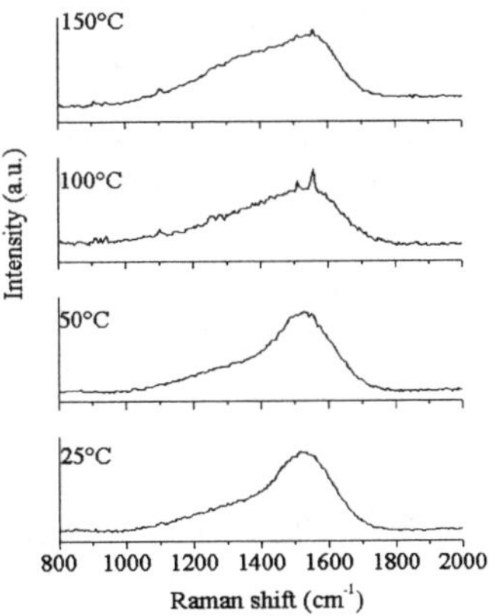

Fig.7 Raman spectra of films deposited in vacuum at different temperatures

Tab.1 Ratio I_D/I_G and D peak FWHM values for different deposition temperatures

T_d (°C)	I_D/I_G	FWHM (cm⁻¹)
580	0.8	103
450	1.2	239
300	0.8	263
180	0.8	317
25	0.3	230

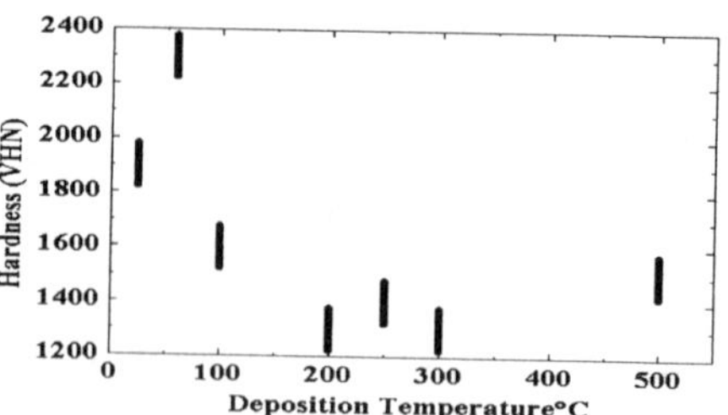

Fig.8 Hardness vs. deposition temperature

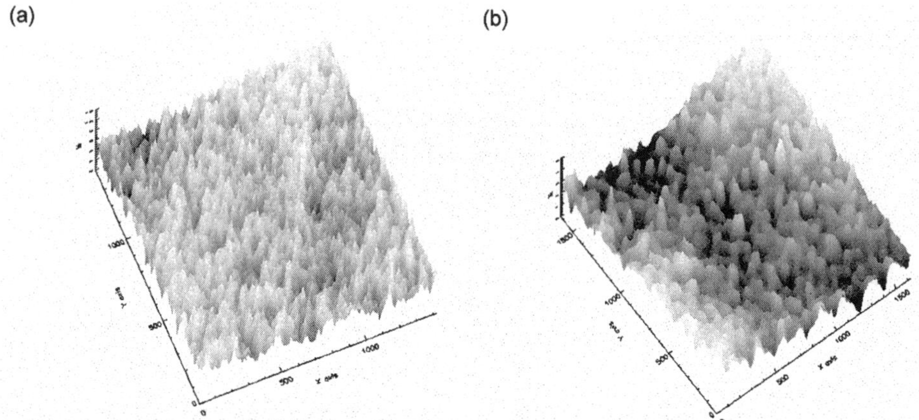

(a)

(b)

Fig.9 The AFM image of the film amorphous carbon film surface deposited in vacuum at 60°C using laser radiation of 308 nm (a), using laser radiation at 248 nm (b).

In fig. 10 it is possible to compare the Raman spectra of two amorphous carbon films, 350 nm thick, deposited using laser radiation at 308 nm spectrum (a) and laser radiation at 248 nm, spectrum (b), the fluence was 200 mJ/pulse. The spectrum (b) has more symmetric shape than the spectrum (a); this is due to the lower intensity of the D peak at 1350 nm. The Si peak at 950 nm is more evident for the film deposited using the 248 nm radiation, indicating a higher film transparency: this two characteristics are connected to a higher diamond-quality of the film due to the shorter wavelength radiation used for the deposition, in agreement with the work of Voevodin et al. [60].

In fig. 11 are shown the spectra of a hydrogen-free sample (a) before and (b) after irradiation performed at a fluence of 6.5 J/cm² by one laser pulse. In the Raman spectrum of the irradiated film there are two distinct peaks at 1335 and 1580 cm⁻¹.

The Raman spectra show that the laser etching of the amorphous carbon is a two-step process. In the first step the ablation creates a graphite layer on top of the film. After this highly absorbing layer has been formed the etching may be characterised by the

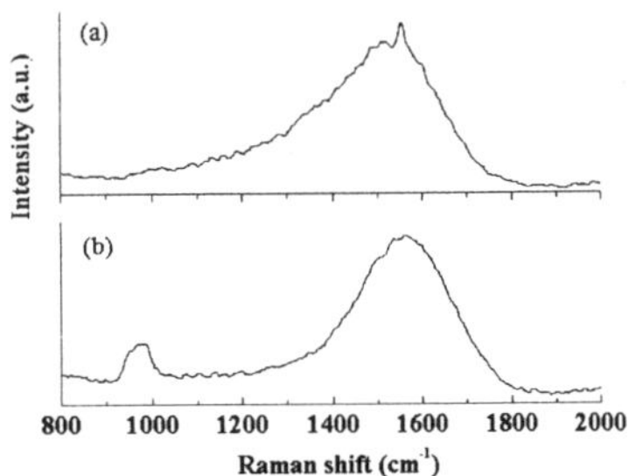

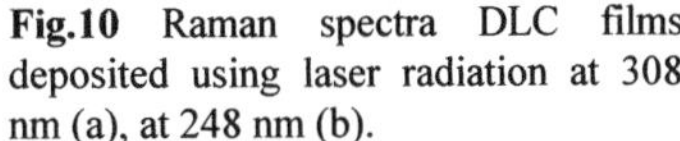

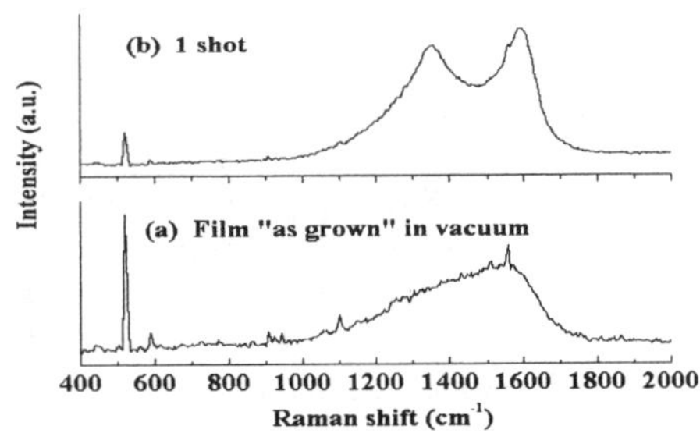

Fig.10 Raman spectra DLC films deposited using laser radiation at 308 nm (a), at 248 nm (b).

Fig.11 Raman spectra of a hydrogen-free sample (a) before and (b) after laser irradiation

dependence of ablation rate on the laser fluence. With this fluence we measured an etch rate of about 400 nm/pulse. The etch rate increases more than linearly with the laser fluence. For hydrogen-free films we found at 0.5 J/cm^2 a fluence threshold for the etching mechanism. To compare the etch-rate for hydrogenated films we irradiated at 0.5J/cm^2 a film grown in hydrogen; and we found an etch-rate of about 30 nm/pulse.

6. Conclusions

We have shown that the layer by layer growth by Pulsed Laser Deposition can be an effective tool in the search for new high T_c materials. By this technique it is possible to engineer structures which cannot be synthesized even by the high pressure technique, for instance increasing the number of CuO_2 planes in the IL block or stacking in sequence several subunits having different structures. Furthermore, while samples obtained by high pressure solid state reaction often contain together different compounds of the same homologous family, a specific component can be grown by PLD without apparent contamination from the other members of the same family. A further important advantage is that samples obtained by layer by layer deposition are monocrystalline epitaxial films. This feature is of particular relevance both with respect to the investigation of the fundamental properties of these new materials (such as anisotropy, critical currents, pinning,...) and to their possible practical applications. In conclusion, crystalline c-axis oriented $(BaCuO_x)_2/(CaCuO_2)_2$ superlattice thin films have been successfully grown on (001) $SrTiO_3$ substrate by PLD and by Laser MBE and the growth process of the films investigated by RHEED. It was shown that RHEED intensity oscillations, even for superlattices of such complex oxides, could be used for a qualitative phase locking of the growth. A detailed *a posteriori* x-ray investigation also demonstrates that $(BaCuO_x)_2/(CaCuO_2)_2$ superlattices grown by Laser MBE are only affected by a small amount of controlled disorder, due to the non exact integer number of unit layers deposited in each iteration, while the random disorder component is negligible. Superconductivity was found only for superlattices grown at oxygen pressures higher than 0.6 mbar. The transition temperature of the Ca containing superlattices was found to be (under high growth oxygen pressure >0.6 mbar) to be as high as 70 K (zero resistance temperature).

We also deposited by PLD amorphous carbon films. We found that: (a) the amorphous carbon films deposited in vacuum have an evident diamond-like quality, (b) the optimal deposition temperature is around 60°C, (c) the film quality increases using a shorter laser wavelength (248 nm), (d) the surface roughness of the deposited films decreases for higher diamond-like quality. We performed investigations of the amorphous carbon excimer laser patterning in air. The modification of the Raman spectra after irradiation is explained in terms of a double step process. We observe the existence of a threshold fluence for etching mechanism which depends on film quality and on the nitrogen content.

References

[1] "Pulsed Laser Deposition of thin films", eds. D.B.Chrisey and G.K. Hubler, John Wiley & sons, Inc., New York 1994

[2] J. J. Harris, B.A. Joyce and P. J. Dobson, Surf. Sci. Lett. **103**, L90 (1981).

[3] G. Balestrino, S. Martellucci, P. G. Medaglia, A. Paoletti, G. Petrocelli, Physica C **302**, 78 (1998).

[4] A. Crisan, G. Balestrino, S. Lavanga, P. G. Medaglia, E. Milani, Physica C **313**, 70 (1999).

[5] H. M. Christen, E. D. Specht, D. P. Norton, M. F. Chrisholm, L. A. Boatner, Appl. Phys. Lett. **72**, 2535 (1998).

[6] M. Joseph, H. Tabata, T. Kawai, Appl. Phys. Lett. **74**, 2534 (1999).

[7] M. Tyunina, J. Levoska, S. Leppavuori, Journ. of Appl. Phys. **83**, 5489 (1998).

[8] M. Izumi, Y. Konishi, T. Nishihara, S. Hayashi, M. Shinohara, M. Kawasaki, Y. Tokura, Appl. Phys. Lett. **73**, 2497 (1998).

[9] K. Ghosh, S. B. Ogale, S. P. Pai, M. Robson, E. Li, I. Jin, Z. Dong, R. L Greene, R. Ramesh, T. Venkatesan, M. Johnson, Appl. Phys. Lett. **73**, 689 (1998).

[10] Y. Shibata, K. Kaya, K. Akashi, M. Kanai, T. Kawai, S. Kawai, Appl. Phys. Lett. **61**, 1000 (1992).

[11] L. R. Tagirov, Phys. Rev. Lett. **83**, 2058 (1999).

[12] H.Shaked, P.M.Keane, J.C.Rodriguez, F.F.Owen, R.L.H.Herman and J.D.Jorgensen, "Crystal structures of the High Tc superconducting copper-oxides" (Elsevier, Amsterdam 1994).

[13] J.D.Jorgensen, D.G.Hinks, O.Chmaissen, D.N.Argyriou, J.F.Mitchell and B.Dabrowski, Proceedings of the first Polish-US Conference on HTS, Wroclaw-Duszniki Zdr., Poland (Springer-Verlag)

[14] H.Ihara, K.Tokiwa, H.Ozawa, M.Hirabayashi and Y.S.Song, Jnp. J. Appl. Phys. **33** L503 (1994).

[15] M.A.Alario-Franco, C.Chaillout, J.J.Capponi, J.L.Tholence and B.Souletie, Physica C **222** (1994) 52.

[16] C.Q.Jin, S.Adachi, X.J.Wu, H.Yamauchi and S.Tanaka, Physica C **223** (1994) 238

[17] H.Ihara, K.Tokiwa, A.Iyo, M.Irabayashi, N.Terada, M.Tokumoto and Y.S.Song, Physica C235-240 (1994) **981**

[18] H.Kumakura, K.Togano, T.Kawashima and Takayama-Muromachi, Physica C **226** (1994) 222.

[19] J.N.Eckstein et al., Appl.Phys.Lett. **57**, 931 (1990).

[20] A.Gupta et al., J.Solid State Chem, **112**, 113 (1994).

[21] G.Balestrino, R.Desfeux, S.Martellucci, A.Paoletti, G.Petrocelli, A.Tebano, B.Mercey and M.Hervieu J. of Mater. Chem. **5**, 1879 (1995).

[22] Gupta, T.M.Shaw, M.Y.Chern, B.W.Hussey, A.M.Guloy, and B.A.Scott, J. Sol. State Chem. **114**, 190 (1995).

[23] C.Aruta, G.Balestrino, R.Desfeux, S.Martellucci, A.Paoletti, G.Petrocelli, Appl. Phys. Lett. **68**, 926 (1996).

[24] T.Maeda, M.Yoshimoto, K.Shimozono, H.Koinuma, Physica C **247**, 142 (1995)

[25] C.Aruta, G.Balestrino, S.Martellucci, A.Paoletti, and G.Petrocelli, J.Appl. Phys. **81**, 220 (1997).

[26] A.Gupta, T. Shaw, M.Y. Chern, B.W. Hussey, A.M. Guloy and B.A. Scott., Journ. Sol. State Chem. **111**, 1 (1994).

[27] G. Koster, G.J.H.M. Rijnders, D.H.A. Blank and H. Rogalla, Appl. Phys. Lett. **74**, 3729, (1999)

[28] G. Balestrino, A.Crisan, S. Lavanga, P.G. Medaglia, Petrocelli, A.A. Varlamov, Phys. Rev. B **60**, 10505 (1999).

[29] J.D.Jorgensen, Physics Today, June 1991, p.34.

[30] J.D.Jorgensen, P.Lightfoot, and S.Pei, Supercond. Sci. Technol. **4**, S11 (1991).

[31] M. Kawasaky, K. Takahashi, T. Maeda, R. Tsuchiya, M. Shinohara, O. Lshiyama, T. Yonezawa, M. Yoshimoto, and H. Koinuma, Science **266**, 1400 (1989).

[32] S. Colonna, F. Arciprete, A. Balzarotti, G. Balestrino, P. G. Medaglia, G. Petrocelli, Physica C **334**, 64 (2000).

[33] G. Balestrino, G. Pasquini, A. Tebano, Phys. Rev. B **62**, 1421, (2000).

[34] I. K. Shuller, M. Grimsditch, F. Chambers, G. Devane, H. Vanderstraeten, D. Neerinck, J. P. Loquet and Y. Bruynseraede, Phys. Rev. Lett. **65**, 1235 (1990).

[35] F. Arciprete, G.Balestrino, S. Martellucci, P.G. Medaglia, A. Paoletti, G. Petrocelli, Appl. Phys. Lett. **71**, 959 (1997).

[36] S. Aisenberg and R. Chabot, J. Appl. Phys., **42** (1971) 2953

[37] H. Tsai, D. B. Bogy, J. Vac. Sci. Technol. A, **5** (1987) 3287

[38] J. Robertson, Adv. Phys., **35** (1986) 317

[39] F.J.Himpsel, J.A.Knapp, J.A.vanVechten, D.E.Eastman, Phys.Rev.B **20** 624 (1979)]

[40] G.A.J.Amaratunga, S.R.P.Silva, Appl.Phys.Lett. **68**, 2529 (1996)

[41] O.Groning, O.M.Kuttel, P.Groning, L.Schlapbach, Appl.Surf.Sci. **111**, 135, (1997)

[42] K.Okano, S.Koizumi, S.R.P.Silva, G.A.J.Amaratunga, Nature **381**, 140 (1996)

[43] M.W.Geis, J.C.Twichell, J.Macaulay, K.Okano, Appl.Phys.Lett. **67**, 1328 (1995)].

[44] M.W.Geis, J.C.Twichell, T.M.Lyszcarz, J.Vac.Sci.Technol. B**14**, 2060 (1996)

[45] Z.H.Huang, P.H.Cutler, N.M.Miskovsky, T.E.Sullivan, Appl.Phys.Lett **65**, 2562 (1994)],

[46] S.R.Silva, J.Robertson, G.Amaratunga et al. J.Appl.Phys.**81**, 2626 (1997)

[47] D.F.Franceschini, C.A.Achete and F.L.Freire Jr., Appl.Phys.Lett., **60** (1992) 3229

[48] F.L.Freire Jr. and D.F.Franceschini, Thin Solid films, **293** (1997) 236]

[49] Y.Lifshitz, Diamond Relat.Mater **3-5** (1996) 388

[50] Y.Lifshitz, G.D.Lempert, E.Grossman, I.Avigal, C.Uzansaguy, R.Kalish, J.Kulik, D.Marton and J.W.Rabalais, Diamond Relat.Mater., **4** (1995) 318

[51] Y.Lifshitz, S.R.Kasi, J.W.Rabelais, Mater.Sci.Forum, **52-53**, (1990) 237

[52] Y.Lifshitz, S.R.Kasi, J.W.Rabelais, Phys.Rev.Lett., **62** (1989), 1290

[53] A.Cricenti, R.Generosi, Rev.Sci.Instrum., **66**, (1995) 4

[54] J.J.Pouch and S.A.Alterovitz (eds.), "Properties and characterization of amorphous carbon films", Mater.Sci.Forum, **52-53** (1990)

[55] R.E.Clausing, J.C.Angus, L.L.Horton, P.Koidl (eds.), "Diamond and diamond-like films and coatings", Proc. of NATO Advances Study Inst., July 22 -August 3, 1990, Castelvecchio Pascoli, Italy, Plenum Press, N.Y., 1991

[56] S.Prawer, K.W.Nugent, Y.Lifshitz, G.D.Lempert, E.Grossman, J.Kulik, I.Avigal, R.Kalish, Diam.Relat.Mater. **5**, 443 (1996)

[57] D.R.McKenzie et al., Diamond Relat.Mater., **1** (1991) 51

[58] P.J.Fallon, V.S.Veerasamy, C.A.Davis, J.Robertson, G.A.J.Amaratunga, W.I.Milne and J.Koskinen, Phys.Rev.B, **48** (1993) 4777

[59] D.R.McKenzie, D.Muller and B.A.Pailthorpe, Phys.Rev.Lett., **67** (1991) 773

[60] A.Voevodin, S.J.P.Laube, S.D.Walck, J.S.Solomon, M.S.Donley, J.S.Zabinski, J.Appl.Phys. **78**, (6) 1995

GNSR 2001
G. Messina and S. Santangelo (Eds.)
IOS Press, 2002

Quality indicators for CVD diamond films: a Raman study

M.G. Donato[a], G. Faggio[a], M. Marinelli[b], G. Messina[a], E. Milani[b],
S. Santangelo[a], A. Tucciarone[b], G. Verona Rinati[b]

(a) INFM, Dipartimento di Meccanica e Materiali
Università diReggio Calabria, 89060 Reggio Calabria, Italy

(b) INFM, Dipartimento di Scienze e Tecnologie Fisiche ed
Energetiche, Università di Roma Tor Vergata, 00133 Roma, Italy

Abstract. A systematic study of the fine structure of Raman spectra in Microwave Plasma Enhanced Chemical Vapour Deposition (MWPECVD) diamond films is carried out. The films are grown with a CH_4-CO_2 gas mixture, at CH_4 concentration ranging between 47.4 and 52% and at two different substrate temperatures (750 and 850 °C). The preferential orientation of the samples is studied by X-ray diffraction (XRD) measurements. At both substrate temperatures, well-structured peaks, indicators of strong residual stress, are observed only in samples grown at low methane concentrations, where the preferential orientation is the result of a competition between various growth sectors. The splitting of the diamond Raman peak is greater close to the grain boundaries and, thus, at the interface with the substrate: this finding evidences a highly inhomogeneous stress distribution in the films. The spreading of the micro-Raman diamond peak positions gives rise to an inhomogeneous broadening of the macro-Raman line whose width may not be, thus, a reliable defect-density indicator in the case of CVD films subjected to large anisotropic stress.

1 Introduction

By means of the Chemical Vapour Deposition (CVD) technique, continuous polycrystalline diamond films can be grown exhibiting the outstanding properties of natural diamond (a large band gap of 5.5 eV, high electric resistance, high thermal conductivity, strong resistance to mechanical and chemical attack). Although they may be applied in various fields, CVD diamond films are particularly suitable to applications in highly aggressive environments. Such applications require a very high quality of the synthesised films: in fact, the presence of impurities and defects (non-diamond carbon inclusions, vacancies, dislocations, grain boundaries) and the production of internal residual stress during the growth process represent a serious limit to the film full exploitation.

Raman and photoluminescence (PL) spectroscopy allow the study of the quality of diamond films. Raman spectroscopy readily distinguishes crystalline diamond phase from other non-diamond carbon phases [1]. In fact, the Raman "fingerprint" of natural diamond is a peak centred at 1332 cm^{-1} whose full width at half maximum (Γ) is usually of 2 cm^{-1}, while a band centred at about 1550 cm^{-1} is the signal of non-diamond carbon phases. PL spectroscopy individuates the presence of impurities, because these, generating some energy levels in the diamond band gap, become PL-active under excitation energies below 5.5 eV [2].

The Raman technique is a valid tool for optimising the deposition process in order to obtain diamond films having high crystalline quality and phase purity. To this aim, the full width half maximum of the diamond Raman peak and the amplitude ratio (I_{ND}/I_D) of the non-diamond band at ~1550 cm^{-1} to the diamond peak are the quality indicators more commonly utilised.

As already noted, CVD diamond films are generally subjected to internal residual stress. It has a thermal component, due to the difference between the thermal expansion coefficients of diamond and the substrate, and an intrinsic component, due to the impurities and defects incorporated in the film. As the presence of residual stress into the films induces a shift in the diamond peak position and, sometimes, the splitting of the peak in two or three components, high-resolution Raman measurements are commonly utilised to investigate the degree of stress to which the diamond films are subjected [3-5].

In this work, Raman measurements, aimed at studying the global quality of CVD diamond films in terms of crystalline quality, phase purity and residual distribution of stress, are presented, with respect to the changes in the substrate temperature (T_S) and the CH_4 concentration in the growth mixture. Demonstration is given that the linewidth of diamond Raman peak, usually regarded as a quality indicator for estimating the crystalline quality of the films, is heavily influenced by the residual stress distribution; for this reason, in the presence of a highly inhomogeneous stress-distribution, it may result an unreliable parameter.

2. Experimental

The investigated diamond films have been grown on 5x5 mm^2 (100) polished p-type single crystal silicon substrates by MWPECVD using a CH_4-CO_2 gas mixture [6] with CH_4 concentrations ranging between 47 and 52%. In order to study the effect of the substrate temperature on the film quality, two sets of samples have been deposited at different substrate temperatures (set A: T_S=750 °C, set B: T_S=850 °C). To obtain information on the film preferential orientation, XRD measurements have been performed using the Cu K_α line of a Bragg-Brentano diffractometer as incident radiation. The Raman spectra have been recorded at room temperature by an Instrument S.A. Ramanor U1000 double monochromator. The excitation source has been the 514.5 nm line of a Coherent Innova 70 Ar+ ion laser. Macro-Raman measurements have been performed using a beam diameter of about 100 μm and a focus depth extending over the whole film thickness. High spatial resolution micro-Raman measurements have been performed, at room temperature, by utilising the X100 objective of an Olympus BX40 microscope. In this way, a spot size of about 1 μm in diameter has been obtained, with a depth resolution of about 2 μm in confocal mode. A spectral slit width of ~1 cm^{-1} (slit width = 100 μm) has been used, so that the slit-induced line broadening substantially does not modify the intrinsic width of the Raman lines to be measured. The laser power has been less than 10 mW at the sample surface. The detection system has consisted of an electrically cooled Hamamatsu R943-02 photomultiplier operating in photon-counting mode. In order to measure linewidth and shift of the diamond peak, the Raman spectra, after subtraction of the luminescence background, have been reproduced by using one or two Lorentzian lineshapes for the diamond line and Gaussian lines for the remaining spectral features.

3. Results

3.1. Macro-Raman measurements

In order to investigate the crystalline quality and the phase purity of the deposited films, the systematic acquisition of their Raman spectra has been performed. In Fig. 1 the values of Γ (FWHM), obtained in the macro-configuration, are shown as a function of the CH_4 concentration at the two substrate temperatures explored. As can be seen, the decrease of T_S leads to an improvement of the crystalline film quality, as indicated by the lower Γ-values measured on the samples of series A.

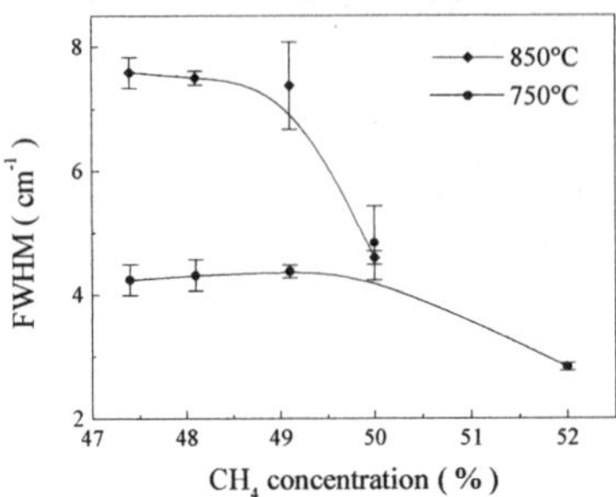

Fig.1 FWHM of macro-Raman diamond peaks as a function of the methane concentration at 750 and 850 °C substrate temperatures (lines are drawn as visual help).

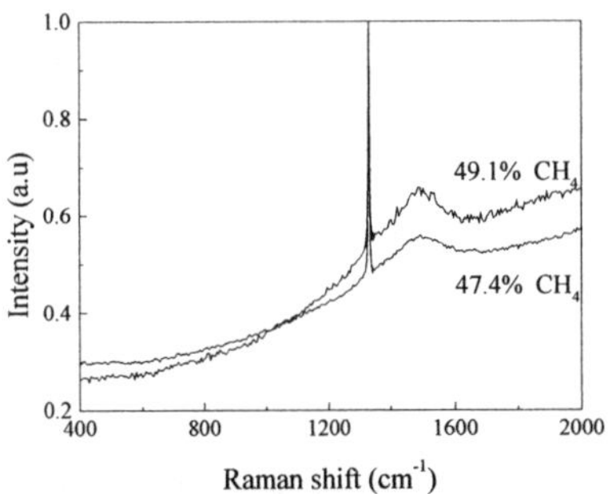

Fig.2 Effect of methane concentration on macro-Raman spectra of samples grown at 750 °C.

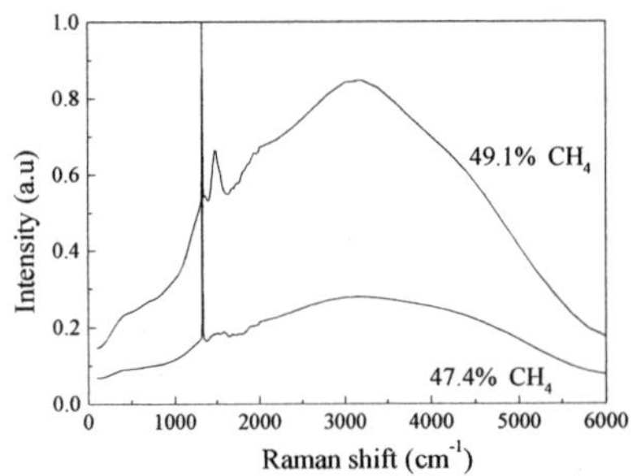

Fig.3 Evolution, produced by the changes in methane concentration, of PL spectra of samples grown at 750 °C.

At both the substrate temperatures, with increasing the CH_4 concentration, Γ remains approximately constant, within the experimental error, until abruptly decreases when the concentration exceeds 49.1% and 50% in the samples grown at 850°C and 750°C, respectively. However, when the CH_4-concentration increases, the phase purity of the samples gets worse, as witnessed by the higher intensity of the ~1550 cm^{-1} non-diamond band (Fig.2). This trend, commonly observed in literature [7,8], is quantitatively confirmed also by the fit-parameters of the spectra shown in Tab.1.

Tab.1 Fit parameters obtained for the samples of Fig.2.

CH_4 (%)	47.4	49.1
FWHM (cm^{-1})	4.24	4.37
I_{ND}/I_D (ad. un.)	0.15	0.27

The worsening of the phase purity induced by the higher CH_4 concentration is also supported by the results of the PL measurements (Fig.3). Films grown with higher methane concentrations exhibit a stronger photoluminescence background, that is usually associated to a higher defect density in the film [9].

3.2 Micro-Raman measurements

In order to clarify the relation between crystalline quality and defect density of the samples, the film residual stress distribution has been investigated by using high spatial resolution Raman spectroscopy, which has given information on the local quality of the films.

To individuate possible correlations between crystalline quality and texturing of the samples, Raman measurements have been supported by X-ray diffraction (XRD) investigations. Figure 4 compares the evolution of the Raman peak at the sample surface (top) and of the corresponding XRD pattern (bottom) produced by the changes in methane concentration at 750 °C. The film preferential orientation changes from <110> to <100> with increasing CH_4 concentration. As for micro-Raman analysis, a narrow and slightly shifted diamond peak is measured only in the <100>-textured sample grown at high CH_4-concentration, meanwhile, at lower methane content, the diamond peak is clearly structured in two components. An analogous situation is revealed for sample series B.

As previously noted, the splitting of the diamond Raman peak in two or three components indicates the occurrence of a strain distribution in the film. For a perfect diamond crystal, without any stress, the first order Raman line at $\omega_0=1332$ cm^{-1} corresponds to inelastic scattering of the laser light by the triply degenerate optical phonon at the Brillouin zone centre. In presence of strain, the cubic symmetry of the crystal is lowered, the triply degeneracy lifted and the diamond peak consequently splitted. Depending on the direction of the stress relative to crystallographic directions, the degeneracy may be partially or completely lifted. The lifting of the degeneracy occurs for uniaxial and biaxial crystallographic strains, but not for hydrostatic strains [3,10,11]. The magnitude of both splitting and shift of Raman line depends on level and nature of the stress, the film is subjected to. For these reasons, the structured lineshape measured on non-textured samples indicates a distribution of strong residual stress, while the non-structured lineshape, measured in <100>-textured samples, is index of low residual stress. The residual-stress evolution in the film, evidenced by micro-Raman measurements at varying CH_4-concentration, is thought as due to the competition between the different growth sectors, leading to the <100>-texturing at high CH_4 contents.

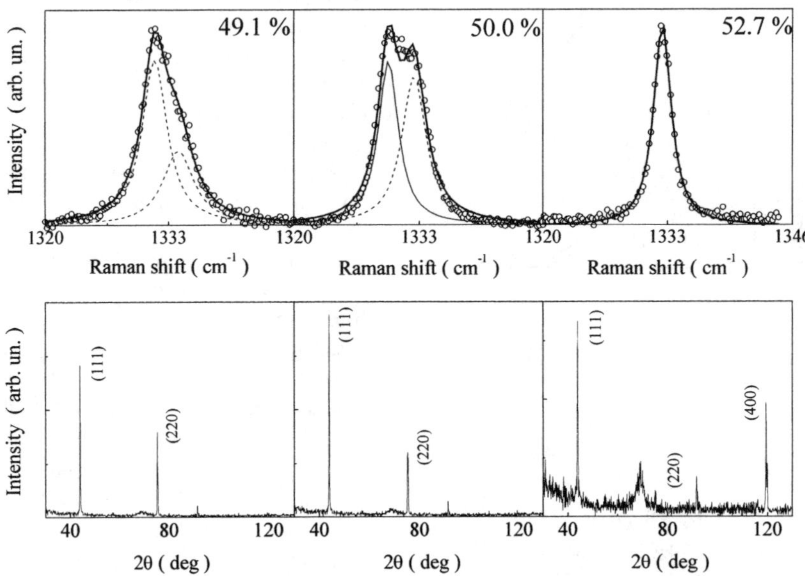

Fig.4 Evolution of the lineshape of the diamond micro-Raman peak produced by the change of CH_4 content in the gas mixture in films deposited at 750°C (top). The circles represent the experimental data; the continuous and dashed lines respectively correspond to the fitted peak and its stress-separated Lorentzian components. The corresponding XRD patterns are also shown (bottom).

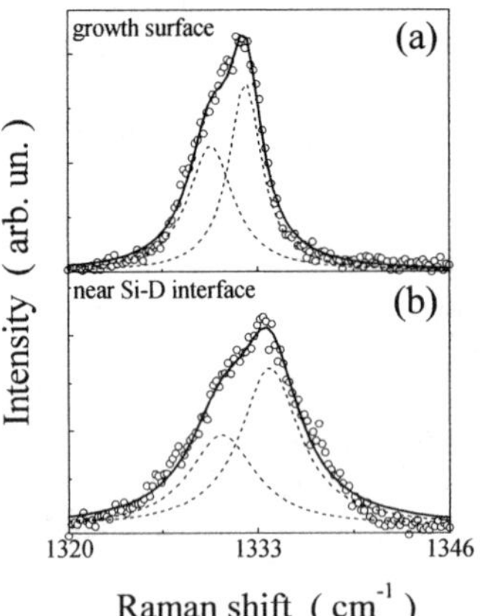

Fig.5 Comparison between micro-Raman diamond lines measured at growth surface (a) and near the film-substrate interface (b). The case of the sample deposited at 750°C utilising 50.0% CH$_4$ percentage is shown. The same symbols as in Fig.4 are used.

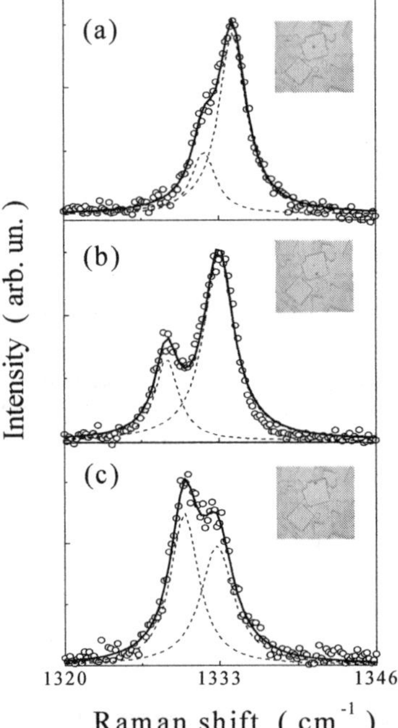

Fig.6 Evolution of the lineshape of the diamond peak observed when the laser probe is moved from the centre (a) to the boundaries (b, c) within the same grain. The micro-Raman peaks shown were measured at the surface of the sample grown at 750°C with 50.0% CH$_4$ concentration. The same symbols as in Fig.4 are used. The optical micrographs, showing the laser spot position, are reported in the inset.

To obtain information about the spatial stress distribution, which the films are subjected to, the diamond peaks measured by focusing at the growth surface and the substrate interface have been compared. Figure 5 shows the diamond peaks measured on a sample grown at 750 °C by using 50.0% CH$_4$ concentration. As can be seen, the peak measured at the substrate interface (Fig.5b) is more heavily stress-affected than the peak measured at the surface (Fig.5a). This fact suggests that the higher density of grain boundaries at the film/substrate interface is responsible, as already reported [4], for the generation of highly inhomogeneous distribution of stress in the film.

The inhomogeneity of the stress distribution in the film, induced by the grain boundaries, is evident also on a smaller scale, when the diamond peaks measured at the sample surface, by varying the laser spot position on the same grain, are compared. The diamond peak recorded at the grain centre (Fig.6a) is less structured than diamond peaks measured by focusing at grain boundaries (Fig.s 6b and 6c), thus indicating that the stress distribution can change not only between different grains, but also on the same grain.

Micro-Raman measurements recorded on a large number of grains can be used to quantitatively estimate the film residual stress level. The problem of deducing quantitative information about the importance of stress in CVD diamond films has been faced by several authors. Ager e Drory [3], under the hypothesis of biaxial stress, calculated the residual stress by using the frequency positions, ω_s and ω_d, of both the "singlet" and "doublet" components in which the diamond peak splits. By averaging these expressions, Ralchenko et al. [12] calculated the average residual stress with the following expression:

$$\sigma = -0.567\,(\omega_m - \omega_0) \qquad (1)$$

where ω_0 stands for the stress-free peak position, σ represents the biaxial stress, expressed in GPa, and ω_m can be calculated by the following expression:

$$\omega_m = 0.5(\omega_s + \omega_d) \qquad (2)$$

According to Eq.1, σ can assume either negative or positive values, corresponding to a compressive or tensile stress, respectively.

By assuming, as usual, that a film deposited on a substrate is in a biaxial in-plane stress state, the changes of the stress-level produced by the varying growth conditions have been evaluated by applying the Ralchenko's method. The ω_s and ω_d values have been calculated by averaging out the frequency positions of the singlet and the doublet components of the diamond peaks relative to several micro-Raman measurements.

Figure 7 reports the biaxial stress, measured on both the series of samples. The negative values of σ indicate that the films studied in this work are subjected to a compressive stress. The set of samples grown at higher temperature is always subjected to a higher level of stress. Furthermore, at a given substrate temperature, the stress level progressively decreases with increasing CH_4 concentration, and relaxes in <100>-textured samples.

4. Discussion

Micro-Raman measurements have evidenced a complex stress-distribution scenario: stress induced modifications on diamond peak position and lineshape are more evident in low CH_4 concentration films, where the <110> and <100> growth sectors compete to settle the film preferential orientation. Moreover, the stress distribution is strongly variable, not only between different grains, but also on the same grain. The great variability of diamond-peak position and lineshape, as evidenced from micro-Raman analysis, can explain the surprisingly high Γ-values measured in macro-Raman configuration on the same films.

In fact, as the laser probe samples a circular area (~100 μm diameter) enclosing a great deal of grains, the single and nearly symmetric diamond peak observed actually arises from the convolution of the responses coming from the individual grains. As different grains are under different stress levels, their corresponding Raman lines are consequently centred at different positions. Thus, the width of the macro-Raman diamond line is determined by the inhomogeneity of the local stress distribution within the volume sampled by the laser probe.

In this frame, the large linewidths observed at low CH_4 concentrations (Fig. 1) are understood as not exclusively due to a lower crystalline perfection of the samples (namely, higher defect density), but as caused by large anisotropic stresses inhomogeneuosly distributed within the film. More specifically, the small Γ-values measured in the samples grown at high CH_4 contents (exhibiting a <100>-texturing) are actually representative of a quite good crystalline-quality. Contrarily, the remarkable broadening of the macro-Raman diamond line detected in non-

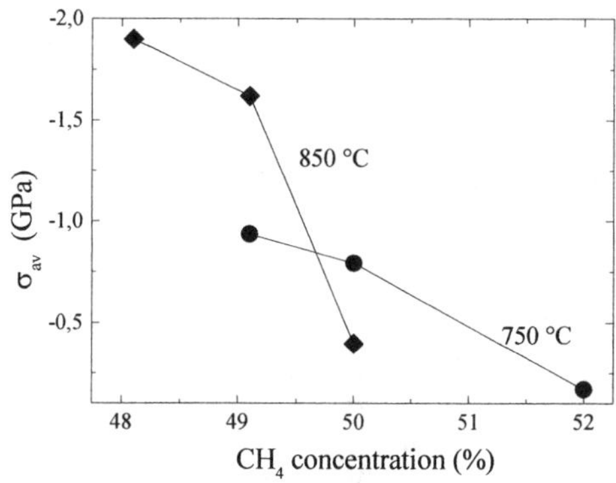

Fig.7 Average stress-level (σ_{av}), as estimated from Eq.1. The dependence of σ_{av} on the CH_4 concentration at the two-substrate temperatures is shown.

textured samples arises from the cooperative effect of both the splitting and the position spreading of the micro-Raman lines [13].

In the light of the above considerations, regarding the Raman linewidth as a defect density indicator can be misleading in the presence of large anisotropic stresses and, rather, result in an under-estimation of the phase-purity of the investigated films.

5. Conclusions

CVD diamond samples, grown by a CH_4-CO_2 gas mixture at two different substrate temperatures and at various CH_4 concentrations, have been studied by means of micro- and macro- Raman measurements, aimed at assessing their crystalline quality and defect density. The crystalline quality, as resulting from macro-Raman measurements, is higher in the samples grown at low temperature. However, it seems to improve with the increase of the methane concentration, even if the phase purity gets worse. This apparent contradiction is resolved if micro-Raman measurements are taken in account. In fact, they evidence a highly inhomogeneous stress distribution in films grown at lower methane concentrations, further accentuated in films grown at higher substrate temperature. The film stress distribution is generated by the competition between the <110> and the <100> growth sectors in settling the film preferential orientation, and relaxes at the highest methane concentration, when the <100>-texturing is achieved. Both the spreading in the diamond peak position and its splitting in two components, evidenced at a local scale, hinder the estimation of the crystalline quality by macro-Raman measurements, because induce an inhomogeneous broadening of the Raman line. This effect is stronger in non-textured films, where the level of stress is higher. Thus, the large linewidth evidenced in films grown at low methane concentration are understood as not exclusively due to a higher density of defects, but to the inhomogeneous broadening caused by the large anisotropic stress within the films.

On the basis of these considerations, it is possible to conclude that, in presence of a highly inhomogeneous stress distribution, the Raman linewidth of the diamond peak may not be a reliable defect-density indicator of CVD diamond films.

References

[1]   D.S. Knight, W.B. White, *J. Mater. Res.* **4** (1989) 385
[2]   W. Zhu, *"Defects in Diamond"*, in *"Diamond : Electronic Properties and Applications"*, edited by L.S. Pan, D.R. Kania, Kluwer Academic Publishers (1995) Boston, pag. 175
[3]   J.W. Ager III, M-D. Drory, *Phys. Rev. B* **48** (1993) 2691
[4]   Y. von Kaenel, J. Stiegler, J. Michler, E. Blank, *J. Appl. Phys.* **81** (1997) 1726
[5]   Q.H. Fan, A. Fernandes, E. Pereira, J. Gracio, *Diam. Relat. Mater.* **8** (1999) 645
[6]   M. Marinelli, E. Milani, A. Paoletti, A. Tucciarone, G. Verona Rinati, N. Randazzo, R. Potenza, M. Pillon, M. Angelone, *Diam. Relat. Mater.* **7** (1998) 519
[7]   Y.K. Kim, K.Y. Lee, J.Y. Lee, *Thin Solid Films,* **272** (1996) 64
[8]   C. Jany, A. Tardieu, A. Gicquel, P. Bergonzo, F. Foulon, *Diam. Relat. Mater.* **9** (2000) 1086
[9]   L. Bergman, M.T. McClure, J.T. Glass, and R.J. Nemanich, *J. Appl. Phys.* **76** (1994) 3020
[10] E. Anastassakis, A. Pinczuk, E. Burstein, F.H. Pollak, M. Cardona, *Solid State Commun.* **8** (1970) 133
[11] M.H. Grimsditch, E. Anastassakis, M. Cardona, *Phys. Rev. B* **18** (1978) 901
[12] V.G. Ralchenko, A.A. Smolin, V.G. Pereverzev, E.D. Obraztsova, K.G. Korotoushenko, V.I. Konov, Yu.V. Lakhotkin, E.N. Loubnin, *Diam. Relat. Mater.* **4** (1995) 754
[13] S.J. Harris, A.M. Weiner, S. Prawer, K. Nugent, *J. Appl. Phys.* **80** (1996) 2187

GNSR 2001
G. Messina and S. Santangelo (Eds.)
IOS Press, 2002

Raman analysis of CVD diamond: influence of the growth parameters

M.G.Donato[a], G.Faggio[a], M.Marinelli[b], G.Messina[a], E.Milani[b], S.Santangelo[a], A.Tucciarone[b], G. Verona Rinati[b]

[a]*INFM, Dipartimento di Meccanica e Materiali, Facoltà di Ingegneria dell'Università, Località Feo di Vito, 89060 Reggio Calabria, Italy*

[b]*INFM, Dipartimento di Scienze e Tecnologie Fisiche ed Energetiche, Università di Roma Tor Vergata, via di Tor Vergata, 00133 Roma, Italy*

Abstract. Synthetic diamond films are deposited on silicon substrates by using Microwave Plasma Enhanced Chemical Vapour Deposition (MWPECVD). The influence of the growth parameters on diamond film properties is systematically investigated. Substrate temperature (ranging between 750 and 850 °C), gas mixture (CH_4-CO_2 or CH_4-H_2) and its methane content (varying in the ranges $47.4 \div 52.7\,\%$ and $0.6 \div 2.2\,\%$, respectively) are shown to be key parameters to improve the quality of the films. The indications emerging from the accurate Raman and photoluminescence (PL) characterisation studies performed suggest the optimal condition for the deposition of very high quality films (namely, substrate temperature fixed at 750 °C and 0.6 % methane in the CH_4-H_2 mixture).

1. Introduction

Diamond is considered an ideal material for many applications *[1]*. The constant progress obtained in the Chemical Vapour Deposition (CVD) processes has allowed to realize low-cost synthetic diamond films with physical and chemical properties approaching or even overcoming, in some aspects, those of natural diamond. CVD diamond is considered a very attractive material because of its great variety of potential applications in microelectronics, optics, micromechanics and, more generally, in all those fields where advanced devices are required to operate in harsh environments.

Unfortunately, compared to natural diamond or to single-crystal synthetic diamond obtained by high-pressure/high-temperature methods, CVD diamond films grown on non-diamond substrates are polycrystalline and generally contain a relative high concentration of impurities and crystal defects. These undesired inclusions may significantly degrade optical and electronic properties of CVD diamond, actually limiting the development of new diamond-based devices. In this context, it is evident how the possibility of producing high-quality films is strictly connected to the minimisation of the defect formation during the film deposition, that is pursued through a feedback procedure, driven by the indications coming out from the results of the characterisation studies, whose ultimate aim is the growth-process optimisation.

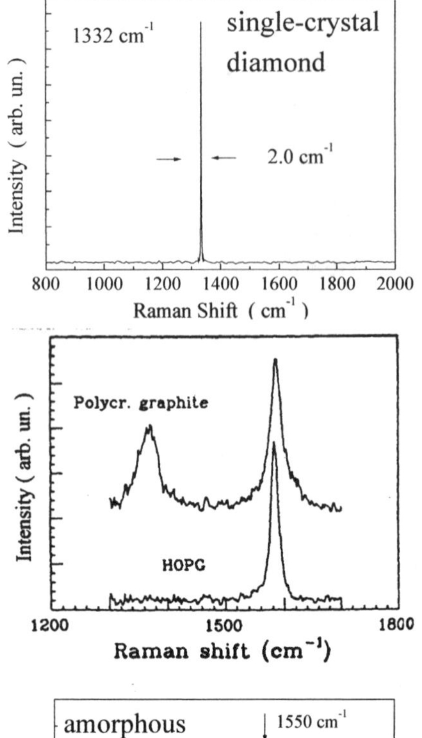

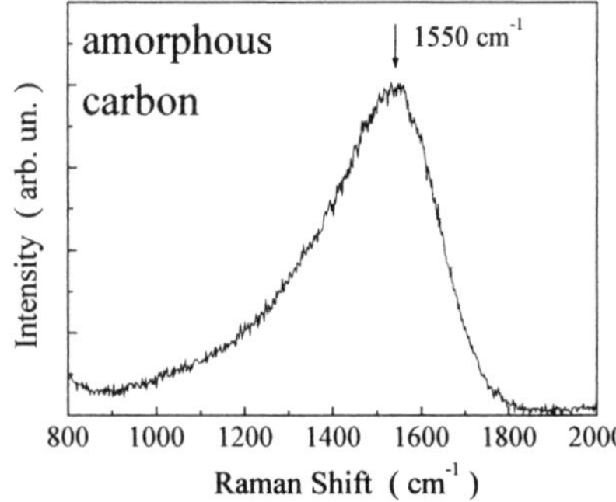

Fig.1 Raman spectra of (top) single-crystal diamond, (centre) polycrystaline graphite and of Highly Oriented Pyrolytic Graphite (HOPG) and (bottom) amorphous carbon.

In the present work, the results are shown of a detailed Raman and PL analysis aimed at clarifying the influence of the growth parameters on the global quality of diamond films grown MWPECVD. In the light of the results of the characterisation of two sample sets grown using a CH_4-CO_2 gas mixture, at different substrate temperatures, the experimental set-up is remarkably improved and a third sample set is then deposited under optimised growth conditions (namely, using a CH_4-H_2 gas mixture at 750 °C substrate temperature).

Micro-Raman measurements, performed at the surface of the latter CVD films, evidence diamond lines narrower than 2.4 cm⁻¹, thus witnessing the intrinsic quality of the synthetic films to be comparable to that of the best natural diamonds.

2. Raman and photoluminescence spectroscopy

Optical techniques, such as PL and Raman spectroscopy are powerful and non-destructive tools widely utilised for evaluating the quality of diamond samples. In particular, the former technique results very useful for studying nature and distribution of crystal defects and impurities having energy levels within the diamond band gap, while the latter, allowing a ready and accurate identification of different carbon allotropes present in the analysed sample *[1-3]*, represents a reliable diagnostic tool for the diamond film characterisation.

The characteristic Raman spectroscopic signal of a natural single crystal of diamond consists in a sharp line at 1332 cm⁻¹, having a full width half maximum (FWHM) of about 2 cm⁻¹ (top of **Fig.1**). Natural single-crystal graphite shows a single Raman peak at 1580 cm⁻¹ ("G"-peak), while an additional band, activated by the disorder caused by symmetry breaking due to the finite crystalline size, appears at about 1355 cm⁻¹ ("D"-peak) in polycrystalline graphite (centre of **Fig.1**) *[4,5]*. Amorphous carbon finally exhibits a broad asymmetric Raman band centred at about 1500 cm⁻¹ (bottom of **Fig.1**) *[1]*.

In CVD diamond films, due to the presence of crystal defects or non-homogeneous stress distribution, the diamond Raman peak is usually much broader than in natural diamond. In addition, upon the occurrence of hydrostatic stress, a shift with respect to 1332 cm^{-1} diamond peak position can be detected *[6-10]*. Non-diamond carbon Raman lines are then usually further detected superimposed to a generally-strong PL background *[1,3,11]*. Therefore, crystalline quality and phase purity of diamond films are conventionally evaluated by measuring the width of the 1332 cm^{-1} line and by comparing the relative intensity of the diamond and non-diamond carbon features, respectively.

3. Experimental

Diamond films were deposited on 5x5mm^2 (100) silicon substrates by using MWPECVD process. In order to promote diamond nucleation on the Si surface, the conventional scratching procedure was adopted. The substrate temperature, T_S, was measured by an optical pyrometer.

Aiming at clarifying the effect of the methane concentration, as well as, the influence of substrate temperature on the film quality, two sets of samples were firstly deposited by using a CH_4-CO_2 gas mixture, with a CH_4 content varying from 47.4 to 52.7 % and substrate temperatures fixed at 750 and 850 °C, respectively.

Being the quality of these films globally not satisfactory, a careful optimisation of the deposition system was carried out before depositing the third sample set. In particular, the reactor geometry was modified in order to optimise the coupling of the microwave power to the plasma, by avoiding power losses in its vertical section. In order to increase the plasma density without raising the substrate temperature, an appropriate water cooling of the substrate holder was necessary.

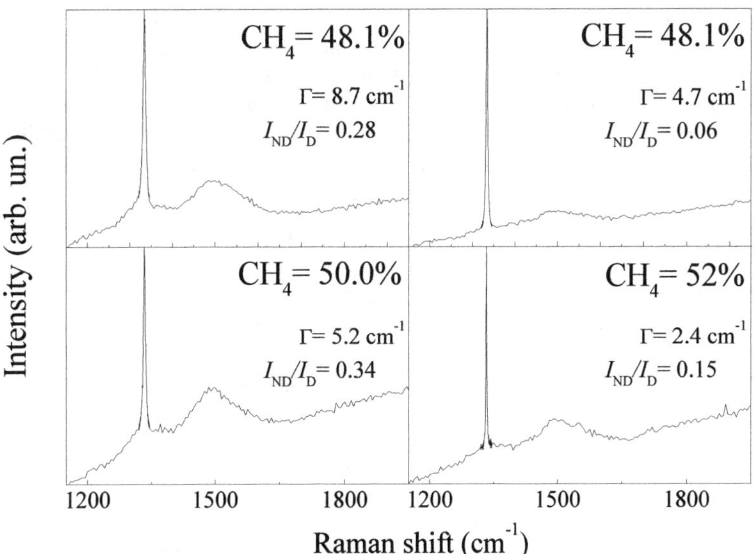

Fig.2 Micro-Raman spectra at the growth surface of diamond samples grown with the CH_4-CO_2 gas mixture. Spectra shown in left (right) column refer to sample set grown at 850 (750) °C. The Γ and I_{ND}/I_D changes produced by the CH_4 concentration variation are also indicated.

The indications, emerged by monitoring the quality of the films grown with the CH_4-CO_2 gas mixture, were utilised to choice the conditions under which were deposited the samples after the optimisation of the experimental set-up.

The third sample set was grown in a CH_4-H_2 atmosphere, that allowed achieving much lower deposition rates than with the CH_4-CO_2 mixture (~0.9 μm/h against ~4 μm/h). The substrate temperature was fixed at 750 °C. In order to systematically change the film morphology, preferential orientation and crystal quality, the CH_4 concentration was varied in the range $0.6 \div 2.2$ %.

In all the above cases, the samples were grown having approximately the same thickness (40 μm).

The Raman and micro-photoluminescence measurements were carried out, at room temperature, on an Instruments S.A. Ramanor U1000 double monochromator, equipped with a microscope Olympus BX40 for micro-Raman sampling and with an electrically cooled Hamamatsu R943-02 photomultiplier for photon-counting detection. Using an X100 objective, the laser beam was focused to a diameter of about 1 μm. A depth resolution of about 2 μm was obtained with a confocal aperture of 100 μm. A spectral slit width of ~0.9 cm^{-1} (slit width = 100 μm) was used. Micro-photoluminescence spectra were recorded in the region $1.6 \div 2.4$ eV (100-6000 cm^{-1} in a relative scale). In both the measurements, the 514.5 nm (2.41 eV) line of a (Coherent Innova 70) argon-ion laser was used as an excitation source.

4. Results and discussion

4.1 CVD films synthesised in CH_4-CO_2 atmosphere

The effect of the methane concentration, as well as, the influence of substrate temperature on the film quality, is established by studying the properties of the two sets of samples deposited in CH_4-CO_2 atmosphere. As shown in **Fig.2**, Raman and PL spectroscopy results clearly demonstrate that both deposition temperature and methane concentration play a quite important role in ultimately determining the CVD diamond film quality.

The full width half maximum, Γ, of the 1332 cm^{-1} diamond Raman peak and the

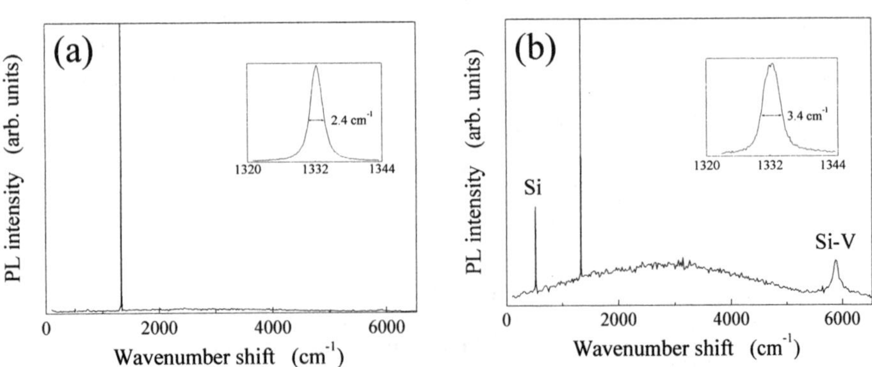

Fig.3 Micro-PL spectra at (a) the growth surface and (b) substrate interface of a sample deposited, at 750 °C, using CH_4-H_2 gas mixture at 0.6% methane concentration (the sharp peak at about 520 cm^{-1} is due to the Raman scattering from the Si substrate).

intensity ratio, I_{ND}/I_D, of the 1500 cm^{-1} non-diamond carbon band to the 1332 cm^{-1} diamond line are, as usual *[12]*, utilised in order to get information about phase purity and crystalline quality of the investigated samples. What emerges is that i) decreasing the substrate temperature from 850 to 750 °C results in a narrowing of the diamond line and ii) a lowering of the large non-diamond band around 1500 cm^{-1}. These findings indicate that, at lower deposition temperature, crystalline quality improves and phase purity increases.

In addition, as, at both temperatures, the I_{ND}/I_D intensity ratio reduces with lowering methane concentration, as expected *[13,14]*, the incorporation of non-diamond phases within the individual grains is deduced to be progressively lesser at lower CH$_4$ content.

Nevertheless, the quality of these films is globally not satisfactory. This suggests the necessity of a careful optimisation of the deposition system.

4.2 CVD films synthesised in CH$_4$-H$_2$ atmosphere

4.2.1 Micro-Raman and micro-PL analysis results

The third sample set is deposited, after having noticeably improved the experimental set-up (*Sect.3*), in growth conditions under which, in the light of the above indications, an optimisation is expected of the film properties. Films are hence deposited at 750 °C substrate temperature, using low CH$_4$ contents. In addition, aiming at achieving lower deposition rates, a CH$_4$-H$_2$ (rather than a CH$_4$-CO$_2$) gas mixture is preferred.

As a general result, the films grown under these conditions exhibit quite better phase purity and crystalline quality. This is demonstrated in **Fig.3a**, where the micro-PL spectrum measured at the growth surface of a diamond sample grown at 0.6 % CH$_4$ is shown. Practically, only the sharp diamond Raman line is detected. Furthermore, the FWHM of the line at ~1332 cm^{-1} is only 2.4 cm^{-1}, thus comparable to that measured in an industrially-prepared single-crystal diamond used as reference sample (line position 1332 cm^{-1}, with $\Gamma \cong 2$ cm^{-1}). A so small Γ-value assesses the very good quality of the sample *[*]*.

Comparable diamond line widths have been recently measured in other CVD diamond films *[15-18]*. In particular, Coe et al. *[16]* report on De Beers CVD diamond films (thickness $\approx$ 1 mm) with FWHM of approximately 3 cm^{-1}; Jany et al. *[17]* show FWHM values down to 2.6 cm^{-1} both in $\approx$20 and $\approx$100μm thick films. In the light of the literature data, the FWHM reported in the present work look quite interesting, especially if the relatively low thickness of our samples is considered.

The excellent phase purity of the individual grains at the growth surface is witnessed by the complete absence of any non-diamond carbon feature around 1500 cm^{-1} and by the quite weak intensity of the PL band centred at ~ 2900 cm^{-1} (2.05 eV).

Moving towards the diamond-Si interface, a slight worsening in the crystalline quality occurs. This is evidenced in **Fig.3b**, where the micro-Raman spectrum, recorded on the same grain as in **Fig.3a** by focusing the laser spot on the interface region, is shown. The larger FWHM at the diamond line reveals a poorer crystalline quality.

At the same time, the intensity of the broad PL at ~2.05 eV is found to increase. Although its origin is still uncertain, the band has been hypothesised *[18-20]* to be due to the optical transitions from a continuous distribution of gap-states introduced by the sp^2 phase. Such an attribution could explain the intensity increase observed going towards the interface. Grain boundaries, in fact, are nucleation sites for sp^2 inclusions and, due to the

---

[*] It is worthwhile noticing that measurements carried out on different grains show line widths ranging from 2.1 to 3.0 cm^{-1} and line position between 1332 and 1333 cm^{-1}, thus confirming the high quality of the film, nearly-unaffected by residual stress.

columnar nature of CVD diamond growth, the most defective film layer is localised mainly at the substrate interface, where the grain boundary density is higher.

Moreover, a PL feature, typically detected in CVD diamond films grown on Si substrates *[21-23]*, is distinctly observed at ~5880 cm^{-1} (1.68 eV). Its appearance is commonly associated with the presence of substitutional or interstitial Si-vacancy complex, incorporated in the growing film owing to the plasma-etching of the Si substrate during the early deposition stage *[24-28]*.

4.2.2 Macro-Raman and macro-PL analysis results

So far, micro-Raman and micro-PL measurements provide information on intrinsic quality of individual grains and on spatial distribution of defects. Unfortunately, due to the highly inhomogeneous nature of the polycrystalline CVD diamond films, the results obtained undergo even strong variations, moving from point to point within the sample *[7]*. Such a variability actually constitutes a serious limitation when different samples, grown under very similar experimental conditions, are systematically studied in order to be finally compared. In principle, the data relative to a great deal of different grains should be properly averaged. However, the problem is generally easily overcome by performing a macro-Raman investigation, since the whole film thickness, over a circular area of about 100 μm diameter, is sampled.

Figure 4 shows the macro-PL spectrum of the sample, which the micro-PL spectra of **Fig.3** refer to. By comparison, the increase of intensity of both ~2900 cm^{-1} PL band and ~1500 cm^{-1} non-diamond carbon band can be noted. Moreover, a width of about 3.4 cm^{-1} is measured at the diamond line. The observed behaviour is readily understood considering that the spectrum recorded in the macro configuration receives contribution from the high-quality layer at the film growth surface, as well as, from the more defective region located at the interface with the substrate. This is the reason why the macro-Raman line width is larger than as measured, in micro configuration, at the sample surface. Nevertheless, the macro-Raman FWHM values remain very small, indicating a quite satisfactory global-quality of the deposited films.

The influence of the methane content in the gas mixture is also investigated. The results are reported in **Fig.5**, where the dependence of the intensity ratio, I_{PL}/I_D, of the ~2.05 eV

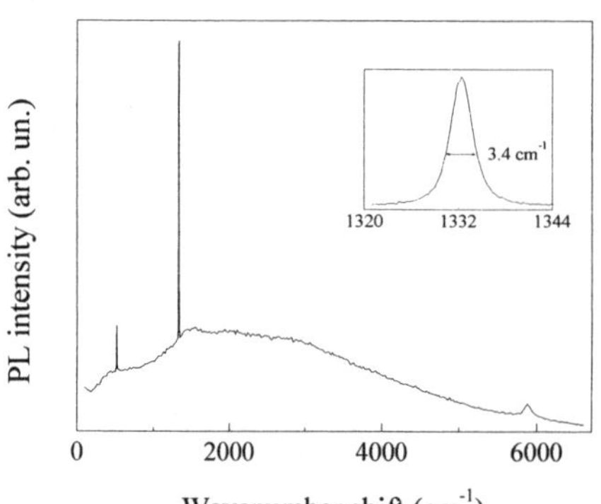

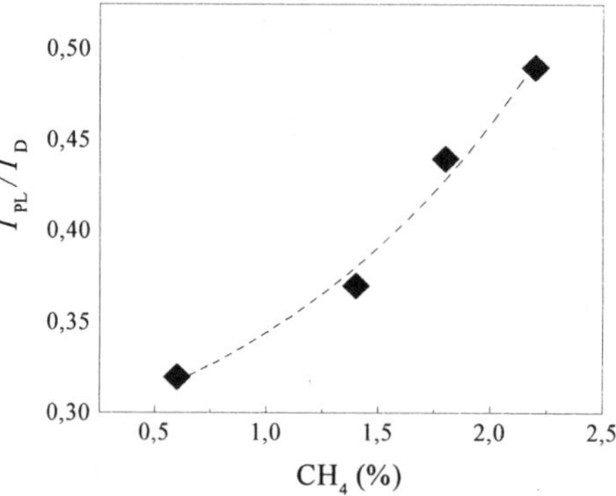

Fig.4 Macro-PL spectrum of a sample deposited, at 750 °C, using CH$_4$-H$_2$ gas mixture at 0.6% methane concentration.

Fig.5 I_{PL}/I_D intensity ratio as a function of the CH$_4$ concentration in the CH$_4$-H$_2$ gas mixture (a line is draw as visual help).

PL band to the ~1332 cm^{-1} diamond line on the CH_4 concentration is shown. The found monotonic increase of I_{PL}/I_D clearly evidences the progressive deterioration, in terms of phase purity, occurring with increasing CH_4 concentration, because of the faster deposition rate achieved at higher methane concentrations *[13,14]*. Hence, the inverse, I_D/I_{PL}, is here assumed as an indicator of the phase purity of the investigated samples.

It is finally worthwhile noticing that, consistently, the I_{ND}/I_D intensity ratio (although its numerical values, obviously lower, are more scattered) exhibits a trend analogous to that of I_{PL}/I_D.

5. Conclusions

In this paper, the results are presented of a systematic Raman and photoluminescence investigation aimed at clarifying the influence of the growth parameters on the physical properties of synthetic diamond films grown by Microwave Plasma Enhanced Chemical Vapour Deposition. The global quality of deposited films is shown to strongly depend on substrate temperature and gas mixture -nature and -composition.

The effect of substrate temperature and gas mixture composition on the film quality, is established by studying the properties of the two sets of samples deposited in CH_4-CO_2 atmosphere (with methane content varying in the range 47.4 ÷ 52.7 %) at 750 and 850 °C, respectively. Although, as a general trend, better results are found in films grown, at lower temperatures, in a CH_4-poorer atmosphere, the quality of these samples is globally not satisfactory.

The fundamental role played by the nature of the gas mixture is thus evidenced. A surprisingly high quality is observed in the third sample set, grown, after the experimental set-up optimisation, in a CH_4-H_2 atmosphere. Due to the much lower deposition rate attained with the CH_4-H_2 gas mixture, films deposited at 750 °C with CH_4 concentration varying in the range 0.6 ÷ 2.2 %, exhibit outstanding crystalline quality and phase purity. Actually, the width of the diamond line, that, in films with low residual stress, can be regarded as a reliable indicator of the sample intrinsic crystalline quality, is comparable to that of the best natural diamonds; meanwhile the intensity ratio, I_D/I_{PL}, of the ~1332 cm^{-1} diamond line to the ~2.05 eV PL band, assumed as an indicator of the film phase purity, shows satisfactorily high values, slowly decreasing at higher methane concentrations.

References

[1] D.S. Knight, W.B. White, *J. Mater. Res.* **4** (1989) 385

[2] G. Faggio, M. Marinelli, G. Messina, E. Milani, A. Paoletti, S. Santangelo, A. Tucciarone, G. Verona Rinati, *Diam. Relat. Mater.* **8** (1999) 640

[3] R.J.Nemanich, J.T.Glass, G. Lucovsky, R.E. Shroder, J.Vac. Sci Technol. A6 (1988)1783-1787

[4] A.M. Bonnot, *Phys. Rev. B* **41** (1990) 6040

[5] A.C.Ferrari, J.Robertson, *Phys. Rev.* B **61** (2000) 14095

[6] M.G. Donato, G. Faggio, M. Marinelli, G. Messina, E. Milani, A. Paoletti, S. Santangelo, A. Tucciarone, G. Verona Rinati, *Diam. Relat. Mater.* **10** (2001) 1535

[7] Y. von Kaenel, J. Stiegler, J. Michler, E. Blank, *J. Appl. Phys.* **81** (1997) 1726

[8] J.W. Ager III, M.D. Drory, *Phys. Rew. B* **48** (1993) 2691

[9] S.A. Stuart, S. Prawer, P.S. Weiser, *Appl. Phys. Lett.* **62** (1993) 1227

[10] M.C.Rossi, *Appl. Phys. Lett.* **73** (1998)1203

[11] P.K. Bachmann, D.U. Wiechert, *Diam. Relat. Mater.* **1** (1992) 422

[12] M.G. Donato, G. Faggio, M. Marinelli, G. Messina, E. Milani, A. Paoletti, S. Santangelo, A. Tucciarone, G. Verona Rinati, *Eur. Phys. J.* **B 20** (2001) 133

[13] Y.K. Kim, K.Y. Lee, J.Y. Lee, *Thin Solid Films,* **272** (1996) 64

[14] S.M. Leeds, T.J. Davis, P.W. May, C.D.O. Pickard, M.N.R. Ashfold, *Diam. Relat. Mater.* **7** (1998) 233

[15] K. Iakoubovskii, A. Stesmans, G.J. Adriaenssens, R. Provoost, R.E. Silverans, and V. Raiko, *Phys. Stat. Sol. (a)* **174** (1999) 137

[16] S.E. Coe, R.S. Sussmann, *Diam. Relat. Mater.* **9** (2000) 1726

[17] C. Jany, A. Tardieu, A. Gicquel, P. Bergonzo, F. Foulon, *Diam. Relat. Mater.* **9** (2000) 1086

[18] D. Meier, *"Diamond Detectors for Particle Detection and Tracking"*, PhD thesis, University of Heidelberg, (1999)

[19] L. Bergman, B.R. Stoner, K.F. Turner, J.T. Glass, and R.J. Nemanich, *J. Appl. Phys.* **73** (1993) 3951

[20] L. Bergman, M.T. McClure, J.T. Glass, and R.J. Nemanich, *J. Appl. Phys.* **76** (1994) 3020

[21] J. Rosa, J. Pangrac, M. Vanecek, V. Vorlicek, M. Nesladek, K. Meykens, C. Quaeyhaegens, L.M. Stals, *Diam. Relat. Mater.* **7** (1998) 1048

[22] T. Feng, B. D. Schwartz, *J. Appl. Phys* **73** (1993) 1415

[23] M.C. Rossi, S. Salvatori, F. Galluzzi, *Diam. Relat. Mater.* **6** (1997) 712

[24] V. S. Vavilov, A. A. Gippius, A. M. Zaitsev, B. V. Deryagin, B. V. Spitsyn, A. E. Aleksenko, *Sov. Phys. Semicond.* **14** (1980) 1078

[25] P. J. Lin-Chung, *Phys. Rev. B* **50** (1994) 16905

[26] C.D. Clark, H. Kanda, I. Kiflawi, G. Sittas, *Phys. Rev. B* **51** (1995) 16681

[27] J.P. Goss, R. Jones, S.J. Breuer, P.R. Briddon, S.Öberg, *Phys. Rev. Lett.* **77** (1996) 3041

[28] L. Bergman, R.J. Nemanich, *J. Appl. Phys.* **78** (1995) 6709 and references therein

Raman and impedance spectroscopic investigation of PEO-lithium triflate films

S.Capoleoni[a], T. Caruso[a], E.Cazzanelli[a], S.Passerini[b], P.Villano[b]

[a]*Dipartimento di Fisica, Università della Calabria and Unità INFM Cosenza; 87036- Arcavacata di Rende, Cosenza, ITALY*
[b]*ENEA, C.R. Casaccia, R&D Electrochemical Energy Conversion Section*

Abstract. PEO (polyethylene oxide) films containing lithium triflate ($LiCF_3SO_3$) were prepared, with a ceramic filler (γ-Al_2O_3) added to improve the electrochemical and mechanical characteristic of the membrane. Such conducting polymers are of particular interest as possible improvements of lithium-battery technology.
Impedance spectroscopy allowed an estimate of the activation energy of the ionic conduction process for the various EO: CF_3SO_3 ratios studied. A useful additional information about the ion transport is the degree of association of the mobile ions, obtained by variable temperature micro-Raman spectroscopy, both in the amorphous and in the crystalline phase, and across the phase transition. In addition the micro-Raman mapping allows for an experimental investigations of the structural and chemical homogeneity of the films.

1. Introduction

Polymeric electrolytes made of an inorganic salt dispersed in a solid polymeric matrix have attracted some attention lately because of their ionic transport properties and of their possible use as solid separators in batteries.

Moreover, the study of these materials offers a chance to investigate and understand fundamental anion-cation interactions and the influence of the polymeric matrix in ionic association and conduction.

Ceramic fillers are routinely used in these systems to improve mechanical and electrochemical characteristics [1], mainly increasing system amorphicity and trapping impurities.

Films obtained from PEO (Polyethylene oxide, from Union Carbide) and lithium triflate ($LiCF_3SO_3$ from 3M, thereafter LiTf) with γ–alumina added (γ–$LiAlO_2$ from Cyprous) were studied. The set of samples under study was $P(EO)_nLiCF_3SO_3$:γ-$LiAlO_2$ (16,7% w), with n=4,8,12,20 and 50, while the molecular weight of the polymeric matrix was 3×10^5. The thickness of the films was about 100 μm. The detailed procedure has been described elsewhere [2]; however, we expose it briefly.

Sample powders have been dried under vacuum for 2 days and then have been by ball-milling in the desired PEO-salt ratio. The powders have then been melted at 90°C under 3 bar of pressure for 2 minutes to obtain the films, which have finally been cold-calendered and then stabilized thermically. The whole procedure was carried out in a dry room (Room Humidity less than 0,1%).

2. DSC Analysis

A differential scanning calorimeter (DSC 2910 from TA Instruments) has been used for DSC analysis in the -70 to 250°C range, with a scan speed of 10°C/min under nitrogen flux. The results have been reported in Figure 1.

It is easy to observe two phase transitions: the first one, at about 60°C, can be assigned to crystalline PEO fusion, while the second one (at a temperature in the range 125-180°C, depending on the sample) can be assigned to the melting of the complex $P(EO)_3LiCF_3SO_3$.

At low temperatures (-40°C) another phase transition can be seen, namely the glass transition. From such measurements, a phase diagram of the system has been plotted (Figure 2). Our results are similar to those in literature [3].

3. Ionic conductivity measurements

The ionic conductivity has been measured for every sample using two copper electrodes, with the cell under vacuum. Temperature has been raised with 10°C steps, with a stabilization time of 24h for each temperature setting. The instruments was a computer-controlled Solartron FRA 1260. The data has been reported in Figure 3.

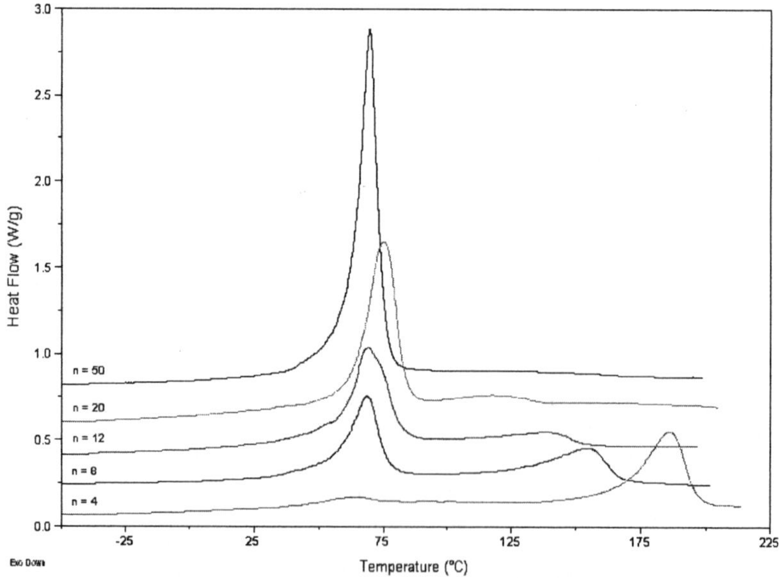

Fig.1 DSC of the samples $P(EO)_nLiCF_3SO_3$:γ-LiAlO$_2$.

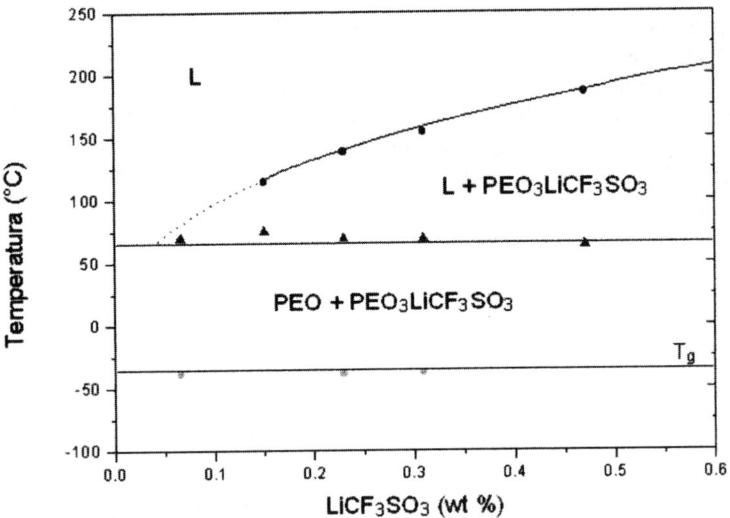

Fig.2 Phase diagram of the system P(EO)$_n$ – LiTf - γLIAlO$_2$.

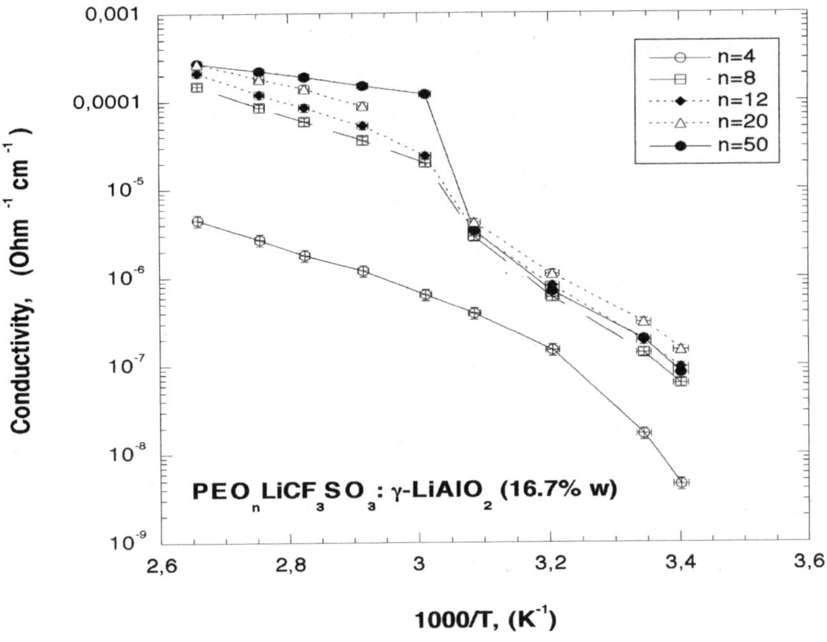

Fig.3 Arrhenius plot of the conductivity of the system.

In Arrhenius theory, conductivity has an exponential relation to the inverse of the temperature. Such a relation holds true for all our samples, with the exception of the $P(EO)_4$-LiTf, provided that we consider two different temperature ranges, and we suppose that Arrhenius law is true for each of them separately. These two zones are divided by the phase transition earlier mentioned, namely the melting of the $P(EO)_n$, at about 60°C.

At temperatures higher than the transition the conductivity increases of about two orders of magnitude, since the sample are crystalline up to about 50°C, thus impeding ionic conductivity; after the transition to the amorphous phase the ionic mobility is obviously much greater.

4. Raman Mapping

A preliminary Raman mapping examination of the samples was carried out, to evaluate their uniformity. For this purpose the Raman instrument (later described) was equipped with an XY positioning stage (lateral resolution better than 1μm) by ISA and a 20x Olympus objective.

We here report the mapping of a 120 μm x 120 area in the samples $P(EO)_{50}$ - Triflate and, with the minimum and maximum salt concentration available, respectively.

The parameter used to evaluate the sample homogeneousness is the intensity ratio of the Raman peak relative to the triflate and the one relative to the P(EO) (supposing that Raman activity depends only on concentration). The two mappings are shown in Figures 4 and 5.

In the first case (P 4) the standard deviation is within 4% of the mean value, whilst in the second case ($P(EO)_{50}$ – Triflate, in Figure 5) it is contained within 7% of the mean value. Even if the area examined is small when compared to the dimension of the sample, the mappings are a proof that our preparation procedure yields samples with a good homogeneity, at least on a sub-millimetric scale.

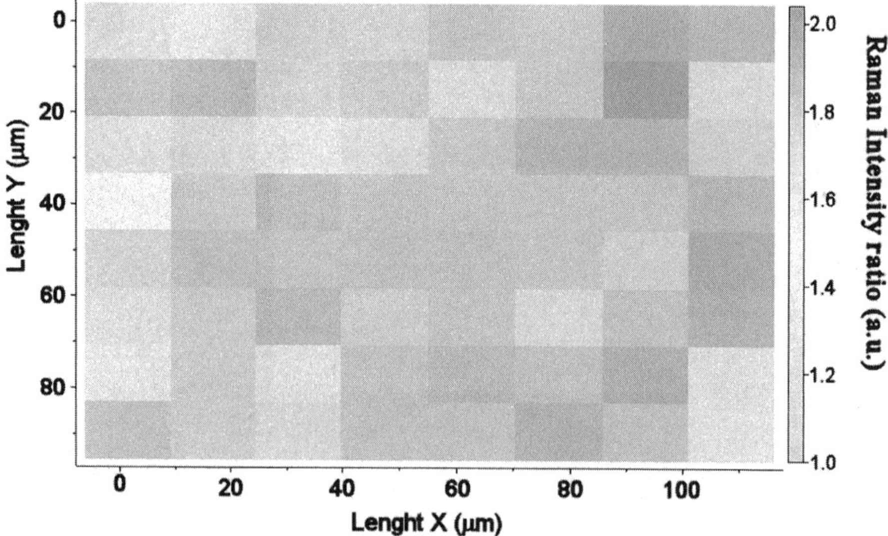

Fig.4 Mapping of the ratio of the Raman intensity of the triflate/P(EO) peaks for the sample $P(EO)_4$ –LiTf - γLiAlO₂.

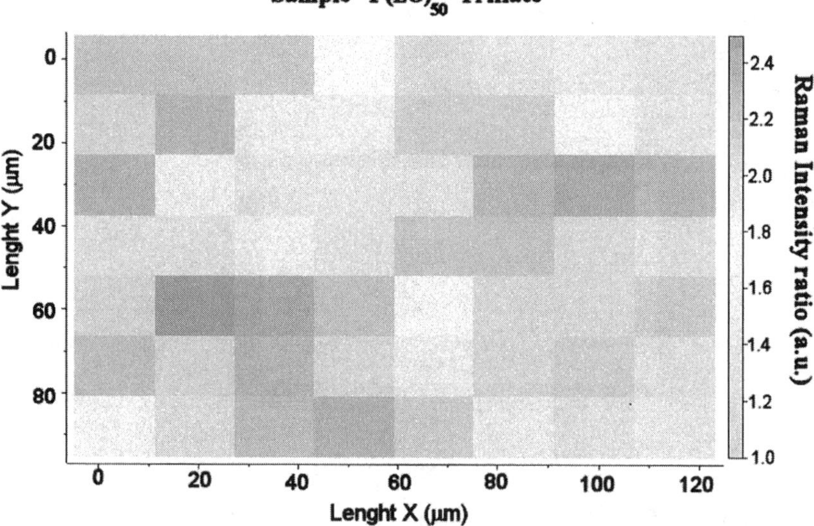

Fig.5 Mapping of the ratio of the Raman intensity of the triflate/P(EO) peaks for the sample P(EO)$_{50}$ –LiTf - γLiAlO$_2$.

5. Raman Spectra

The instrument used for the Raman measurements is the micro-Raman LABRAM from ISA, computer-controlled via the software Labspec 3.01. The illuminating laser is a 17 mW power HeNe laser, an the objective used was in this case a 50x from Olympus. For the variable-temperature spectra a heating cell THMS 500 from Linkam was used.

Spectra have been recorded from samples (not exposed to air) after a stabilization of 45 min for each temperature. The spectra of the pure γLi-AlO$_2$ powder, of the pure PEO and of the LiTf salt have also been recorder as a reference, so to have a more exact interpretation of each peak in the Raman spectrum.

We analyzed the variation of the Raman peaks in the region centered at 1040 cm^{-1}, at various temperatures for our samples. Such region contains peaks which can be assigned to the symmetric vibrations of the SO$_3^-$. These peaks are different from the ones which have been measured in a water solution [4] (which shows only one Raman mode at 1032 cm^{-1}). The higher frequency modes (1041 cm^{-1} and 1055 cm^{-1}) can be assigned correspondingly to ion pairs and ion aggregates [5]. We note that the latter are only present in the samples with a higher salt concentration.

If we suppose that Raman activity of the modes under exam does not vary appreciably in the various aggregation states, we can then link the Raman intensity ratio to the concentration ratios of the various complexes of the triflate anion. This in turn allows us to observe the variations of the relative concentrations of free ions, ion pairs and ion aggregates, as the temperature varies

In Figure 6 we show the details of such analysis for one of the samples, namely the P(EO)$_{20}$-Triflate. In the Figures 7,8 the results for the whole series of samples at two different temperatures (above and below the phase change point) are shown.

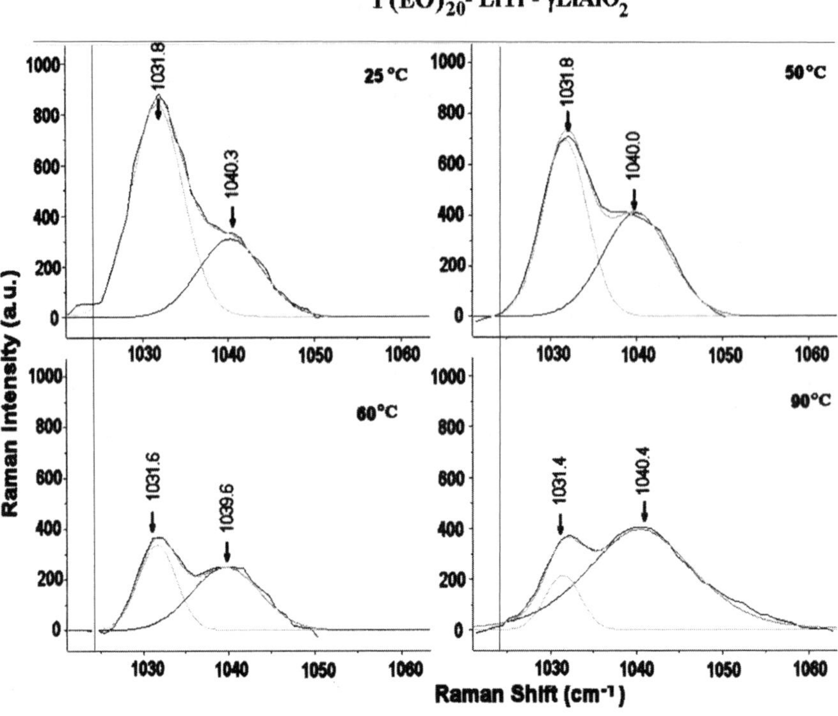

Fig.6 Analysis of the peaks assigned to the free ions (1031 cm^{-1}) and ion pairs (1040 cm^{-1}) in the sample P(EO)$_{20}$ - Triflate. The concentration of ion aggregates is below the experimental uncertainty.

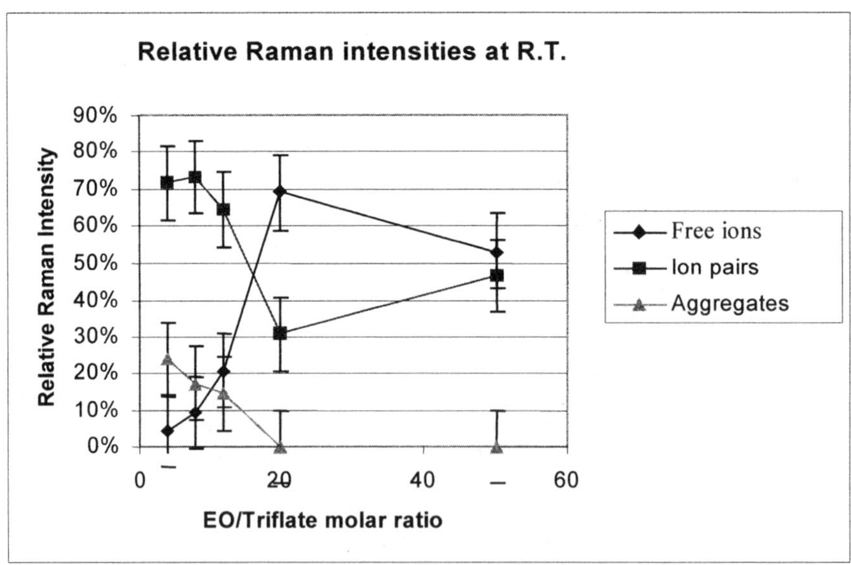

Fig.7 Relative concentration of free ions, ion pairs and ion aggregates (where detectable) for our samples at 20°C.

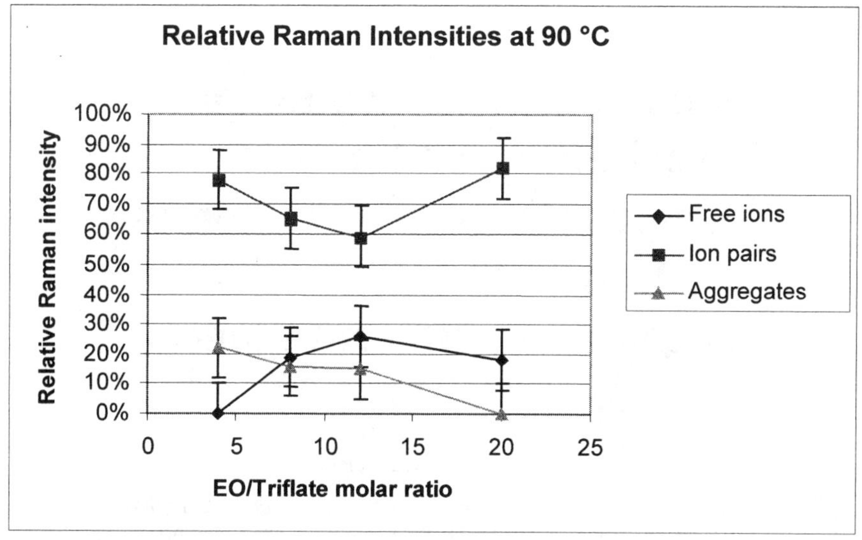

Fig.8 Relative concentration of free ions, ion pairs and ion aggregates (where detectable) for our samples at 90°C., above the amorphous phase transition at 50°C.

For each temperature, the variations of the relative concentration of the free ions at various EO/Triflate molar ratios are similar to that of the conductivity (Figure 3).

The conductivity of our samples is actually ionic and depends mainly on the free ions, whose concentration increases (not in a linear relation) as the salt concentration decreases.

As the temperature raises over the amorphous phase point, we can note that the concentration of free ions decreases markedly, as they form ion pairs. However, such decrease is matched by a large rise in conductivity. We can then assume that the mobility of the ions increases dramatically in the amorphous phase, by about 10^3 with respect to the conductivity in the crystalline phase of P(EO).

References

[1] B.Choi, Y.Kim, K. Shin *J. Power Source* **1997** *68*, 350
[2] P.P. Prosini, S. Passerini, R. Vellone and W.H. Smyrl, *J. Power Sources* **1998**, *75*, 73.
[3] J.F. Moulin, P.Damman, M.Dosière *Polymer* **1999**, *40*, 5843
[4] S.Schantz, *J.Chem.Phys.* **1991**, *94*, 6296
[5] L.M.Torell, P.Jacobbson,G.Petersen, *Pol. Adv. Tech.* **1993**, *4*, 152

GNSR 2001
G. Messina and S. Santangelo (Eds.)
IOS Press, 2002

Raman spectra of amorphous carbon-based thin films: a comparative discussion on the analysis of the 1000-1800 cm^{-1} region by different models

G.Messina, S.Santangelo[*]

INFM, Dipartimento di Meccanica e Materiali, Facoltà di Ingegneria, Università "Mediterranea", località Feo di Vito, 89060 Reggio Calabria, Italy
** E-mail: santange@ing.unirc.it tel.: +39.(0)965.875305; fax:+39.(0)965.875201*

A.Tebano, A.Tucciarone

INFM, Dipartimento di Scienze e Tecnologie Fisiche ed Energetiche, Università di Roma "Tor Vergata", via Tor Vergata 110, 00133 Roma, Italy

Abstract The example of amorphous-carbon coatings grown by a conventional pulsed KrF-laser deposition system operating at 248 nm is utilised in order to evidence the possibility, for the indications deduced from the different approaches to the analysis of the 1000-1800 cm^{-1} region of the Raman spectra of disordered carbon-based materials, of being contradictory. The region dominated by the sp^2-C arising modes is analysed and the results attained are comparatively discussed in the light of the currently-utilised models involving the well-known D- and G- bands, as well as of the recently-emerging interpretations contemplating vibrations of even- and odd- fold C-rings. Novel parameters are therewith introduced, able to monitor the average optical-transparency, stress-level and structural-order degree of the films.

1. Introduction

Raman spectroscopy is the most-widely utilised non-destructive film-diagnostic tool *[1-13]*. The optimisation of the growth conditions, as well as the tailoring of the material physical-properties is commonly accomplished by studying the evolution of the Raman spectra. In disordered carbon-based materials, the analysis is usually focused on the 1000-1800 cm^{-1} region. Frequency-position, width and intensity of the bands dominating this spectral region, in fact, strongly depend on the structural and bonding characteristics of the films.

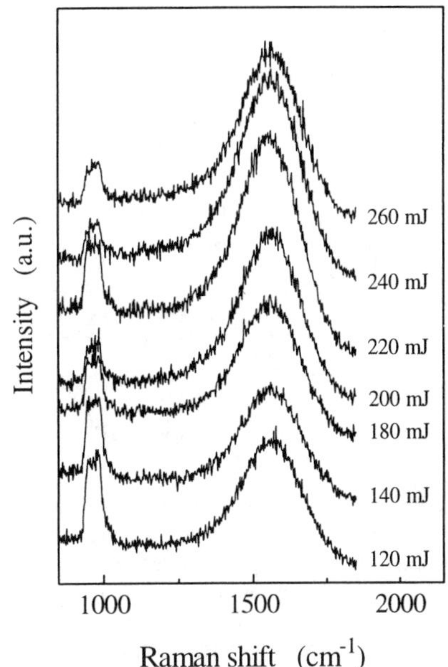

Fig.1 As-measured Raman spectra of the PLD grown a-C films investigated. Excitation wavelength is 514.5 nm. The shape evolution produced by the different KrF-laser pulse-energy is shown.

However, since there is no a priori reason to choose a particular function to fit the spectrum, some practical problem may arise in comparing the descriptive parameters attained by different approaches. The indications emerging by applying to different models and the related fitting procedures may appear contradictory and, as a consequence, the results of the Raman analysis be not always of ready-comprehension.

This is here evidenced by considering the example of amorphous-carbon (a-C) coatings grown by a conventional pulsed KrF-laser deposition system operating at 248 nm. The results, attained by analysing the spectral region of the sp^2-C arising modes, are comparatively discussed in the light of the currently-utilised models involving the well- known D- and G- bands *[1-4]*, as well as of the recently-emerging interpretations contemplating vibrations of even- and odd- fold C-rings *[5-7]*. In the discussion, particular emphasis is given to material characteristics, such as transparency-degree and stress-level, of crucial importance in view of the film utilisation for newly-arising applications. The different approaches are shown to provide complementary information about the local structure of the sp^2-C clusters of investigated films. The descriptive parameters attained by the different models and the related fitting procedures are compared and the origin of the eventual differences clarified. Novel parameters are therewith introduced, able to provide a reliable description of the average film-strain level originating from the inclusion of curved-rings in C clusters and the average ordering-degree of the film resulting from the formation of planar C-ring clusters.

2. Experimental details

A KrF excimer laser operating at the 248 nm wavelength is utilsed to ablate amorphous carbon films from a (99.99 % purity, 20 cm in diameter) pyrolitic-graphite target in UHV chamber at base pressure $<10^{-7}$ Torr. Carbon deposits onto rotating (~ 1 cm^2 surface area) B-doped (100) Si substrates, 5 cm distant from the target, mantained at 60°C. In order to prevent excessive cratering, laser pulses (duration ~ 10 ns, repetition-rate 10 Hz) are focused onto a rectangular spot with an incidence angle of 45° off-normal with an area of ~ 0.03 cm^2. Further details about the growth process are reported elsewhere *[8]*.

The Raman spectra are recorded, at room temperature, in the 200-2000 cm^{-1} region by using a Jobin Yvon Ramanor U-1000 double monochromator equipped with an electrically cooled Hamamatsu R943-02 photo-multiplier as a detector and photon counting electronics. A Coherent Innova 70 Ar$^+$ laser, operating at 514.5 nm wavelength, is utilised as excitation source. The S/N ratio is improved by recording multiple scans and a power density of ~20 W/mm^2 at the sample surface is utilised in order to prevent sample annealing.

3. Raman analysis

Raman spectroscopy is the most-widely utilised non-destructive tool for the characterisation of crystalline, nanocrystalline and amorphous carbon-based films. The Raman spectrum of large carbon single-crystals consists in a sharp line located at 1332 cm^{-1} or 1580 cm^{-1} depending on the bond symmetry. The former corresponds to the zone-centre T_{2g} triply-degenerate optical phonon (k=0) originating from tetrahedrally coordinated C atoms *[9,10]*; the latter, known as G-line, corresponds to the zone-centre in-plane E_{2g} C-C stretching mode vibrations of trigonally bonded C atoms *[11,12]*. In polycrystalline graphite an additional structure, known as D-peak, can be observed at about 1355 cm^{-1}. Such a feature is attributed to an A_{1g} breathing mode at K point activated by the relaxation of the k=0 wave-vector selection rule *[11]* or related to the maxima in the VDOS of graphite at M and K points *[12,13]*.

Since there is no a priori reason to choose a particular function to fit the spectrum, different lineshapes are utilised in order to reproduce the sp^2-C mode arising features detected in the 1000-1800 cm^{-1} region of the Raman spectra of disordered carbon-based materials. At times, on the basis of the graphite VDOS asymmetry, a Breit-Wigner-Fano (BWF) line-shape is used to reproduce the G-band *[2-4]*. Actually, in this case, excellent fits are obtained for a-C films with low sp^2 content, with no need to include any contribution at 1355 cm^{-1} *[2-4]*. The addition of a Lorentzian D-band in films with higher sp^2 content eventually evidences

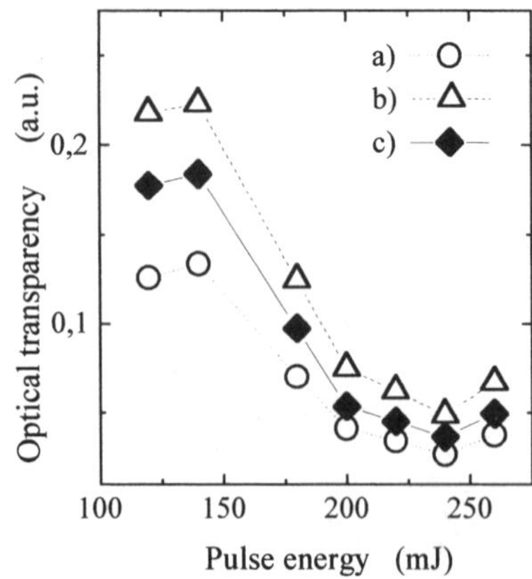

Fig.2 The film optical-transparency degree, as conventionally measured by the second-order Si/G-band intensity ratio, $^{II}I_{Si}/I_G$, as function of the pulse energy. Both the cases of (a) BWF- and (b) Gaussian- lineshape reproducing the G-band are shown, and compared with the second-order Si/sp^2-C intensity ratio $^{II}I_{Si}/I_{sp2-C}$ (c).

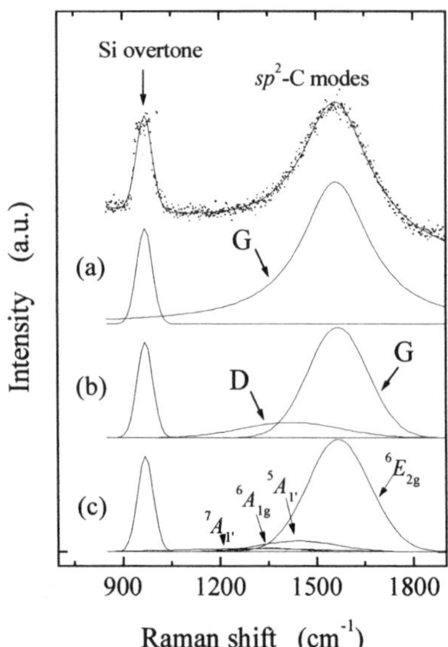

Fig.3 As-measured and fitted spectrum of the sample deposited at 120 mJ pulse-energy. Comparison among the spectral features obtained on the basis of the conventional D-and G- band approach, both in cases of (a) BWF- and (b) Gaussian- G-band, and according to the idealised 5-, 6- and 7- fold vibrating C-ring model.

the presence of small graphitic crystallites [2-4]. Often, since a Gaussian lineshape is expected for a random distribution of phonon lifetimes in disordered materials, D- and G-bands are reproduced by a double Gaussian fit [1]. Recently, four Gaussian bands, corresponding to the stretching and breathing modes (E- and A- type, respectively) of idealised 5-, 6- and 7- membered π-bonded C rings [5], have been utilised to reproduce the 1000-1900 cm^{-1} region in Raman spectra of a-C based materials [6,7].

4. Results and discussion

The Raman spectra of the investigated a-C films are shown in Fig.1. The shape evolution is the effect of the laser pulse energy variation during the deposition of the considered sample set. It is worthwhile noticing that the samples, here utilised in order to discuss the results attained on the basis of the models quoted in the previous Section, are deposited under non-optimal energy-densities ($\sim$4-9 J/cm^2). Our interest, far from the growth-process optimisation, in fact, focuses mainly on comparing the descriptive parameters achieved by applying the above-mentioned fitting procedures, eventually clarifying the origin of the differences observed. In the discussion, particular emphasis is then given to the material characteristics of crucial importance in view of the film utilisation for newly-arising applications, requiring transparency and better substrate adhesion.

4.1 Film transparency-degree

In order to monitor the variations, produced by the changing growth conditions, in the optical transparency of the deposited films, the ratio is often considered of the (first- or) second- order Si-peak to the G-band the integrated intensities. The results obtained for the investigated a-C films are plotted in Fig.2 against the pulse energy. The film transparency-degree, as measured by the 2nd-order Si-peak/G-band intensity-ratio, $^{II}I_{Si}/I_{G}$, decreases with increasing the pulse energy, evidencing the gradual deterioration of the optical film characteristics, occurring due to the progressive moving-away from the optimal deposition-conditions [6]. As a major result, the film transparency-degree is found to exhibit qualitatively the same trend, independently of the specific fitting procedure applied.

Going into more details, the comparison between the cases of a BWF and a Gaussian G-band evidences that, quantitavely, lower $^{II}I_{Si}/I_{G}$ values are obtained in the former case. This is because, as shown in Fig.3a, the BFW lineshape suffices alone (i.e. with no need of additional Lorentzian D-band) to fit the entire sp^2-C portion of the Raman spectra of the a-C samples considered. The division of a same $^{II}I_{Si}$- by a consequently larger I_{G}- value results in a ratio lower than that attained, correspondingly, in case of a Gaussian G-band. If a different approach is chosen for analysing the region of interest in Raman spectra of a-C based materials and the spectral features there detected reproduced by the four Gaussian bands (Fig.3c) corresponding to the A- and E-modes of idealised 5-, 6- and 7- membered C rings [5], the film transparency-degree is evaluated as the 2nd-order Si-peak/sp^2-C band intensity-ratio, $^{II}I_{Si}/I_{sp2-C}$ [6,7]. In such a case, as shown in Fig.2 for the considered sample set, intermediate $^{II}I_{Si}/I_{sp2-C}$ values are found. The different normalisation in the ratio measuring the film optical-transparency causes the latter values to be lower than those achieved, via the D- and G- band approach, in case of a Gaussian G-band; meanwhile, the use of the BWF lineshape (Fig.3a), making part of the spectrum, handled as background in the other two cases (Fig.3b,c), to be included into the integrated-intensity of the asymmetrical G-band, is

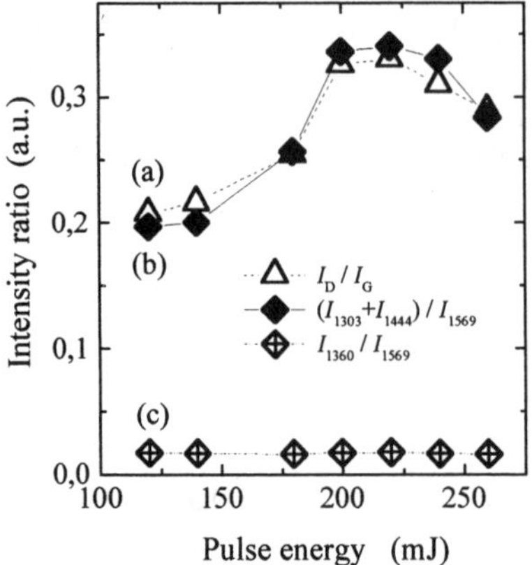

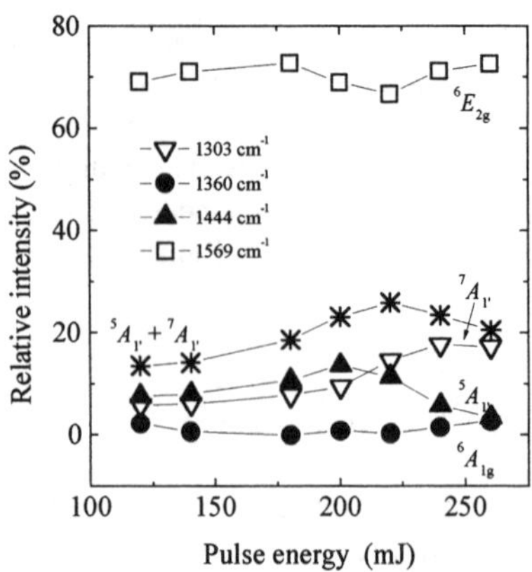

Fig.4 The film stress-level, as conventionally measured by the D/G intensity ratio, I_D/I_G (a), as function of the pulse energy. The parameters R_{dist} (b) and R_{plan} (c), respectively accounting for the deviation from the optimal bond-angle and the film ordering-degree, are also reported.

Fig.5 Idealised 5-, 6- and 7- fold vibrating C-ring model. Relative contribution of 1303 cm^{-1} ($^7A_{1'}$), 1360 cm^{-1} ($^6A_{1g}$), 1444 cm^{-1} ($^5A_{1'}$) and 1569 cm^{-1} ($^6E_{2g}$) vibrational-frequencies to the entire first-order C-related portion of the Raman spectra.

responsible for the achievement of relatively higher $^{II}I_{Si}/I_{sp2-C}$ values. Nevertheless, the present analysis suggests that, if the transparency-degree of the sample is evaluated by means of the $^{II}I_{Si}/I_{1569}$ ratio (i.e. dividing by I_{1569} in place of I_{sp2-C}), values nearly-coincident with those of $^{II}I_{Si}/I_G$ for Gaussian D-bands are found.

Thus, if the changes of the optical properties of a same sample set has to be monitored in order to optimise the growth process by evidencing the differences, produced, sample by sample, by the variations in the deposition conditions, the choice of a specific fitting procedure to reproduce the spectra is unimportant, because, as shown, the different models provide, qualitatively, the same information. The knowledge of which approach has been followed becomes important when a quantitative comparison among different sample sets is requested.

4.2 Film stress-level

As well known, a stress lowering generally results in improved mechanical film charateristics, as well as in bettered substrate adhesion *[8,14-16]*. However, actually some problem may arise in monitoring and comparing the average stress-level, achieved under different deposition conditions, because of the contradictory indications emerging, even in the ambit of a same model, from the different interpretations of the Raman analysis results.

According to the conventional D- and G- band approach, the film stress-level is generally measured by the D/G intensity ratio, I_D/I_G, and the disappearance of the D-band, usually associated to an improvement in the mechanical properties, is currently assumed to signal a stress-diminishing *[8,14-16]*. In the light of this interpretation, on one hand, the I_D/I_G changes of the a-C samples considered, plotted, in Fig.4, against the pulse energy, would evidence a structural-disorder and film-strain enhancement around 220 mJ; while, on the other hand, the absence of any spectral feature at ~1355 cm^{-1} in case of the BWF G-band would oppositely show a quite-low structural-order level at any pulse energy.

However, the development of a D-band has been recently related to the increase of aromatic C-ring number and suggested to be, hence, indicative of ordering *[17]*. Therefore, the reduction of flat 6-fold rings, in favour of distorted structures, would determine a corresponding higher average structural-disorder degree and stress-level. In this scenery, the absence of a D-band in case of the BWF G-band, in contrast with the indications emerging from the D/G intensity ratio, would signal the occurrence of the ideal situation of films with unsignificant strain-level.

In order to get more insight into the problem, eventually claryfing the origin of the discrepancy between different interpretations, the relative contribution of vibration-modes arising from even- and odd- fold C-rings to the entire sp^2-bond region is plotted in Fig.5. As can be seen, the stress-increase, evidenced in Fig.4 by the I_D/I_G variations, occurs in correspondence to the raise of the relative contribution of the $^5A_{1'}$- and $^7A_{1'}$- modes to the entire C-related spectrum portion; meanwhile, the negligible fractional contribution of the $^6A_{1g}$-mode reminds the Lorentzian D-band absence. These findings suggest that current and more-recent interpretations give complementary information and, as an effect, the parameter conventionally utilised to monitor film-strains does not suffice to provide a clear picture of the film structural-properties and related stress-levels. Hence, in order to measure the average film-strain level, originating from the inclusion of curved-rings in C clusters, and the structural-order degree, deriving from the formation of flat aromatic rings within the film, the introduction of two novel parameters is here proposed. The former, here called R_{dist}, is defined as the ratio of the integrated intensities of the 1303 and 1444 cm^{-1} to the 1569 cm^{-1} centred bands, $(I_{1303}+I_{1444})/I_{1569}$, while the latter, named R_{plan}, is defined as the ratio of the integrated intensities of the 1360 cm^{-1} to the 1569 cm^{-1} centred bands, I_{1360}/I_{1569}. The results relative to the sample here discussed are shown in Fig.4. The very small R_{plan} values are consistent with the low ordering-degree evidenced, on the basis of the recent interpretations, by the absence of any feature at 1355 cm^{-1} in case of a BFW G-band; while the R_{dist} variations, giving a measure of the deviation from the optimal bond-angle, indicate the occurrence of the distorsions, responsible for the film-stress changes signalled, in the light of the current interpretation, by the D/G intensity ratio.

5. Conclusion

The example of a-C films, grown by a conventional pulsed KrF-laser deposition system operating at 248 nm, is considered in order to discuss the results of the analysis of the spectral region of the sp^2-C arising modes, in the light of the currently-utilised models involving the D- and G- bands, as well as of the recently-emerging interpretations contemplating vibrations of even- and odd- membered C-rings. Particular emphasis is given to the material characteristics of crucial importance in view of the film utilisation as transparent coatings. The different approaches are shown to provide complementary information about local structure of the sp^2-C clusters, transparency-degree and stress-level of the investigated films. The origin of the eventual failure of the conventionally-utilised parameters is clarified and the introduction is therewith proposed of novel parameters, able to provide a reliable description of the average film-strain level, originating from the inclusion of curved-rings in C clusters, and ordering-degree of the film, resulting from the formation of planar C-ring clusters.

References

[1] M.A.Tamor, W.C.Vassel, *J. Appl. Phys.* **76** (1994) 3823

[2] D.G.McCullock, S.Prawer, A.Hoffman, *Phys. Rev.* B **50** (1994) 5905

[3] Y.Lifshitz, *Diamond Relat. Mater.* **5** (1996) 388

[4] S.Prawer, K.W.Nugent, Y.Lifshitz, G.D.Lempert, E.Grossman, J.Kulik, I.Avigal, R.Kalish, *Diamond Relat. Mater.* **5** (1996) 433

[5] T.E.Doyle, J.R.Dennison, *Phys. Rev.* B**51** (1995) 196

[6] M.P.Siegal, P.N.Provencio, D.R.Tallant, R.L.Simpson, *Appl. Phys. Lett.* **76** (2000) 2047

[7] M.P.Siegal, D.R.Tallant, L.J.Martinez-Miranda, J.C.Barbour, R.L.Simpson, D.L.Overmyer, *Phys. Rev.* B **61** (2000) 10541

[8] G.Messina, A.Paoletti, S.Santangelo, A.Tebano, A.Tucciarone, *Microsystem Technol.* **26** (1999) 30

[9] S.A.Solin, A.K.Ramdas, *Phys. Rev.* B**1** (1970) 1687

[10] D.S.Knight, W.B.White, *J. Mater. Res.* **4** (1989) 385

[11] F.Tuinstra, J.L.Koenig, *J. Chem. Phys.* **53** (1970) 1126

[12] R.J.Nemanich, S.A.Solin, *Phys. Rev.* B**20** (1979) 392

[13] R.Al Jishi, G.Dresselhaus, *Phys. Rev.* B**26** (1982) 4514

[14] B.Schultrich, H.J.Scheibe, G.Grandremy, D.Schneider, *Phys. Stat. Sol.* a**145** (1994) 385

[15] D.Schneider, C.F.Meyer, H.Mai, B.Schöneich, H.Ziegele, H.J.Scheibe, Y.Lifshitz, *Diamond Relat. Mater.* **7** (1998) 973

[16] M.K.Fung, W.C.Chan, Z.Q.Gao, I.Bello, C.S.Lee, S.T.Lee, *Diamond Relat. Mater.* **8** (1999) 472

[17] A.C.Ferrari, J.Robertson, *Phys. Rev.* B**61** (2000) 14095

GNSR 2001
G. Messina and S. Santangelo (Eds.)
IOS Press, 2002

Relaxational dynamics of water in porous glasses

Vincenza Crupi[*], Domenico Majolino, Placido Migliardo and
Valentina Venuti

*Dipartimento di Fisica dell'Università di Messina and INFM Unità di
Ricerca di Messina,
C.da Papardo, S.ta Sperone 31, 98166 Messina, Italy*
Phone: +39-090-391478, Fax: +39-090-395004, E-mail: crupi@dsme01.messina.infm.it

Abstract. We present a detailed dynamical analysis performed by Raman spectroscopy on water in the bulk state and confined inside a sol-gel porous matrix with 26 Å interconnected pores at different hydration levels. The confinement seems to induce strong destructive effects in the interfacial water, where the hydration phenomena cause different environments related to new spectral features and originate the lost of tetrahedrical arrangement (icelike), typical of pure water. This last evidence is clearly evidenced in the VDOS spectra carried out by incoherent neutron scattering.

1. Introduction

The modification, relatively to the bulk state, of the behaviour of liquids in restricted geometries and especially at interfaces is a field of rapid growing interest because of the close connection with relevant technological and biophysical problems [1]. It concerns, in fact, a great variety of physical situations occurring in surfaces, lamellar systems, amorphous systems, ionic polymers and biomolecular systems. The topic is quite complex because the specific differences in structural and dynamical properties depend on the particular interaction between liquid and substrate, the size of the particles, the size of the confining region. Nevertheless, some general consideration can be made. In fact a fundamental role, in studying the relaxational dynamics of a liquid that diffuse and reorient in a pore, is played by two competitive processes: surface interaction (chemical effect) and geometric restrictions (physical effect) [2]. The first one takes into account the degree of the adsorption of the molecules on the active surface sites, which is expected to slow down the dynamics; the second is the confinement effect that can lead to an increase in the free volume of a molecule, which then results in accelerated dynamics together with a decrease of the glass transition temperature relative to the one of the bulk phase. The interplay of these two effects depends strongly on the particle density and the size of the confining system, and represents a not completely solved problem [3].

In particular, the study of properties of interfacial water as a function of temperature or hydration level is important and appeals for both experimental and theoretical approaches [4]. In fact, the strong dipole of water molecules, and the ability to form hydrogen bonds between molecules or with some substrate have an important influence on the behaviour of water confined in a small volume or at the vicinity of an interface. Such situations are very frequent in a lot of systems of interest to biology, chemistry and geophysics. Those properties are

particularly relevant in understanding phenomena like the mobility of water in biological channels or the dynamics of hydrated proteins [5,6]. From a theoretical point of view, molecular mobility of water in confined geometry has been studied by computer simulation [7]. This represents a very suitable tool especially in the supercooled regime, without the limitation of the nucleation process, which takes place in the real experiment. The collective and single-particle dynamics of water confined in nanopores have also been studied by different experimental techniques such as Rayleigh wing and Raman scattering, Incoherent Quasi Elastic and Inelastic neutron scattering (IQENS and IINS, respectively), neutron diffraction and nuclear magnetic resonance (NMR) [8]. It is still difficult, however, to find general trends, systematic studies of the dynamics of confined water have not been attempted until now. Nevertheless, the majority of theoretical and experimental studies [9,10] on water in porous silica supports the existence of a thin layer (<10Å) of surface water differing from bulk water for:
i) a lower mobility (a lower diffusion coefficient, higher viscosity), ii) an enhanced hydrogen bonding, iii) a lower density, iv) a lower nucleation temperature. In our case, being the pore diameter very small (26 Å), the effects induced by the surface forces on the water dynamics are relevant.

In this paper we present result of an experimental analysis, performed by Rayleigh wing, Raman and Neutron scattering, on water within a GelSil glass. As we'll see, the confinement of water strongly influences both the low-frequency quasi-elastic and inelastic contributions and the fundamental intramolecular OH stretching mode region.

2. Experimental set-up and data handling

We analysed Rayleigh wing, Raman and Incoherent Inelastic Neutron Scattering (IINS) spectra of water at different hydration percentages N/N_0 and different temperatures.
Experimental polarized (VV) and depolarised (VH) Rayleigh wing and Raman spectra were collected, at room temperature, for bulk and confined water at N/N_0=100%, 96.7%, 95.6%, 15.4% and 5.9%, with a good signal-to-noise ratio by a high-resolution, fully computerized Spex-Ramalog 5 triple monochromator, used in a 90° scattering geometry [11]. It was coupled with an INNOVA70 Ar^+-Kr^+ Ion laser with a mean exciting power of 1W and working with the vertically polarized 6471Å laser line. This wavelength induced a low-fluorescence contribution from the glass. Following a well established procedure, a spectral resolution of 0.05 cm^{-1} HWHM in the region ranging from –20 to 20 cm^{-1}, of 0.25cm^{-1} in the one between –50 to 50 cm^{-1} and finally of 2.5 cm^{-1} in the –100 to 500 cm^{-1} was used. The spectra with different resolution were subsequently numerically matched and properly normalized by means of the detailed balance law. In the high frequency region, VV and VH spectra were taken in the 2800÷3700 cm^{-1} range, with a resolution of 4 cm^{-1}.

We used cylindrical glasses (10mm diameter, 5mm thick for light scattering measurements and 5mm diameter, 1mm thick for neutron scattering measurements), produced by the sol-gel technology and purchased from GelTech Co. with nominal pores of diameters of 26Å (5% standard deviation), pore volume fractions of 0.39, surface areas of 609 m^2/g of the glass, and bulk density of 1.2 g/cm^3. The pores in the glass are highly branched and interconnected in a fractal-like geometry. These glasses contain, on the surface, a great number of SiOH groups, strong active sites for H-bond with the OH groups of the water, allowing the existence also of double HBs between surface and adsorbed water [12]. In order to get measurements at various water content, GelSil was firstly full filled by immersing it in water inside an optical cell for a sufficient time (34 hours) and then outgassed by a vacuum pump, at room temperature for different outgassing times, in the range $60 \div 2.02 \times 10^5$ sec. For estimating the water amount dispersed within the pores, we have simply connected the Raman scattered intensity I^{OH} (integrated in the $2800 \div 3700$ cm^{-1} range) to the number N of water scatterers in the scattering volume. It is to be noticed that simple calculations allow to get the value of N/N_0 of water adsorbed to the Gelsil surface in the first layer (considering the molecular diameter of water ~3Å), N/N_0=41.6%. This means that the data taken for N/N_0=15.4% and N/N_0=5.9% refer to water bonded to the glass inner surface only in the first layer.

As far as IINS measurements are concerned, they were performed on water in bulk and confined into sol-gel glasses, at N/N_0 = 100%, 95%, 5% and T = -35°C, -12°C, 20°C, 40°C.

The porous glasses have the same characteristics of the ones used for the light scattering measurements, except for the diameter (5 mm) and the thickness (1 mm).

The data were collected on the IN6 t.o.f. spectrometer at Institut Laue-Langevin (ILL) in Grenoble. The incident neutron wavelength was 5.12 Å, the elastic energy resolution (hwhm) 50μeV, determined by a vanadium standard, and the exchanged momentum range from 0.28 to 1.94 Å^{-1}. We obtained the one-phonon amplitude-weighted proton vibrational density of states (VDOS), $Z(\omega)$, following a standard reduction method as applied in [13], by evaluating and subtracting the multiphonon and background contribution from the experimentally obtained generalized frequency distribution $P(\alpha, \beta)$. For all the data reduction we used appropriate IINS programs available in the ILL software library.

3. Results and discussion

As far as Raman measurements are concerned, we analysed the imtramolecular O-H stretching spectral region (2800 cm^{-1} < ω < 3700 cm^{-1}) in order to get information on the effects of the porous matrix on the HB network of water.

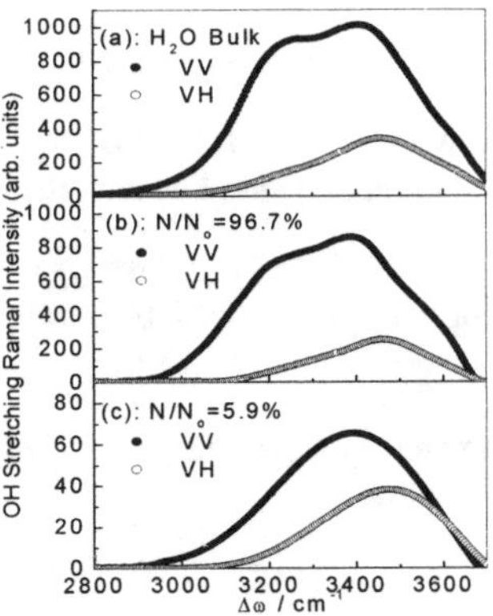

Fig.1 OH stretching Raman intensity for VV and VH configurations for water in the bulk state (a) and confined at N/N_0=96.7% (b) and 5.9% (c).

We followed the spectral stripping procedure of the polarised O-H stretching band suggested, some years ago, by Green, Lacey and Sceats (GLS) [14]: the low-frequency shoulder in the I_{VV} spectrum of H_2O arises from "collective" modes of the OH groups, and this collective band is characterised by a depolarisation ratio different from that of uncoupled OH oscillators. The relative intensity of this mode can be obtained Assuming that the I_{VH} spectrum basically looks like the scaled-down version of the I_{VV} spectrum without the collective band I_C, the relative intensity of this mode can be obtained after an appropriate data reduction, by subtracting I_{VH} from I_{VV}. In pure water it has been proved that I_C linearly increases with falling temperature, and approaches the value of ice at the singularity temperature of water ($T_S = 218K$). On the basis of this occurrence, GLS considered this collective band as arising from the collective in phase stretching motion of the water molecules in the fully bonded tetrahedral network, to which liquid water tends as it is supercooled. In Fig. 1 we show VV and VH Raman spectra in bulk and confined water.

As can be clearly seen, we observed the falling down of the collective contribution, whose centre-frequency is evidenced by the dashed arrow, when the water content is decreasing. Fig. reports, as result of the stripping procedure, the collective band for various values of N/N_0.

In the case of bulk water, this contribution is quite similar to the "open" water contribution, shown in [15], already attributed by the authors to the O-H vibration in tetrabonded H_2O molecules having an "intact bond". We can hence conclude that the falling down of this contribution indicates a destructuring effect of the Gelsil matrix on the tetrahedral H-bond network of pure water.

Concerning Rayleigh wing data, in Fig. 2 we report the spectra of water in the bulk state and confined in GelSil at $N/N_0=5.9\%$, together with the best-fit obtained by the law: $I_{VH}(\omega)=Res(\omega)+H.N.(\omega)+Vgt_1(\omega)+Vgt_2(\omega)+Bkg$, in which $Res(\omega)$ is the resolution enlarged Gaussian contribution, $Vgt_1(\omega)$ and $Vgt_2(\omega)$ are Voigt profiles, Bkg is a flat background, and $H.N.(\omega)$ represents the imaginary part of the Havriliak Negami profile. This last [16] is mathematically described as:

$$H.N.(\omega) = -(1/\omega)\mathrm{Im}\left[1 + (i\dot{\omega}\tau_{HN})^\alpha\right]^{-\gamma} \tag{1}$$

where τ_{HN} is a characteristic relaxation time, α and γ ($0 < \alpha, \gamma < 1$) are shape parameters. This function reflects the presence of a distribution of relaxation times, and plays, in the ω-domain, the same role played by the Kolrausch – Williams – Watt profile,

$$KWW(t) = \exp\left[-(t/\tau_{KWW})^\beta\right] \tag{2}$$

in the t-domain. As can be seen, when $\alpha = \gamma = 1$, the Lorentzian profile is recovered.

In literature [16], a mathematical relationship between the relevant parameters of these two profiles has been found:

$$\log_{10}\left[\tau_{HN}/\tau_{KWW}\right] = 2.6(1-\beta)^{0.5}\exp(-3\beta) \tag{3}$$

in which $\alpha\gamma=\beta^{1.23}$.

Starting from the fitting parameters (τ_{HN}, α and γ) it is possible to define a mean relaxation time by:

$$\langle\tau\rangle = (\tau_{KWW}/\beta)\Gamma(1/\beta) \tag{4}$$

where Γ is the Gamma function.

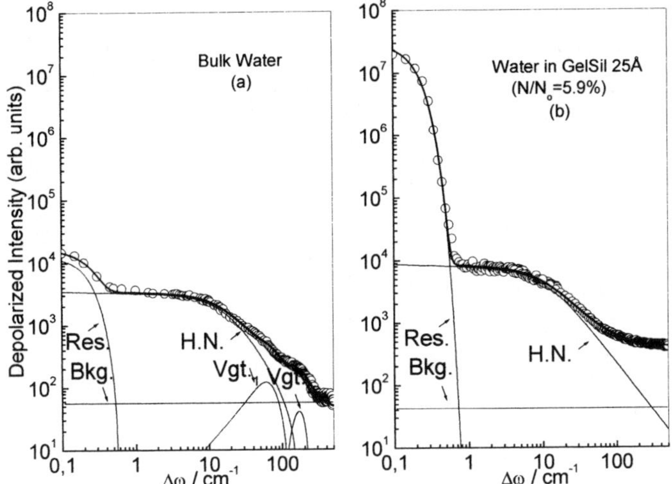

Fig.2 Rayleigh wing spectra of water in the bulk state (a) and confined at N/N_0=5.9% (b), together with the theoretical fit /continuous line) and the deconvolution components (dashed lines).

From an inspection of Tab. 1, in which all the best-fit parameters for all the investigated samples are reported, we observe, first of all, that in the case of bulk water the H.N.(ω) strictly recall a Lorentzian profile.

We can interpret the origin of this contribution as due to the motion of water into the various transient cages H-bond imposed. These latter give rise to a multiplicity of the primary elemental
rates, i.e. to the obtained distribution of relaxation times. We revealed, in agreement with previous measurements [9], an increase of $<\tau>$ with the decreasing of the amount of water, and these results have been related to the presence of both chemical and physical traps [17]. On the other hand, when N/N_0 diminishes, the width of the distribution, β, decreases from 0.88 to 0.64. From the evolution of these two parameters we can hypothesize that, firstly, the H-bond between the glass surface and water is energetically stronger with respect to the one between water molecules. In addition, the existence of an interfacial H-bond imposes the presence of a large number of new transient species respect to the ones characteristic of bulk water [12]. The two Voigt bands, centred respectively at 60 cm^{-1} and 170 cm^{-1}, already observed in pure water, correspond to the H-bond bending and stretching collective lattice modes. They disappear starting from N/N_0=95.6%.

For explaining the spectra in the restricted translational region, we calculated the effective susceptibility

$$\chi''(\omega) = I_{VH}(\omega) \cdot (n(\omega,T)+1)^{-1} \qquad (5)$$

in which n(ω,T)+1 represents the well known Bose-Einstein population factor. From $\chi''(\omega)$ we obtained the Raman effective vibrational density of states (VDOS), $g_{eff}^{R}(\omega)$, given by:

$$g_{eff}^{R}(\omega) = \omega \cdot \chi''(\omega) = \sum_{b} P_b(\omega) g_b(\omega) \qquad (6)$$

Tab.1 Havriliak Negami best-fit parameter for all the analysed systems.

$N/N_0(\%)$	$\tau_{HN}(psec)$	α	γ	β	$<\tau>$
H_2O	0.31	0.97	0.88	0.88	0.31
100	0.42	0.72	0.77	0.62	0.48
96.7	0.44	0.80	0.77	0.67	0.49
95.6	0.50	0.93	0.75	0.75	0.51
15.4	0.53	0.93	0.65	0.66	0.58
5.9	0.55	0.96	0.60	0.64	0.61

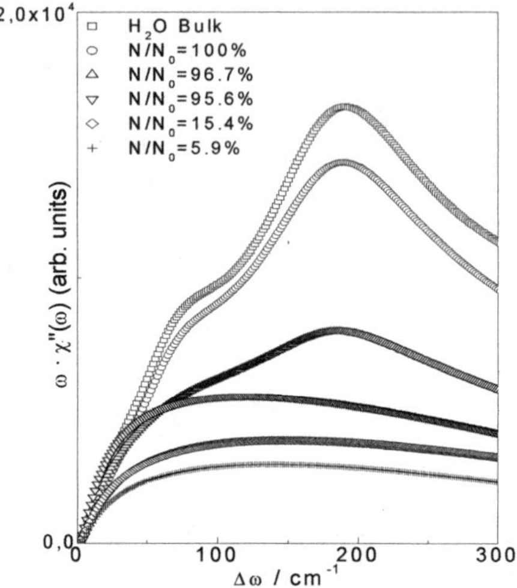

Fig.3 Raman effective VDOS for bulk and confined water at all the analysed hydration levels.

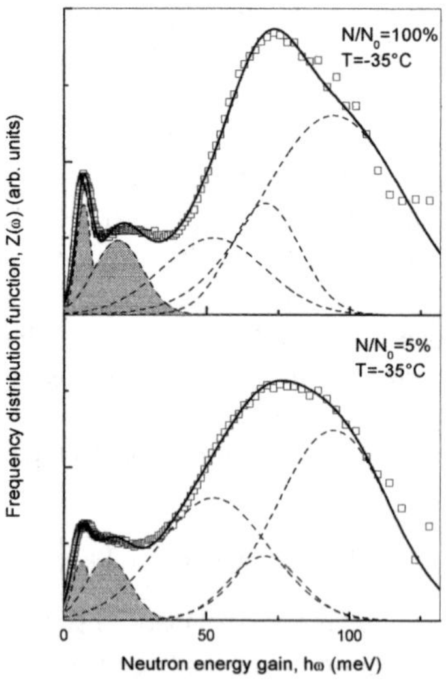

Fig. 4: One phonon proton effective VDOS at different hydration levels at T=-35°C. Dashed lines represent the theoretical fit with Gaussian symmetric profiles.

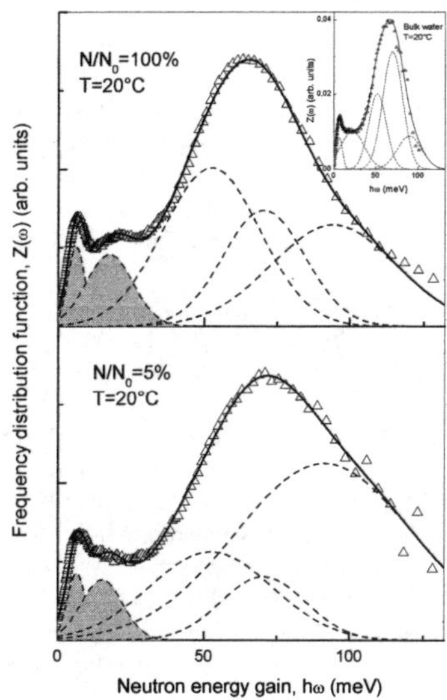

Fig. 5: One phonon proton effective VDOS at different hydration levels at room temperature

where the summation is extended to the various sub-bands b, $P_b(\omega)$ is the electron-vibrational coupling factor and $g_b(\omega)$ is the "true" amplitude weighted VDOS, as obtained in an IINS experiment.

As can be seen by an inspection of Fig. 3, in which we showed $g_{eff}^R(\omega)$ for all the analysed samples, in the case of bulk water the two characteristic bumps, the first one convoluted with the acoustic-like contribution. When the hydration percentage decreases, we observe a flattening of the $g_{eff}^R(\omega)$ spectral profile, in which the symmetries typical of pure water are lost, since the interactions among water molecules and active Si-OH surface groups destroy the intermolecular network of water.

For what IINS data are concerned, we show in Figs. 4 and 5, as an example, the evolution of the one-phonon asmplitude-weighted VDOS Z(ω) for different samples, deconvoluted in Gaussian bands. Two of them (centred at about 8 meV and 20 meV) are related to translational modes, according to the two-peak structure observed by MD in supercooled water [18], the other three (about 52 meV, 70 meV and 91 meV of centre-frequency) are connected to the three ibrational modes. The most relevant results can be summarized in the following point: a) flattening and attenuation of the hindered translational modes (8 and 20 meV) with respect to bulk water, indicating a strong destructuring effect in the interfacial water; b) enhancement, with respect to bulk water, of the librational mode centred at about 91 meV, mainly connected with the hindered rotation of water bonded with the two Si-OH surface groups.

References

[1]   G. Carini, V. Crupi, G. D'Angelo, D. Majolino, P. Migliardo and Y. B. Mel'nichenko, *J. Chem. Phys.* **107** (1997) 2292.

[2]   W. D. Dozier, J. M. Drake and J. Klafter, *Phys. Rev. Lett.* **56** (1986) 197.

[3]   M. Arndt, R. Stannarius, H. Groothues, E. Hempel and F. Kremer, *Phys. Rev. Lett.* **79** (1997) 2077.

[4]   R. M. Lynden-Bell and J. C. Rasaiah, *J. Chem. Phys.* **105** (1996) 9266.

[5]   M. S. P. Sansom, I. K. Kerr, J. Breed and R. Sankararamakrishnan, *Biophys. J.* **70** (1996) 693.

[6]   M. Settles and W. Doster, *Faraday Discussion of the Chem. Soc.* **103** (1996) 269.

[7]   T. Fehr and H. Löwen, *Phys. Rev. E* **52** (1995) 4016.

[8]   See, as an example, J. M. Zanotti, M. C. Bellissent-Funel and S. H. Chen, *Phys. Rev. E* **59** (1999) 3084, and references therein.

[9]   M. C. Bellissent-Funel, S. H. Chen and J. M. Zanotti, *Phys. Rev. E* **51** (1995) 4558.

[10]   D. C. Steytler, J. C. Dore and C. J. Wright, *Mol. Phys.* **48** (1983) 1031.

[11]   F. Aliotta, C. Vasi, G. Maisano, D. Majolino, F. Mallamace and P. Migliardo, *J. Chem. Phys.* **84** (1986) 4731.

[12]   P. G. Hall, A. Pidduck and C. J. Wright, *J. of Coll. and Inter. Sc* **79** (1981) 339.

[13]   A. Fontana, F. Rocca, M. P. Fontana, B. Rosi, A. J. Dianoux, *Phys. Rev. B* **41** (1990) 3778.

[14]   J. L. Green, A. R. Lacey and M. G. Sceats, *J. Phys. Chem.* **90** (1986) 3958.

[15]   S. Magazù, G. Maisano, D. Majolino and P. Migliardo in: H. J. White, J. V. Sengers, D. B. Neumann, J. C. Bellows (eds.), Physical Chemistry of Aqueous Systems. Wallingford, New York, 1995, p. 361.

[16]   F. Alvarez, A. Alegria and J. Colmenero, *Phys. Rev. B* **44** (1991) 7306.

[17]   V. Crupi, G. Maisano, D. Majolino, P. Migliardo and V. Venuti, *J. Chem. Phys.* **109** (1998) 7394.

[18]   S. H. Chen, C. Liao, F. Sciortino, P. Gallo and P. Tartaglia, *Phys. Rev. E* **59** (1999) 6708.

Saturation effects in degenerate four wave mixing lineshape on FeI atomic vapours

L. De Dominicis, M. Di Fino, R. Fantoni, S. Martelli
ENEA Applied Physics Division, Via E. Fermi 45,
00044 Frascati (Rome) - Italy

O. Bomatì Miguel*, S. Veintemillas Verdagüer*
Instituto de Ciencia de Materiales de Madrid,
Cantoblanco 28049 Madrid - Spain

Abstract. Line-splitting phenomena, occurring upon strong saturation conditions at high laser power, were investigated on DFWM lineshape in iron atomic vapours. Measurements have been performed monitoring the a^5D_2-$y^5D^0_2$ atomic transitions. Atomic iron vapours were obtained by thermal decomposition of iron pentacarbonyl. Experimental results have been modelled following the radiative re-normalizzation theoretical approach. A physical mechanism to explain the physical nature of the splitting effect is proposed and discussed.

1. Introduction

Degenerate Four Wave Mixing (DFWM) spectroscopy is a non-linear spectroscopic technique based on light scattering from laser induced gratings. In DFWM experiments two pump laser beams with both the same wavelength λ_P and polarization state, travel through the investigated medium forming each other an angle θ. If λ_P matches with one of the allowed transitions of one or more chemical species in the medium, the generated pattern of optical fringes induces a sinusoidal modulation of the molecular internal energy and a modulation of the refraction complex index. This modulation originates, in turn, a grating able to diffract a fraction of a third probe beam (at degenerate at wavelength λ_P) which is crossing the interaction region at the Bragg angle. DFWM resulted to be a powerful method for quantitative measurements of physical parameters (as for instance gas concentration and temperature) in chemically reacting media (e.g. flames and plasma) [1-3]. In particular it has been demonstrated that DFWM signal is considerably less affected by collisional quenching as compared to LIF, thus making its use preferable when the detection of species at nearly atmospheric or higher pressure is required [4].
DFWM on atomic vapours, in the counterpropagating phase-matching geometry [5], received a great deal of attention mainly for the laser beam saturation effects in the lineshape. As first predicted theoretically and then experimentally observed in sodium vapours [6], highly saturated atomic DFWM lineshapes are characterized by a splitting when the laser is tuned at exact resonance with the probed transitions. Despite the physics

underlying this effect is still not completely understood, this splitting should be carefully accounted for any time DFWM has to be used for quantitative measurements in atomic vapours.

In the present contribution we report the study of saturation effects on neutral atomic iron DFWM lineshape. Neutral atomic iron vapours (FeI) were produced by thermal dissociation of iron pentacarbonyl, $Fe(CO)_5$, a volatile oily complex which decomposes into Fe+5(CO) upon heating above 523°K. The $a^5D_2y^5D_2^0$ transitions of FeI, located at $33067.4cm^{-1}$ was recorded after having heated up to 673°C an optical cell filled with a mixture of $Fe(CO)_5$ vapours diluted in Ar. In order to investigate the saturation effects the $a^5D_2\text{-}y^5D_2^0$ DFWM lineshape has been recorded at different values of pump laser energy. A theoretical simulation, based on the radiative re-normalisation method reported in [7] is used to model the splitting mechanism. In addition a physical mechanism to explain the physical nature of the splitting effect is proposed and discussed.

2. Theoretical background

Saturation effects in Forward DFWM (FDFWM) lineshape were formerly studied by using the radiative re-normalization technique [7]. The relevant parameter S introduced in the model is defined as

$$S = \frac{\Omega^2}{\gamma_0 \gamma_{ab}} \tag{1}$$

where γ_0 and γ_{ab} are respectively the population and collisional relaxation rate of the excited transition, while Ω is the Rabi frequency of the transition:

$$\Omega = \frac{\mu_{ij}E_p}{\hbar} \tag{2}$$

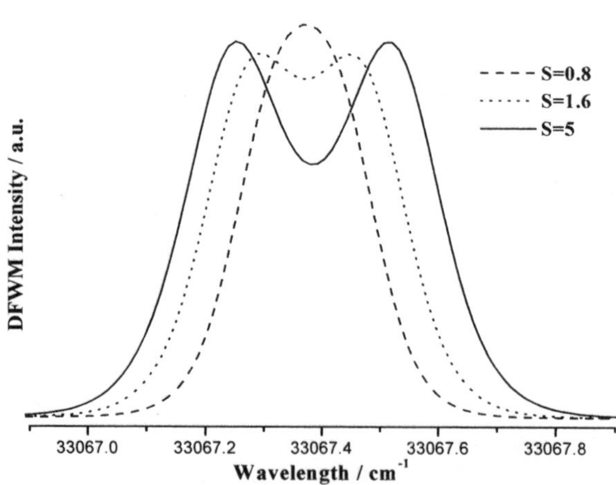

Fig.1 FDWFM spectrum of the a^5D_2 - $y^5D^0_2$ transition in FeI calculated for different values of the saturation parameter S. Details about the radiative re-normalization model used are given in ref.7.

with μ_{ij} induced dipole moment of the probed transition and E_p electric field strength of the pump beam.

From eq. 1 and eq. 2 the parameter S turns out to be a-dimensional. If the partial gas pressures and the temperature are kept fixed, S increases as pump laser power is raised. Any time the pump energy exceeds the saturation threshold level (corresponding to S=1), the line broadening due to Rabi oscillations dominates the collisional width.

The theoretical FDWFM spectrum of the $a^5D_2 - y^5D^0_2$ transition in FeI at different degree of saturation (i.e. different values of S) and calculated using theory reported in ref.7, is reported in fig.1.

As predicted by the model, the fingerprint for the saturation of FDFWM lineshape is the appearance of a dip located at exact resonance with the transition frequency whose depth increases with S. The model is capable to reproduce the line-splitting of saturated DFWM lines observed in the crossing pump beams geometry for the excitation of atoms, including iron, nevertheless a comprehensive description of the physics underlying this phenomenon is still lacking. In the absence of an adequate theory, the modelling of saturation effects must be based on an exhaustive experimental characterization of each atomic species used as active medium, in order to avoid to introduce errors in quantitative FDFWM measurements.

3. Experimental apparatus

The laser radiation for the excitation of the FeI a^5D_J-$y^5D^0_{J'}$ transitions was generated by frequency doubling in a BBOI crystal the output of a narrow-band dye laser operating with Rhodamin B. The dye laser was pumped by a doubled Nd:YAG laser (JK2000) providing up to 250mJ per pulse at 532nm with a repetition rate of 1Hz. In this experimental scheme laser pulses in the spectral range between 300-310nm were produced, with 15ns of time duration and 600μJ of energy. The bandwidth $\Delta\nu_L$ of the dye laser is specified to be 0.2 cm^{-1} by manufacturer. The FDFWM phase matching geometry [7], chosen for an efficient signal generation, required the use of suitable optics to generate three co-propagating beams travelling at the corners of a square before focalisation by a lens (f=1000mm). The two pump beams had the same energy, while the probe beam was four time weaker. The interaction region was a cylinder with 1cm of length along the beams propagation direction and 1mm of diameter in the transversal section..

The FDFWM signal emerged from the interaction region, located at the center of the 60 cm long optical cell, as if originating from the fourth corner of the square, as shown in the scheme in fig.2. The signal, after being spatially filtered, was detected by a photomultiplier tube connected to a Boxcar Averager (SRS245). The dye laser scan and data acquisition was simultaneously controlled by a PC (Epson 386DX). In the recorded spectrum each data point corresponded to the average of signals from 10 subsequent laser shots.

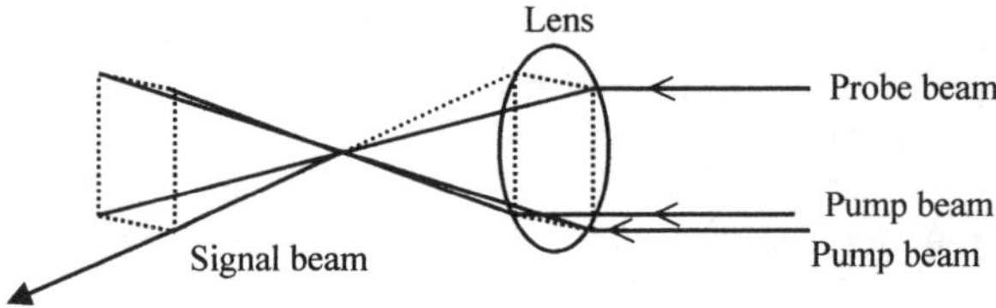

Fig.2 Forward DFWM phase matching geometry

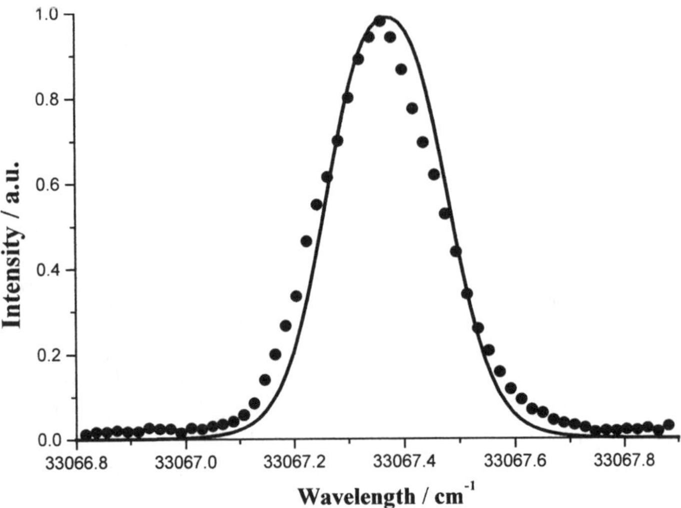

Fig.3 Experimental lineshape (dots) of the a^5D_2 - $y^5D^0_2$ transition in FeI at 673°K. Pump beam energy was 100μJ. The continuous curve is the result of theoretical simulation with S=0.93

FeI atomic vapours was generated by raising up to 673°K the temperature of the cell filled with a mixture of 10mbar of $Fe(CO)_5$ and 190mbar of Ar. The cell was equipped with a Pt/Rh thermocouple to measure the gas temperature.

4. Experimental results

The lineshape associated to the a^5D_2 - $y^5D^0_2$ atomic transition in FeI has been recorded at temperature of 673°K and at different value of the pump beam laser energy I_P. Synthetic lineshape through the experimental curve have been evaluated according to [7], the results are shown in fig.3 for I_P = 100μJ.

The calculated Rabi frequency in this experimental configuration is 0.034cm^{-1}, while μ_{ij}= 0.028D as determined by the relation

$$\mu_{if} = \frac{1}{g_i g_f} \sqrt{\frac{3\hbar e^2 f_{if}}{2m_e \omega_{if}}} \tag{3}$$

where f_{ij}=0.024 and $\hbar\omega_{ij}$=4.09eV are the oscillator strength and the frequency of the transition, respectively and g the involved level degeneracy. The lifetime of the $y^5D^0_2$ level is 6.5ns [9], which gives a value of $4.1\cdot10^{-3}$ cm^{-1} for the excited state population lifetime γ_{aa}. The closest agreement was achieved for γ_{ab}=0.3cm^{-1}, which corresponds to S=0.93. In fig.4 the spectra recorded at I_P = 200μJ and I_P = 250μJ are reported together with the theoretical simulations performed respectively with S=1.86 and S=2.45. The presence of a dip at the center of the experimental FDFWM lineshape is clearly observable. The

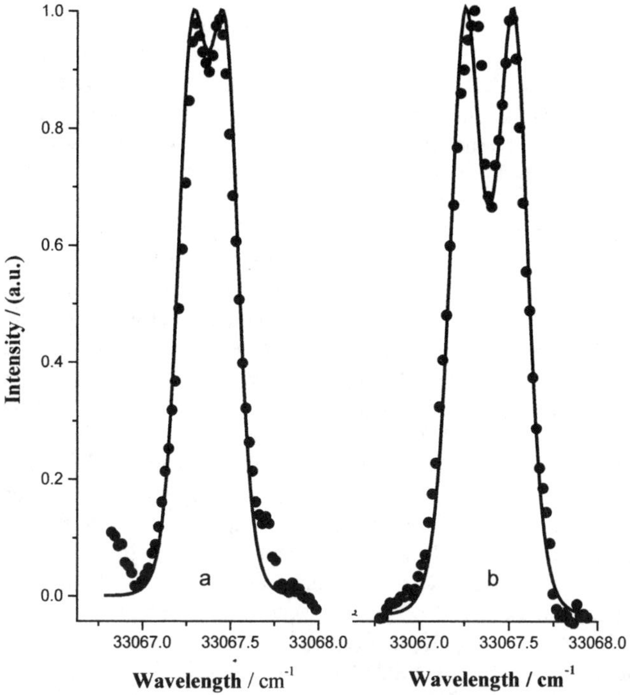

Fig.4 Experimental lineshape (dots) of the a^5D_2 - $y^5D^0_2$ transition in FeI at 673°K. a) Pump beam energy was 200μJ. The continuous curve is the result of theoretical simulation with S=1.86. b) Pump beam energy was 250μJ. The continuous curve is the result of theoretical simulation with S=2.45.

experimental findings indicate that the line splitting becomes more and more evident as S increases. The adopted model reproduces with good accuracy the lineshape in all the investigated experimental conditions, and the apparent discrepancies can be ascribed to the multi-mode axial structure of the laser emission. In fact, due to the non-homogenous energy distribution in the focal volume, space regions contributing to FDFWM signal with different degrees of saturation, may coexist. It comes out that saturation induces strong lineshape distortions in Fe DFWM spectrum, which completely hinder the application of this technique for quantitative measurements of physical parameters at high laser intensity.

5. Discussion and conclusions

The spectrum of the a^5D_2 - $y^5D^0_2$ system of FeI recorded with a non-linear spectroscopic technique (DFWM) based on scattering from laser induced gratings in a gaseous medium is reported for the first time. FeI atomic vapours were produced by heating up to 673°K iron pentacarbonyl vapours. The a^5D_2 - $y^5D^0_2$ transition was chosen to monitor the saturation effect on FeI DFWM lineshape at 673°K since among the recorded lines it is the least affected by pumps beams depletion and signal reabsorption. The evidence of line-splitting due to saturation indicates that atomic vapours are targets which can significantly contribute to shed light on the physical nature of the saturation line-splitting in FDFWM. It is worthwhile to mention that in our previous experiment dealing with saturated DFWM on NO_2 molecule no evidence for such a dip was observed [10]. The induced dipole moment in FeI ($\sim 3 \cdot 10^{-2}$D) is considerable lower than in NO_2 (~ 0.4D), this difference is rather general in comparing atoms with molecules and in our opinion may help towards a deeper comprehension of their different saturation behaviour. The line splitting mechanism can be due to a combination of two different effects: the Dynamic Stark Effect (DSE) and Coherent Trapping of atomic Populations (CTP). When a strong laser beam shines an atom, DSE induces a splitting of each individual atomic level into two levels separated by an amount of energy equal to $h\Omega$. These two non-degenerate levels can be simultaneously coupled, under the excitation of the two pump laser beams, with the upper level of the probed transition if $\Omega < \Delta v_L$, where Δv_L is the laser bandwidth. In this case the condition for the CTP effect are satisfied. The efficiency of the pump beams in transferring atomic populations from the lower to the upper level is then considerably reduced and the splitting in the DFWM lineshape appears. This mechanism can explain why the dip is observed mainly when transitions with low values of the induced dipole moment μ are probed. In fact, being the Rabi frequency a linear function of μ, the condition $\Omega < \Delta v_L$, at the same laser energy for the pump beams , is easily satisfied in transitions with a low values of μ. It should be emphasized that in our experiment being $\Delta v_L = 0.2 \text{cm}^{-1}$ and the maximum value of the Rabi frequency $\Omega_{max} = 0.053 \text{cm}^{-1}$, the condition required for the existence of the dip, according to the proposed mechanism, is well satisfied. Within this scheme the theoretical and experimental evidence that DFWM lineshapes are not splitted for $S < 1$, is related to the hypothesis that, due to the low value of the laser excitation energy, no efficient DSE energy level split takes place. It should be mentioned that, in a previous DFWM experiment [11], line intensity alterations and line position shifts due to saturation have been experimentally observed on two satellite transitions of the $A^2\Sigma$-$X^2\Pi$ system of OH. The experimental data were modelled using the dressed molecule theoretical approach [7]. This observation supports the hypothesis at the basis of the physical mechanism here proposed, that saturation induces strong coupling effects between adjacent transitions which results in strong modification of lineshape and line position. The model proposes to explain the physics underlying the line splitting of saturated FDFWM lineshape account for all experimental evidence reported in this work. Nevertheless it seems in future desirable to going into further details in the data analysis and to perform experiments on other atomic vapours samples.

Acknowledgements

We would like to thank S. Pignataro, B. Attal Tretout and G. Gatti for helpful discussions. The contribution of M.Giorgi at the early stage of the work is gratefully acknowledged.

References

[1] P. Ewart, S.V. O'Leary, *Opt. Lett* **11**, 279 (1986)

[2] T. Dreier, D.J. Rakestraw, *Appl. Phys B* **50**, 479 (1990)

[3] S. Williams, R.N. Zare, L.A. Rahn, *J. Chem Phys.***101**, 1093 (1994)

[4] P. Ljungberg, O. Axner, *Appl. Phys. B* **63**, 69 (1996)

[5] G. Grynberg, M.Pinard, P.Verkerk, *Opt. Comm.* **50**, 261 (1984)

[6] G. Grynberg, M.Pinard, P.Verkerk, *J. Physique* **47**, 617 (1986)

[7] B.Attal Tretout, H. Bervas, J.P. Taran, S. Le Boiteux, P.Kelley, T.K. Gustafson, *J. Phys B* **30**, 497 (1997)

[8] M. Kotzian, N. Rösch, H. Schröder, M.C. Zerner *J. Am. Chem. Soc.* **111,** 7687 (1989)

[9] T.R. O'Brian, M.E. Wickliffe, J.E. Lawler, W. Whaling, J.W. Brault, *J.Opt. Soc. Am. B* **8**,1185 (1991)

[10]R. Fantoni, L. De Dominicis, M. Giorgi, R.B.Williams. *Chem. Phys. Lett.* **259,** 342 (1996)

[11]R. Fantoni, L. De Dominicis, M.Giorgi, D.A. Sidorov Biryukov, M. D'Apice, S. Giammartini, in *Proceedings of Itarus '99* edited by G. Ferrante, M. Vaselli and A. Zheltikov, Intellekt Moscow (2000)

GNSR 2001
G. Messina and S. Santangelo (Eds.)
IOS Press, 2002

SER studies of 1H-1,2,4-triazole on silver sol

Barbara Pergolese, Adriano Bigotto[*]

Department of Chemical Sciences-University of Trieste
Via Licio Giorgieri 1-34127 Trieste
Phone: +39-040-6763950, Fax: +39-040-6763903,
[*]E-mail: bigotto@.univ.trieste.it

Abstract. Surface enhanced Raman spectra of 1H-1,2,4-triazole (Tz4) adsorbed on silver sol were obtained. The SER data were interpreted on the basis of previous vibrational assignments, with the help of the results of *ab-initio* calculations carried out using the 6-31G** basis set. From the comparison of SER and normal Raman spectra, it can be deduced that 1H-1,2,4-triazole is non-dissociatively adsorbed on metal surface and that it interacts with it through the N_4 and the N_2 atoms. The molecular plane assumes a tilted orientation with respect to the silver surface.

1. Introduction

The azoles are extensively used in surface treatment of materials, in particular in the field of protection of metals from corrosion. Therefore, a large number of investigations were carried out in order to clarify the interaction mechanism between these substances and metal surfaces and, for this purpose, the SERS technique has proved very informative.

1H-1,2,4-triazole (Tz4) is a very effective corrosion inhibitor for- copper [1], but the studies of the adsorption behaviour of Tz4 are scarce. In particular SERS investigations on Tz4 were performed only for the molecule adsorbed on microlithographically prepared copper surfaces [2]. However, in this study no conclusions about the molecular sites of interaction with the metal were drawn. Therefore, in order to gain a better understanding of the adsorption behaviour of Tz4 on metal surface, we planned SERS investigations on this molecule adsorbed on silver sols and we present here the preliminary results obtained from them. A correct interpretation of the SER spectra requires a detailed assignment of the vibrational spectra of the investigated substance. The vibrational modes of Tz4 in gaseous and solid phase were previously assigned on firm experimental basis[3]. We rely on these assignments to interpret the results of the SERS measurements. Moreover we have planned ab-initio calculations of the harmonic wavenumbers and of the normal modes of vibration of Tz4 with a view to giving a detailed description of each normal mode in terms of internal coordinates. SER spectra have been interpreted on the basis of the surface selection rules[4,5], making reference to the results of their application to some planar molecules and ions [5,11].

2. Experimental procedure

Tz4 (98%) was obtained from Aldrich. SERS measurements were performed using silver colloids prepared by reduction of $AgNO_3$ with excess $NaBH_4$ accordingly to Creighton *et al.*[12]. Milli-Q water was used in the preparation. All glassware was thoroughly cleaned with HNO_3 and washed with milli-Q water in order to avoid impurities in the colloid preparation. Tz4 was added to the sol in order to obtain a final concentration of about 10^{-3} M . The sols, before and after addition of Tz4, were characterized using UV-visible absorption spectra obtained with a UNICAM Heλios spectrophotometer. In order to enhance the Raman signal, NaCl or KCl were added so that the final concentration of electrolyte was 0.02-0.03 M. Raman spectra were obtained with a SPEX Ramalog instrument. An AT personal computer was used for data acquisition and monochromator control. Excitation was provided by 514.5 nm radiation from a Spectra-Physics 165 argon ion laser laser and the samples were contained in capillary cells. FT-Raman spectra were obtained using a Perkin-Elmer SYSTEM 2000 instrument with excitation provided by 1064 nm radiation from a diode-pumped Nd:YAG laser. The treatment of the spectral data was performed using the Perkin-Elmer IRDM and the Galactic GRAMS386 software.

Table 1 Experimental spectral data, calculated wavenumbers (scaled values) and assignments for Tz4

v_{oss}[a]	v_{calc}	PED[b]
3510	3500 (A')	$100r_{6,1}$
3140	3139 (A')	$95r_{7,3}$
—	3131 (A')	$95r_{8,5}$
1503	1531 (A')	$26r_{3,2}$, $21r_{5,4}$, $11\alpha_{4,3,7}$, $7\alpha_{1,5,8}$, $7\alpha_{2,3,7}$, $5\alpha_{2,1,6}$, $5\alpha_{5,1,6}$, $5\alpha_{4,5,8}$
1430	1429 (A')	$32r_{5,1}$, $26\alpha_{2,1,6}$, $20\alpha_{5,1,6}$, $6\alpha_{4,5,8}$
1360	1362 (A')	$35r_{3,2}$, $21r_{5,4}$, $14\alpha_{1,5,8}$, $9\alpha_{4,5,8}$
1280	1269 (A')	$24r_{5,4}$, $21r_{4,3}$, $10r_{2,1}$, $16\alpha_{2,3,7}$, $7\alpha_{2,3,4}$, $5\alpha_{4,5,8}$
1260	1223 (A')	$8r_{5,4}$, $7r_{3,2}$, $6r_{5,1}$, $24\alpha_{2,3,7}$, $20\alpha_{4,3,7}$, $12\alpha_{4,5,8}$, $9\alpha_{1,5,8}$, $7\alpha_{2,1,6}$, $6\alpha_{5,1,6}$
1155	1115 (A')	$31r_{4,3}$, $19r_{2,1}$, $5r_{5,1}$, $10\alpha_{4,3,7}$, $10\alpha_{1,5,8}$, $5\alpha_{2,1,6}$
1103	1079 (A')	$33r_{5,1}$, $22r_{4,3}$, $5r_{3,2}$, $11\alpha_{5,1,6}$, $10\alpha_{2,1,6}$
1040	1025 (A')	$52r_{2,1}$, $15\alpha_{4,5,8}$, $8\alpha_{1,5,8}$, $6\alpha_{2,1,5}$
970	936 (A')	$7r_{2,1}$, $27\alpha_{1,2,3}$, $22\alpha_{2,1,5}$, $14\alpha_{1,5,4}$, $6\alpha_{3,4,5}$, $6\alpha_{5,1,6}$, $5\alpha_{4,3,7}$
—	916 (A')	$5r_{4,3}$, $32\alpha_{3,4,5}$, $19\alpha_{2,3,4}$, $12\alpha_{1,2,3}$, $10\alpha_{1,5,4}$, $8\alpha_{1,5,8}$
883	910 (A")	$72\gamma_{7,3}$, $8\gamma_{8,5}$, $8\tau_{3,4}$, $7\tau_{3,2}$
842	886 (A")	$75\gamma_{8,5}$, $9\gamma_{7,3}$, $7\tau_{1,5}$, $7\tau_{5,4}$
678	659 (A")	$33\tau_{3,2}$, $29\tau_{3,4}$, $24\tau_{1,2}$, $8\tau_{5,4}$
—	640 (A")	$6\gamma_{6,1}$, $36\tau_{1,5}$, $28\tau_{1,2}$, $19\tau_{5,4}$, $10\tau_{3,4}$
538	520 (A")	$72\gamma_{6,1}$, $12\tau_{1,5}$, $11\tau_{1,2}$

a) fundamental vibrations of gaseous Tz4[3] b) **r**= stretching, α= bending in plane, γ= bending out-of-plane, τ = torsion

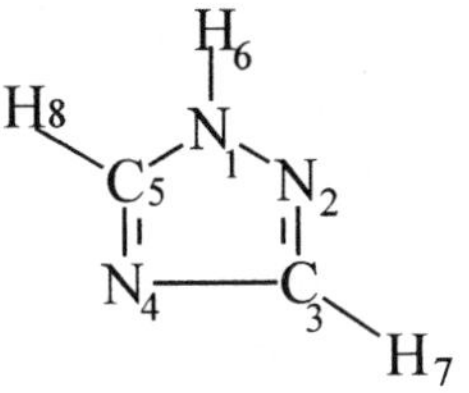

Scheme 1

3. Computational procedure

Calculations were performed using the GAUSSIAN 94 package[13]. The geometry of Tz4 was optimised at the RHF level of theory, using the 6-31G** basis set. A C_s symmetry was imposed to the input geometry. The harmonic wavenumbers and the normal modes were calculated analytically at the same level of approximation. The transformation of *ab-initio* cartesian coordinate force constants to internal coordinate space was performed according to the method of Boatz and Gordon[14]: a redundant set of primitiva internal valence coordinates was used throughout the calculations, following the approach of Baker *et al.*[15]. Torsional coordinates were defined keeping in mind the recommendations of Keresztury *et al.*[16]. The **F** and **G** matrices built in the previous step were used to perform a standard zero-order **GF**-matrix treatment from which the wavenumbers of the vibrational modes and the Potential Energy

Distribution (P.E.D.) were obtained. Four scaling factors were applied to the force constants obtained from the calculations: 0.75 for C-N stretching internal coordinates, 0.79 for the N-H stretching internal coordinate, 0.83 for C-H stretching internal coordinates and 0.76 for the other internal coordinates. The results of calculations are reported in Table 1, together with the observed wavenumbers and the atomic numbering is reported in Scheme 1. It is worth-mentioning that in the first column of the Table 1 the IR data of Tz4 in gaseous phase obtained by Bougeard *et al.* are reported instead of the wavenumbers observed in our Raman spectra in solid state. In fact Tz4 is characterized by strong hydrogen bondings in the solid state which influence the position of the bands. Since the wavenumbers obtained from the calculations related to the isolated molecule are better correlated to the data obtained in gaseous phase than to those in solid phase.

4. Results and discussion

The Raman spectra obtained, both normal (NR in the following) and surface-enhanced, are reported in Fig. 1. The interpretation of the vibrational spectra was carried out making reference to the descriptions of the vibrational normal modes of Tz4 in terms of internal coordinates, as provided by P.E.D..

A comparison between the SER spectra of Tz4 adsorbed on Ag sols and the NR spectra of Tz4 in the solid state reveals that the SER spectra may be well correlated to the normal Raman spectra and that there are no marked differences in the position of the bands on going from the NR to the SER spectra. This observation suggests that the molecule interacts with the silver surface in the non-dissociated form, in agreement with the results obtained from the SER study of Tz4 adsorbed on microlithographically prepared copper surfaces [2].

It can be observed that the most enhanced band in the SER spectrum is that at 979 cm^{-1}. It can be related either with the NR band at 960 cm^{-1} or with the one at 980 cm^{-1}. The latter one is correlated to the calculated normal mode at 936 cm^{-1}, which is described, by means of P.E.D., essentially as a N_1-N_2-C_3 bending mode (27%) mixed with a N_2-N_1-C_5 (22%) bending and a N_1-C_5-N_4 bending (14%). On the other hand, the NR band at 960 cm^{-1} is associated with the normal mode calculated at 916 cm^{-1}, which has a prevailing contribution from the C_3-N_4-C_5 (32%), N_2-C_3-N_4(19%), N_1-N_2-C_3(12%) and N_1-C_5-N_4(10%) bending internal coordinates. Therefore, both the NR bands are assigned to normal modes which can be characterized as ring-breathing modes.

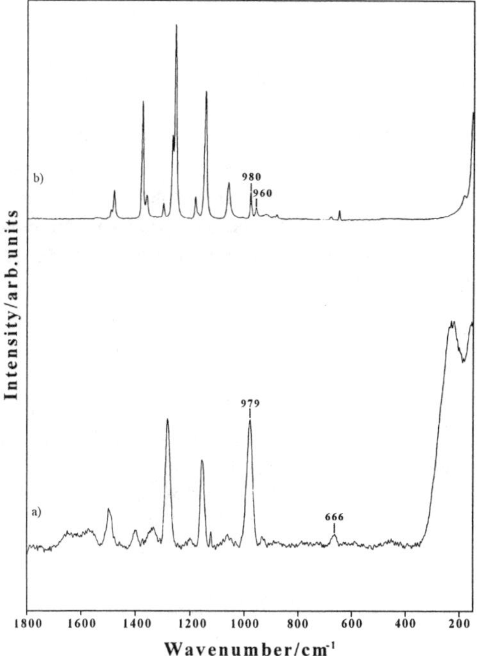

Fig. 1 SERS spectrum of Tz4 on silver sol (a); normal Raman spectrum of solid Tz4 (b)

It is evident that, whichever is the choice, the normal modes include perceptible contributions from internal coordinates involving the N_2 and N_4 atoms. Moreover, the Mulliken charge distribution shows that the density of negative charge is localized mainly at the N_4 atom but also at the N_2 atom. These arguments suggest that Tz4 interacts with the metal surface through the lone pairs of N_4 and N_2 atoms. In order to confirm this hypothesis, it would be interesting to perform DFT calculations of the harmonic wavenumbers and of the related normal modes of vibration of Tz4 bound to at least one silver atom. In fact, from the comparison between the calculated force constants values of the free molecule and those of the molecule bound to metal, it can be obtained further information about the molecular sites of interaction with the metal surface. The results of these calculations and the conclusions drawn from them will be reported in another paper.

Regarding the molecular orientation assumed by Tz4 on the metal surface, it can be observed that the SER spectrum displays an enhanced band at 666 cm⁻¹ which is assigned to an out-of-plane fundamental. According to the surface selection rules of SER spectroscopy[4], the enhancement of an out-of-plane fundamental rules out the possibility of a purely edge-on orientation of the molecular skeleton on the metal surface. Moreover in the region of the C-H stretching modes the SER spectrum displays no bands: this observation is diagnostic of a very titled or even parallel orientation of the molecular skeleton with respect to the metal surface.

5. Conclusion

The preliminary study on the adsorption behaviour of 1H-1,2,4-triazole on silver sols by means of SER spectroscopy is reported. From the comparison of the SER and NR spectra of Tz4 in solid state, it can be deduced that the molecule interacts in the non-dissociated form with the metal surface.

From the description of the fundamentals that undergo the most significant enhancements in the SER spectra, it can be deduced that the molecule interacts with the silver sol with the lone pairs of the N_4 atom and of the N_2 atom, assuming a very tilted orientation with respect to the metal surface.

References

[1]   F.Zucchi, M. Fonsati, G. Trabanelli, Int. Corros. Congr., Proc., 13th, Paper 322/1-Paper 322/9 Australasian Corrosion Association: Clayton, Australia, 1996.

[2]   D. Thierry, C. Leygraf, *J. Electrochem. Soc.* **133** (1986) 2236.

[3]   J.A. Creighton In *Advances in Spectroscopy: Spectroscopy of surfaces*, vol 16, Clark, RJH, Hester RE (eds). John. Wiley & Sons: Chichester (1988) 37.

[4]   M. Moskovitz, J.S. Suh, *J. Phys. Chem.* **88** (1984) 1293.

[5]   M. Moskovitz, J.S. Suh, *J. Phys. Chem.* **88**, (1984) 5526.

[6]   M. Moskovitz, J.S. Suh, *J. Phys. Chem.* **92**, (1988), 6327.

[7]   T.C. Strekas, P.S. Diamandopoulos, *J. Phys. Chem.* **94** (1990) 1986.

[8]   M.L. Patterson, M.J. Weaver, *J.Phys. Chem.* **89** (1985) 5046.

[9]   X. Gao, J.P. Davies, M.P. Weaver, *J. Phys. Chem.* **94**, (1990) 6858.

[10] Y.J. Kwon, D.H. Son, S.J. Ahn, K. Kim, *J. Phys. Chem.* **98** (1994) 8481.

[11] D.Bougeard, N. Le Calvè, B. Saint Roch, A. Novak, *J. Chem. Phys.* **64** (1976) 5152.

[12] A. Creighton, C.G. Blatchford, M.G. Albrecht, *J. Chem. Soc., Faraday Trans.* 2 **75** (1979) 790.

[13] M.J. Frisch, G.W. Trucks, H.B. Schlegel, P.M.W. Gill, B.G. Johnson, M.A. Robb, J.R. Cheeseman, T. Keith, G.A. Petersson, J.A. Montgomery, K. Raghavachari, M.A. Al-Laham, V.G. Zakrzewski, J.V. Ortiz, J.B. Foresman, J. Cioslowski, B.B. Stefanov, A. Nanayakkara, M. Challacombe, C.Y. Peng, P.Y. Ayala, W. Chen, M.W. Wong, J.L. Andres, E.S. Replogle, R. Gomperts, R.L. Martin, D.J. Fox, J.S. Binkley, D.J. Defrees, J. Baker, J.P. Stewart, M. Head-Gordon, C. Gonzalez, J.A. Pople GAUSSIAN 94 Revision E.1., Gaussian Inc., Pittsburgh PA, 1995.

[14] J.A. Boatz, M.S. Gordon, *J.Phys.Chem.* **93** (1989) 1819.

[15] J. Baker, A.A. Jarzeckj, P. Pulay, *J.Phys.Chem. A* **102** (1998) 1412.

[16] G.Keresztury, A.Y.Wang, J.R. Durig, *Spectrochim.Acta* **48A** (1992) 199.

Spatially resolved CARS thermometry and CH LIF detection on laboratory flames

M. D'Apice, M. Marrocco, S. Giammartini, P. Cavazza, L. Crecco,
L. De Dominicis*

*Sez. Tecnologie della Combustione, ENEA C.R. Casaccia, Via Anguillarese 301, 00060
S.M. di Galeria (Roma)*
Phone: 06-30484685, Fax: 06-30484811, Email:massimo.dapice@casaccia.enea.it
**Div. Fisica Applicata, ENEA C.R. Frascati, Via E. Fermi 45, 00044 Frascati (Roma)*
Phone: 06-30484685, Fax: 06-30484811, Email:dedomini@frascati.enea.it

Abstract: During last years several theoretical codes have been developed to simulate the physical and chemical flame behaviour. Due to the complexity of these models, complete and rigorous calculations have been performed only on a limited number of cases. Validation of these simulations requires accurate measurements on real flames where geometrical, physical and chemical parameters are strictly controlled. In particular, temperature and flame front thickness, being flame characteristics strongly depending on burner geometry, reactant flows and chemical kinetics, have to be measured with high accuracy in order to compare experimental results with theoretical predictions.

In this work the results of two different measurement campaigns, performed in our molecular spectroscopy laboratory located at ENEA-Casaccia Research Centre (Rome, Italy), are reported.

During the first campaign we performed a spatially resolved thermometry performed with CARS (Coherent Anti-Stokes Raman Scattering) technique on N_2 molecules [1] in a well-controlled Bunsen flame fed with a premixed air-methane flow. Methane was chosen as fuel due to the relatively simple chemical kinetics of the associate combustion reaction in air. A turbulent flow of reactants, in the range of Reynolds number 800-1500, was investigated. Measurements allowed to visualise the flame front structure while the highest temperature value measured was in good agreement with the theoretical adiabatic value.

The flame front geometry was also investigated in the second campaign by using a different technique based on the LIF (Laser Induced Fluorescence) detection of CH radicals emitted in flame. In fact, this kind of radicals is a good tracer of the flame front where the temperatures are high enough to remarkably enhance the production rates of some important pollutants [2]. LIF technique was applied to a Bunsen flame, fed with liquid Propane gas, in laminar flow condition (Reynolds number <100).

1. Introduction

The main difficulty to face with reactive flow study lies in the strict relation between fluid-dynamics and chemical kinetics which constitute a single thermal and fluid-dynamical problem.

From a theoretical point of view, this problem could be faced directly solving, through a DNS (Direct Numerical Simulation) method, the set of Navier-Stokes, energy and transport equations for each chemical specie involved. This procedure relies on the correct description of all spatial scales, from the largest one (in the order of the whole physical domain) to the smallest (the Kolmogorov scale), all the temporal scales (related to fluid-dynamics and chemical kinetics) and all the chemical species, each one characterised by a specific transport equation.

If we consider that a complete description of the combustion kinetics for an air-methane flow, one of the simplest case, involves several tens of chemical species, we can easily understand why it is impossible to solve any practical problem without assuming any kind of simplification. In this respect, simplified approaches were developed through the RANS (Reynolds Averaged Navier-Stokes) or LES (Large Eddy Simulation) codes, which do not consider all the complexity of the chemical and fluid-dynamical interactions in a combustion process assuming a drastically simplified kinetics. However, in such a way any prediction about some physical variables, especially temperature and chemical concentrations, are basically unreliable.

For this reason, it is so important to perform experimental programs to measure a set of physical variables needed for developing and testing more reliable theoretical codes.

This work reports the results of two experimental campaigns, both performed by non linear spectroscopic techniques in order to measure temperature and CH radical concentration in laboratory reference flames provided by a Bunsen burner. Such kind of burner was chosen due to its widespread use in free flame studies and the consequent large availability of literature data, also derived by different techniques, to be compared and/or integrated with ours. Furthermore, in spite of its construction simplicity, a Bunsen burner is able to provide a rather large number of different flame conditions (laminar, turbulent, diffusive or premixed flow) easily allowing the process control.

2. CARS thermometry

Temperature measurements were performed by CARS technique on a premixed flame provided by a modified Bunsen burner, supplied with an air/Methane gas fuel mixing. CARS experiments were tuned on Nitrogen molecule as its relative abundance in air guarantees a high intensity CARS signal in any test condition.

The burner geometrical simplicity and the Methane choice as fuel allowed to easily run some numerical simulations whose output data could be compared with experimental results. In fact, Methane is the hydrocarbon with the simplest combustion process in air with a kinetics characterised by a few hundreds of intermediate chemical reactions involving some 60 species. For an even simpler process we could study Hydrogen combustion, but in this case we should adopt such a lot of safety cares that would make experiments quite impractical.

Without entering in a detailed CARS technique description, widely reported in literature [3], we can just consider the main characteristics that makes our CARS apparatus preferable respect to more traditional devices (e.g. thermocouples) in flame thermometry:

- Non intrusive: experiments are performed focusing laser beams on a measurement point without any change in the process fluid and thermal dynamics.
- High spatial and temporal resolutions: our experimental system can operate on very small measurement volumes, well below 1 mm^3, integrating the useful signal for a few seconds, or on relatively large volumes (with a maximum dimension of a few millimetres) acquiring signal from a single laser shot (10 ns long).

- Real time acquisition: data acquisition and elaboration are performed in real time by a specialised software based on a neural network approach [4].

This last feature, specifically developed for our CARS system, represents a remarkable improvement to apply CARS technique for on-site full scale industrial devices without the need to work in a controlled experimental area.

2-1. CARS set-up

Our apparatus for CARS thermometry on Nitrogen is shown in fig. 1. A first pulsed Nd:YAG laser, able to generate 1200mJ pulses at 1064nm (near IR) with 10Hz frequency, is duplicated by a non linear KDP crystal to provide 500 mJ pulses at 532nm (green). Pulse time width is typically around 10 ns in the classic Q-switched operating condition.

The green laser beam (500mJ, 532nm) is then split (crossing a beam-splitter) in two beams with 30% and 70% of the total energy:

- the first beam (170mJ, 30% of total energy) is directed to "pump" a dye laser (working with Rhodamina 610) which, in turn, generates a Stokes beam at λs=607 nm (red) with only 30 mJ energy due to the rather low dye laser efficiency (15-20%);
- the second beam (340mJ, 70% of total energy) is sent to cross another beam-splitter to generate two equal beams at 532nm with 50% energy (170mJ). One of these two

EXPERIMENTAL SET-UP (CARS spectroscopy on N_2)

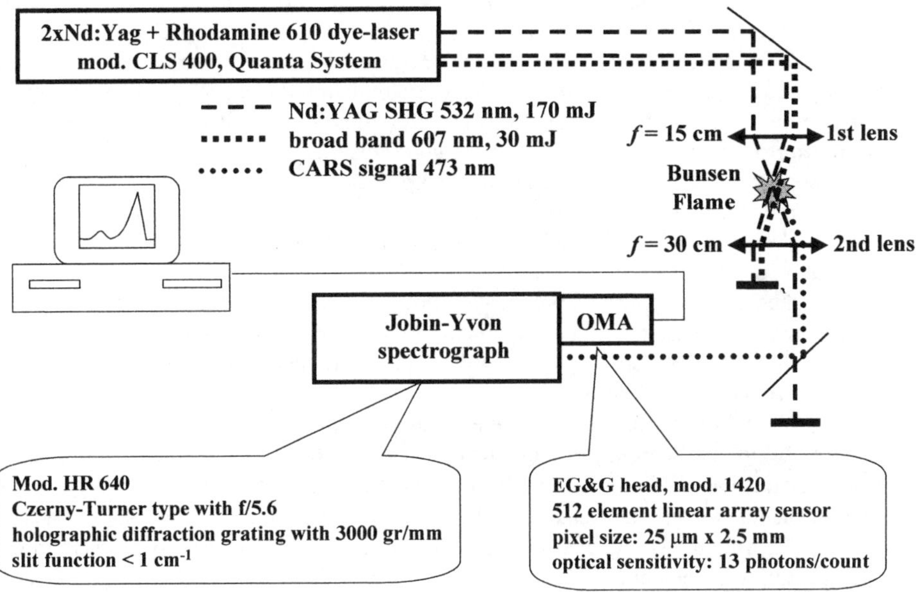

Fig.1 CARS set-up

parallel beams is then superimposed on the Stokes beam to realise the beam phase-matching according to the so called planar-boxcar geometry [3]. In such a way we obtain the physical and geometrical conditions needed to generate the anti-Stokes beam (the useful CARS signal) excluding, at the same time, the cold contribute due to Nitrogen normally present in the air outside the flame measurement zone.

The two so generated laser beams, the single green (532nm, 170 mJ) one and the compound one made of two superimposed red (607 nm, 30 mJ) and green (532nm, 170 mJ), are then focusing, by a first lens, on a point inside the sample flame. The two working wavelengths (λp: 532 nm; λs: 607 nm) are selected to excite the first vibrational Raman transition 0-1 of the Nitrogen molecules inside the flame, so producing an anti-Stokes beam at λas= 457 nm.

In effect, in the multiplex CARS technique we applied, we worked with a broadband dye laser generating a Stokes band, rather than a single line, with a spectral width around 10 nm. In this way also the anti-Stokes signal is generated as a band, with a similar spectral width centred around 457 nm, to be properly analysed.

Such a signal is then collected through a second lens, reflected by a dichroic mirror (optimised for 457 nm), "cleaned" from residual beams at 532 and 607 nm by crossing an interferential filter (CW = 470 nm, FWHM = 50 nm) and, finally, focused by a third lens on the slit of a Jobin-Yvon HR-640 monochromator (3000 grooves/mm). Signal spectrum produced by monochromator is then sent to an EG&G Optical Multichannel Analyzer (OMA), with a 512-element photosensor array, wired to a Personal Computer (PC).

OMA provides a spectral histogram to be compared, by an elaboration software defined as CARSnet, with a set of reference theoretical spectra of Nitrogen molecule, each one referred to a different temperature with 50°K step in the range 300-2500°K. CARSnet is based on a neural network (NN) algorithm and is able to provide a real time temperature evaluation from the best fitting between experimental and reference spectra [4].

A test to establish the reliability of the NN method has been carried out. At different temperatures a theoretical spectrum with noise simulating the experimental conditions was fed into the NN. The comparison between the known theoretical temperature with the temperature extracted by the NN showed a good linearity with uncertainty of about 25 °K. Another test conducted on a C_2H_2/air flame measured the reliability of the NN on temperature predictions against other experimental techniques (namely usual CARS fitting routines, Degenerate Four Wave Mixing on NO molecules, thermocouples). This test gave a deviation of only 3% at about 1600 °K.

An example of a typical experimental run is finally reported in figure 2 where the measured spectrum (dotted line) of the main vibrational line $v=1\rightarrow0$ of N_2 molecules with Raman shift of 2330 cm^{-1} is compared to the theoretical spectrum (continuous line) obtained from a fitting routine.

As we should expect the temperature extraction obtained by the NN gives a value that is very close to the temperature prediction determined by the numerical fitting. The comparison is based on a portion of the total spectral band only, precisely from the $v=1\rightarrow0$ line till the local maximum relative to the position of the hot band (i.e. $v=2\rightarrow1$ transition, still visible in figure 3). The reason of this choice relies on the sensitivity of this portion of the spectrum to temperature changes which affect the width of the main vibrational transition as well as the relative intensity between the $v=1\rightarrow0$ and $v=2\rightarrow1$ line. It would not add any further improvement to extend the analysis to either sides of the spectrum.

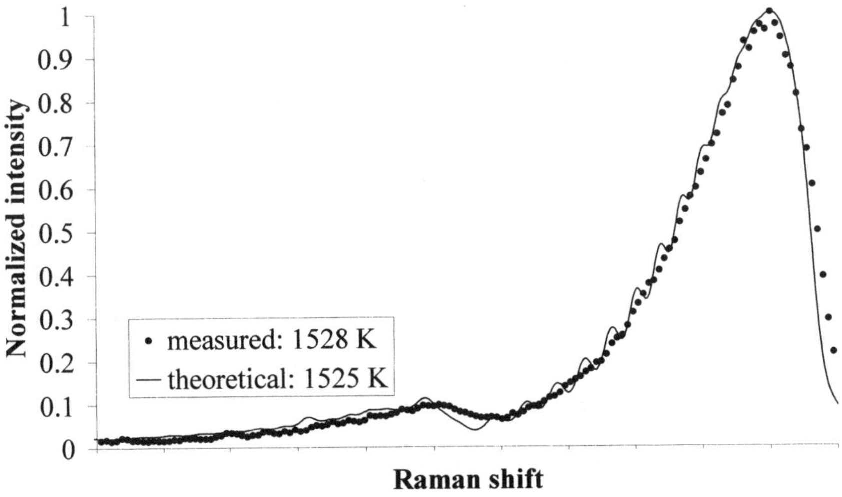

Fig.2 Comparison between numerical (continuous line) and experimental (dotted line) CARS spectra.

The additional effect of temperature in figure 2 is relative to the presence of rotational structures, which reveal themselves on the numerical fitting but not in the experimental data. This practical limitation was due to the insufficient spectral purity of the pump beams and it prevented us from discriminating the rotational Q-band in spite of the availability of a good slit function of the spectrograph. Nevertheless the interplay between the width of the main peak and the relative intensity of the two vibrational lines seems enough to get a good level of prediction on temperature from the NN scheme adopted in the present work.

The burner used in our experiment was a Bunsen type, drawn in fig.3, with a modified outlet nozzle, 137mm long, 21mm in diameter and 1.5mm of wall thickness. The length/diameter ratio, greater than 6, was chosen to obtain a rather even and, mainly, axially symmetrical outlet velocity profile. Moreover, this geometry allows an effective Methane/air premixing before ignition.

The burner was placed on an XYZ translator, step-motor operated and PC controlled, to precisely select the measurement point inside the flame and realise a complete thermal map. Spatial step was 20μm in the horizontal plane (XY) and 50μm along the vertical axis (Z). The three linear translators could assure a span width of 25cm along X axis and 45cm along Y-Z ones, well above the burner dimensions. Furthermore, consider that the maximum dimension of the measure volume, depending on the first lens focal length (15cm), was in the order of tenths of a millimetre. So, spatial resolutions and span widths of our translator system were quite adequate for our purposes.

The Bunsen burner was fed with Methane and a molar mix with 80% in Nitrogen and 20% in Oxygen. Gas flows, provided by two pressure cylinders, were measured through two Venturi meters placed on the ongoing piping to Bunsen.

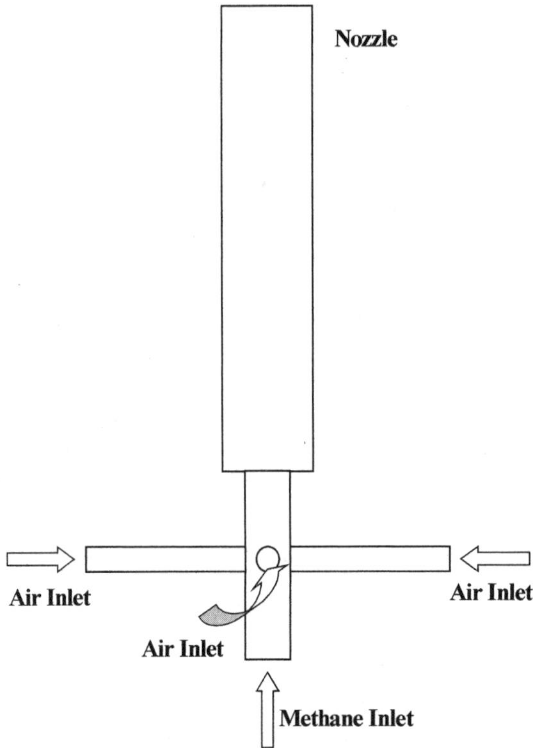

Fig.3 The modified Bunsen burner (simplified scheme)

2-2. CARS experimental results

Our measurement campaign was started with a preliminary run performed to define the best operative conditions for the experiment. For this first run the Bunsen premixed flame was fed with a Methane flow of $4.5 \cdot 10^{-5}$ kg/s and a Nitrogen/Oxygen mixture (80/20) flow of $3.8 \cdot 10^{-4}$ kg/s. In this way we obtained a nozzle outlet mean velocity (u_0) of 1.15 m/s, a Methane/air mixture density (ρ_0, at 25°C) of 1.07 kg/m^3 and a mixture kinematic viscosity (ν_0) of $1.61 \cdot 10^{-5}$ m^2/s. Such values led to the following physical parameters:

$$\phi = 2.1; \qquad Re = u_0\, L_0/\nu_0 = 1500; \qquad Gr = \frac{(T_1 - T_0)gL_0^3}{T_0\nu_0} = 2.28 \cdot 10^6; \qquad Gr/Re^2 = 1$$

Where $\lambda = 1/\phi$ is the air excess ratio (i.e. the actual air/CH$_4$ mass flow rate divided by the stoichiometric air/CH$_4$ mass flow rate); L_0 is a reference dimension (in our case the nozzle inside diameter); T_0 and T_1 are two reference temperatures (in our case the ambient temperature T_0 and the flame adiabatic temperature for Methane combustion in air around 2230 °K); g is the gravity acceleration; Re is the Reynolds number, an index of flow instability (turbulence); Gr is the Grashof number, representative of floating to viscous (or floating to inertial as Gr/Re^2) forces responsible of slow swirl structure formation, in the outer part of the flame, sized in the reference dimension (L_0) order.

Measurements were taken radially moving the check point from burner axis with 1 mm step. In each point from 16 to 21 temperature acquisitions were taken, through which arithmetic mean ($\overline{T}$) and standard deviation values were calculated:

$$\overline{T} = \frac{1}{N}\sum_k T_k \qquad devstd = \sqrt{\frac{\sum_k (T_k - \overline{T})}{N-1}} = \sqrt{\frac{N\sum_k T_k^2 - (\sum_k T_k)^2}{N(N-1)}}$$

In order to guarantee an effective signal to noise ratio and the statistical reliability of measurement, temperature values were estimated through a spectrum averaged on 50 laser shots (one temperature acquisition in 5 sec at 10Hz laser pulse frequency). Radial temperature profiles, so acquired at three different heights (1.5-11-21mm for r/D=0.07-0.52-1) above burner nozzle, showed high temperature gradients, particularly near the Bunsen wall (between 0.5 and 0.57 r/D), demonstrating the fine spatial resolution obtained in our CARS experiment. In order to evaluate the radial width of the thermal field, measurement was stopped when temperature evaluation fell below 400 °K. The results of the first measurement run are reported in fig. 4.

Based on these successful preliminary results, we chose the experimental conditions for the second measurement run, confirming most of the parameter values previously adopted. For this second run the Bunsen premixed flame was fed with a Methane flow of $2.5\cdot10^{-5}$ kg/s and a Nitrogen/Oxygen mixture (80/20) flow of $2.1\cdot10^{-4}$ kg/s. In this way we obtained a nozzle outlet mean velocity (u_0) of 0.64 m/s, a Methane/air mixture density (ρ_0, at 25°C) of 1.07 kg/m^3 and a mixture kinematic viscosity (v_0) of $1.61\cdot10^{-5}$ m^2/s. Such values led to the same adimensional physical parameters realised in the first run:

$$\phi = 2.1; \qquad Re = u_0 L_0/v_0 = 1500; \qquad Gr = \frac{(T_1 - T_0)gL_0^3}{T_0 v_0} = 2.28\cdot10^6; \qquad Gr/Re^2 = 1$$

In this case the Bunsen flame showed two zones of light emission (fig.5): an internal one, with a basically conical shape, nearly 2 nozzle diameter (40mm) high and an external one,

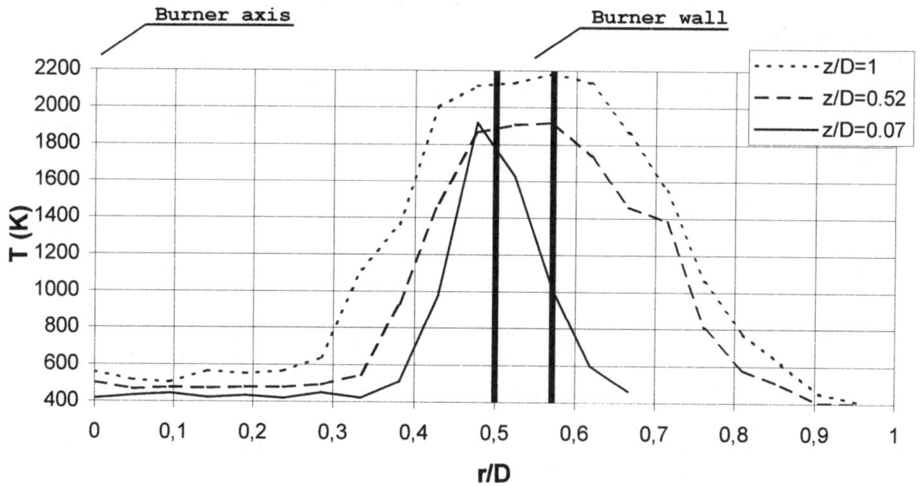

Fig.4 First measurement run results

dimmer than former, nearly 4 nozzle diameter (80mm) high. As the air/fuel mixture is very rich in fuel, only a part of the available Methane was burnt in the internal (premixed) flame, consuming all the Oxygen in the mixture. In the external (diffusive) flame all residual CH_4 was burnt, consuming O_2 from ambient air in nearly stoichiometric conditions. In this zone we observed the maximum temperature (2150°K), very close to the theoretical adiabatic value (around 2230°K).

In each point of the experimental grid 22 temperature evaluations (each based on 50 spectra acquisition) were taken to guarantee the statistical reliability of measurement. Data acquisition was repeated at seven different heights (0.5 to 84mm for r/D=0.02-0.5-1-1.5-2-3-4) above burner nozzle moving radially from Bunsen axis to 20mm (21 points, 1mm apart, for each temperature profile). Again, in each point we calculated:

- the temperature arithmetic mean value : $\overline{T} = \dfrac{1}{N}\sum_k T_k$

- the standard deviation : $std\ dev = \sqrt{\dfrac{\sum_k \left(T_k - \overline{T}\right)}{N-1}}$

- the relative standard deviation : $rel\ std\ dev = std\ dev / \overline{T} = \dfrac{1}{\overline{T}}\sqrt{\dfrac{\sum_k \left(T_k - \overline{T}\right)}{N-1}}$

In fig. 6, 7 and 8 are shown, respectively, the three maps of the mean thermal field, the *std dev* and the *rel std dev*. They all were symmetrically built from data taken on a single side of Bunsen axis. In fig. 6 three hot zones are evident: two anchored to the nozzle rim and the third on the higher part of the flame axis where we registered the maximum temperature of 2150 °K. Such a high temperature can be the result of CO oxidation to CO_2, following a quite slow but highly exothermal reaction, taking place at a certain distance from the flame basis. This reaction provides nearly 40% of the total heat produced in CH_4 combustion process.

In fig. 7 and 8 can be identified the zones with *std dev* highest values, namely where the maximum temperature fluctuations took place. In particular, such fluctuations are evident in the upper zone (z/D between 3 and 4), where the flame showed obvious oscillations even at first sight, and in the area around the nozzle rim, probably due to the presence of whirls released by the Bunsen wall.

Fig. 8 shows a middle region (between z/D=0.5 and 1.5) with *rel std dev* values higher than the surrounding zones. Temperature fluctuations in this region could be somehow related to the experimental acquisition process, more reliable to detect high rather than low temperatures.

3. LIF CH detection

LIF (Laser Induced Fluorescence) spectroscopic technique is one of the most reliable method to measure chemical concentrations in hostile environments [3]. Its main advantage lays in the linearity of the involved process, consisting in molecular relaxation among electronic levels (fluorescence), so that the relative cross section is several orders greater than in other physical phenomena involved in alternative techniques (e.g. Raman diffusion and non linear spectroscopy methods like CARS and DFWM). The LIF minimum sensitivity threshold is in the range of a few ppm (part per million). So, this is a useful

Fig.5 Bunsen flame

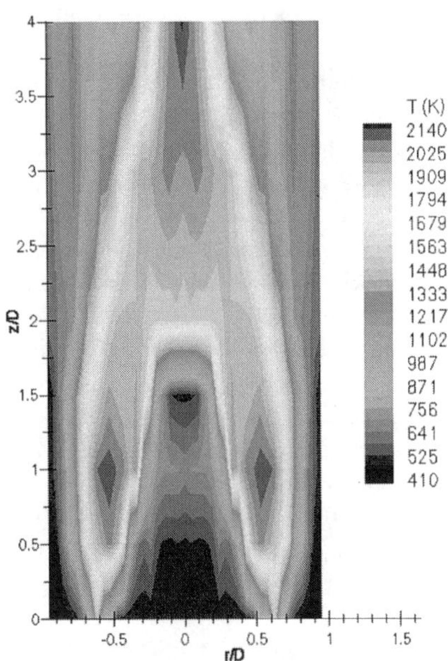

Fig.6 Flame thermal field

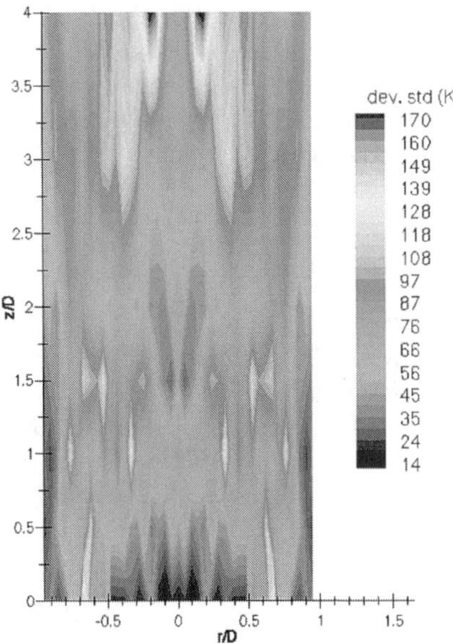

Fig.7 Standard Deviation

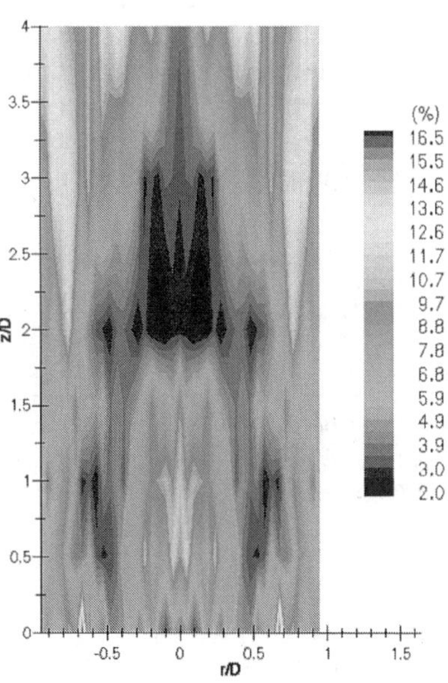

Fig.8 Relative Standard Deviation

technique to detect absolute or relative chemical concentrations, especially for species, like radicals, that
for their intrinsic nature are scarcely present in a flame with a very short life time (due to their high reactivity).

OH is the most commonly studied radical in combustion processes, both for its abundance in any hydrocarbon combustion reaction in air and for the deep available knowledge about its chemical structure and physical properties, including the emission spectrum in the visible and UV bands. Furthermore, it is also a good indicator of structure and thickness of the flame front, where its production is mainly concentrated due to high temperature.

As an alternative indicator to this aim, we can use also the CH radical, another extremely reactive transient molecule, which can be useful to quantify, in pollution studies, the "prompt NOx" production as well. The main advantage in trying to detect the CH radical instead of OH lays in the faster 2-body diffusion of CH produced on the flame front compared with the relatively slow 3-body process of OH diffusion. This means that CH remains highly localised in its formation zone, i.e. the flame front, and immediately disappears just outside due to fast recombination processes. In other words, detecting CH distribution definitely results in a better definition of flame front structure and morphology.

However, there is a relevant drawback too. In fact, from the point of view of LIF detection, the CH fluorescence signal is remarkably weaker than OH one due to different factors. Firstly, CH radical production in CH_4/air combustion is intrinsically well below OH formation rate. Secondly, the higher CH localisation reduces the fluorescence emitting zone. Finally, due to the lighter CH molecular weight inducing a higher collisional mobility, CH quenching processes are much more important than in OH case and produce a strong reduction of radicals effectively involved in CH fluorescence emission [3].

In our work LIF technique was applied in order to detect CH fluorescence signals in a diffusive flame in a laminar regime (Reynolds number below 100), ignited by feeding a standard Bunsen burner with liquid gas Propane in stoichiometric conditions.

3-1. LIF set-up

Our apparatus for LIF detection of CH radicals is shown in fig. 1. The main laser source is the same used for CARS experiment: a pulsed Nd:YAG laser, able to generate 1200mJ pulses at 1064nm (near IR) with 10Hz frequency, is duplicated by a non linear KDP crystal to provide 500 mJ pulses at 532nm (green) with a pulse time width around 10 ns in the classic Q-switched operating condition. The second harmonic generation (SHG) beam (532nm) is then recombined with the fundamental frequency beam (1064nm) in another non linear KDP crystal to produce a third harmonic generation (THG) pulsed beam at 355nm (near UV) with a 35 mJ energy. Finally, this UV laser beam is used to pump a tunable dye laser LDL 105/205 that, through an internal diffraction grating, allows to select the frequency suited to excite CH molecules. To this aim we used the transition system A-X, which is highly diagonal, i.e. there are no roto-vibrational couplings during transients due to laser excitation.

This fact, along with the necessity to avoid strong signal interferences due to elastic scattering, requires the fluorescence signal, derived by excitation of (0,0) band, be acquired through a suited filter, even accepting a signal intensity weakening. Alternatively, it could be possible to acquire the fluorescence signal from band (0,1)

EXPERIMENTAL SET-UP (Linear LIF on CH)

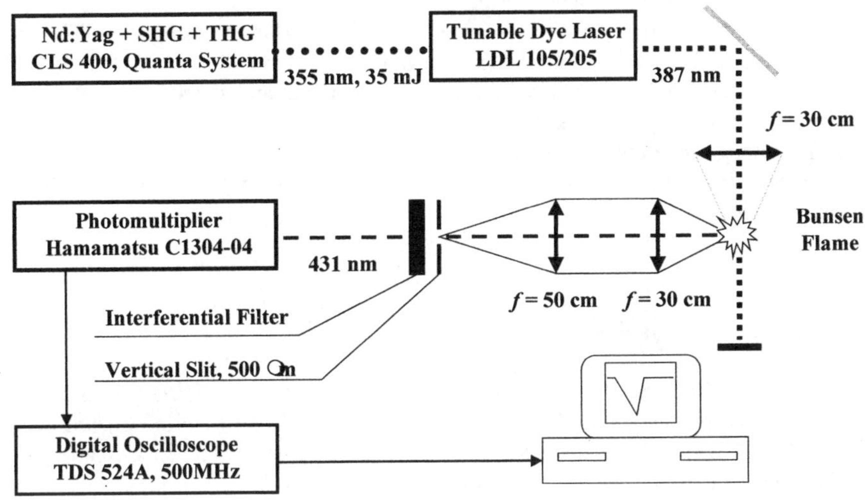

Fig.9 LIF set-up

emission, but this would require a high energy laser source with rather long pulse width. For our work, due to a minimum in dye laser efficiency placed around 416 nm, i.e. rather close to 431.5 nm where is placed one of the most intense CH excitation band, we chose to excite the A-X (1,0) band (see fig.10) at a wavelength of 387 nm and to detect the fluorescence signal at 431 nm, according to the scheme already adopted in [2].

After all these wavelength controls, the excitation beam at 387nm was directed to the burner flame by a mirror and then focused on the measurement zone through an UV Silica lens, 30 cm in focal length. In this way we obtained a reference laser beam of 30 μm in diameter across the whole burner nozzle. A micrometric support system with a resolution of 10 μm was used to accurately focus the lens. LIF signal was acquired perpendicularly to the excitation beam and detected by a Hamamatsu C1304-04 photomultiplier. Two more lenses, respectively 30 and 50 cm in focal length, allowed to focus the fluorescence signal on a vertical slit, 500 μm wide, placed in front of the photosensor to select a thin

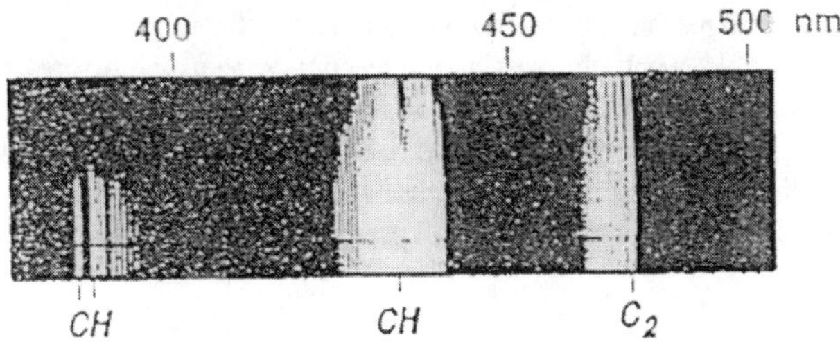

Fig.10 CH excitation spectrum

flame section. Moreover, an interferential filter with CW around 431 nm was inserted in the optical lay out in order to reduce to a minimum the noise signals due to Rayleigh or Mie-Lorentz diffusion phenomena inside the flame.

The fluorescence signal was inspected on a 500 MHz digital oscilloscope TDS 524A, integrating for a 200 ns time interval (synchronised with fluorescence starting, just after a spurious peak due to the above cited diffusion phenomena) and averaging the responses to nearly 300 pulses. The integration was necessary to guarantee a measurement linearity with the fluorescence energy, i.e. with the CH concentration. Before each measurement session, the linear LIF regime was regularly tested through calibrated neutral density filters.

3-2. LIF experimental results

Fig. 11 shows LIF profiles of CH molecule concentration for the six different heights (referred to the burner nozzle diameter D=10mm, h/D=0.9-1.2-1.5-1.8-2.1-2.4) from the Bunsen outlet we tested. In abscissa the distance from Bunsen axis (up to r/D=+/-0.45) is reported while in ordinate the fluorescence signal relative intensity is expressed in arbitrary units.

At the lowest height (h/D=0.9) CH is present in small quantity, as oxidation of heavy hydro-carbons contained in our mixing, especially Propane and Butane, is a rather slow reaction and needs some time, and then space, to be completed. The flame front is very thin, with the two sides nearly 6mm apart.

Rising the height to h/D=1.2, CH production rate remarkably increases, as clearly showed by the fluorescence signal rising, and the flame width falls to 5.5mm as the two sides of the reaction front start approaching each other. Concentration peaks are broadening due to thermal diffusion of the CH, produced in the reaction front, towards the surrounding zones. This aspect results in a relative uncertainty in flame front thickness determination, even if referring to CH concentration peak thickness at half maximum leads to values in agreement with the related available literature data. Based on this criterion, it can be observed that the thickness of a diffusion flame front is not so precisely detectable like in a premixed flame.

The same considerations can be made for the next height (h/D=1.5) where the peak distance falls to 4.5mm.

The concentration profile at h/D=1.8 seems to nearly pass through the reaction zone only, with a minimum value not so far from the maximum one. The non-perfect symmetry of the profile was due to a temporary decrease of laser beam intensity during measurement and must not be considered of any physical evidence. In fact, in our operative conditions, the flame was highly stable and there was no reason to expect a non-symmetric behaviour.

At h/D=2.1 the profile is entirely crossing the reaction zone as the two reaction fronts are now indistinguishable and fused together. The overall flame shape is going to shrink.

Finally at h/D=2.4, the maximum height we tested, CH concentration is highly reduced as the profile is above the reaction zone and is crossing the combustion product zone, where only a small quantity of CH not yet recombined is still present due to molecular diffusion.

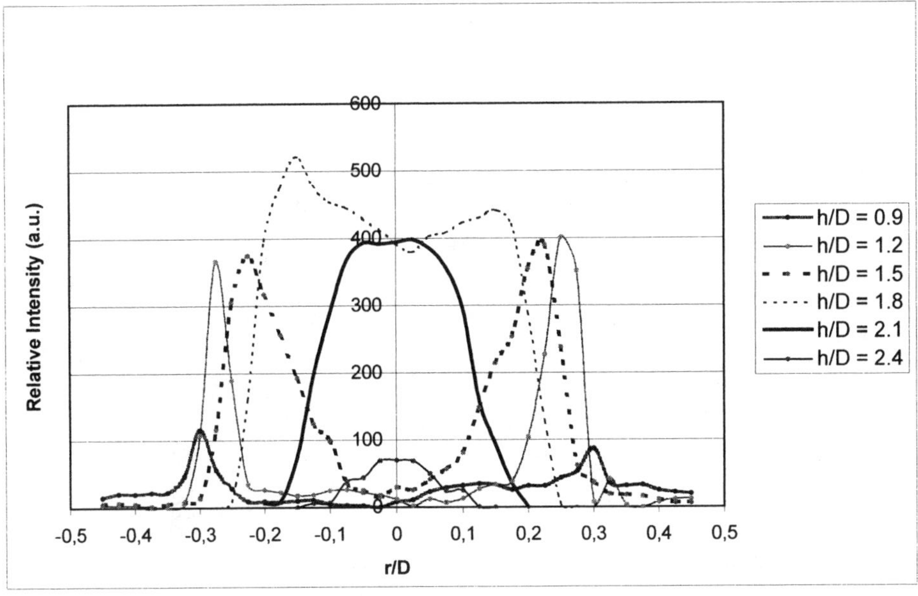

Fig.11 CH LIF profiles

It is interesting to note as the CH concentration gradients are steeper on the external side of the flame than on the internal one. This should be the result of a larger CH diffusion towards the non-reacted fuel inside the flame rather than the ambient air (comburent) and reaction products. From the point of view of pollutant emission, this is an advantage because not all the produced CH can react with the Nitrogen in ambient air (outside the flame) to generate "prompt" NOx as follows:

$$CH + N_2 \rightarrow N + HCN$$
$$N + O_2 \rightarrow NO + O$$

For all LIF signals we measured a maximum relative intensity error within 10%, thanks to the steadiness of signals and the minimum flame fluctuations.

Fig. 12 shows the reaction front profiles (internal and external). Such profiles were built according the the same criterion previously defined, i.e. interpolating the points identified, for each investigated height, on the abscissa corresponding to 50% of the maximum LIF signal intensity.

Fig. 13 reports the same CH concentration profiles but in a 3D plot for better viewing the flame front morphology.

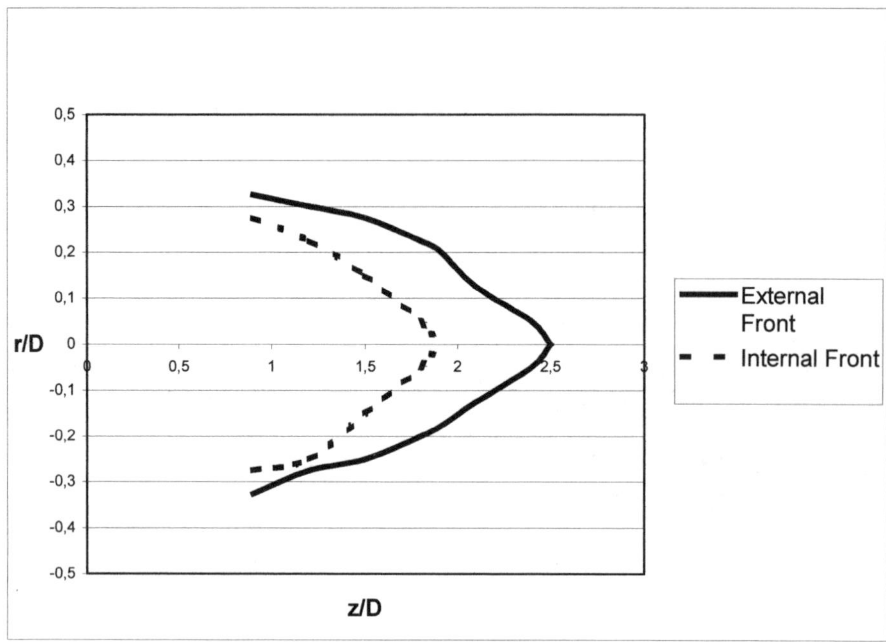

Fig.12 reaction front geometry

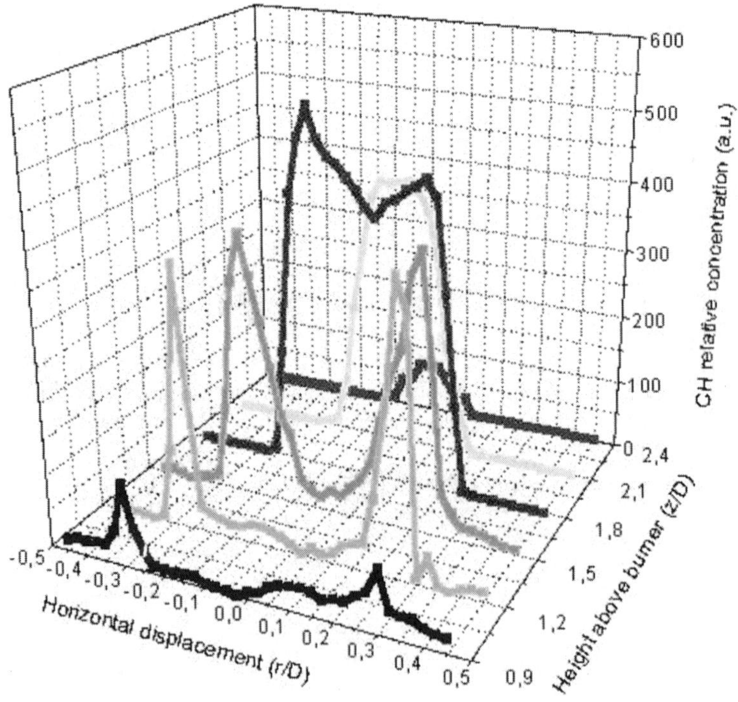

Fig.13 3D CH concentration profiles

4. Conclusive remarks

The experiments described above allowed to investigate the combustion process in different reference conditions:

- diffusive and premixed flames,
- laminar and turbulent flows,
- light (Methane CH_4) and heavy (Propane/Butane C_3H_8/C_4H_{10}) hydrocarbon combustion in air, tracing thermal maps and detecting flame fronts.

All the results obtained on Bunsen burners constitute a reference basis for further interpretative investigations on the more complex combustion phenomenology applied to industrial combustors, especially the ones destined to gas turbine. Such an application presently represents one of the main field of interest for our Combustion Technology Section in ENEA-Casaccia Research Centre (Rome, Italy). In this respect, special efforts are being aimed to studies related to pollutant emission (particularly NOx) from industrial burners in order to define practical solutions to effectively reduce environmental impact.

In our experimental facility COMET (Combustion Experimental Test), some measurement campaigns were already performed on industrial burner prototypes using CARS [5] and DFWM techniques. In the next future we are planning to apply polarization techniques as an alternative way to detect temperature and, mainly, pollutant chemical concentrations in industrial combustor flames.

References

[1]　R.Fantoni et al., J. Raman Spectroscopy 31, 697-701 (2000).

[2]　P.H. Paul and J.E. Dec, *Imaging of reaction zones in hydrocarbon-air flames by use of planar laser-induced fluorescence of CH*, Opt. Lett. 19, 998 (1994).

[3]　A.C. Eckbreth, *Laser Diagnostics for combustion temperature and species*, Gordon and Breach Publishers, 1996.

[4]　H.J. Van der Steen and J.D. Black, *Temperature analysis of coherent anti-Stokes Raman spectra using a neural network approach*, Neural Comput. & Applic. 5, pp.248-257, 1997.

[5]　M.Marrocco et al., *CARS spectroscopy on a dry low NOx methane/air burner*, Proc. SPIE Vol. 4201, pp.82-90, 2000.

GNSR 2001
G. Messina and S. Santangelo (Eds.)
IOS Press, 2002

The G-band frequency-position in Raman spectra of amorphous carbon-nitride based materials: correlation with the chemical composition

G.Messina, S.Santangelo*

INFM, Dipartimento di Meccanica e Materiali, Facoltà di Ingegneria, Università "Mediterranea", località Feo di Vito, 89060 Reggio Calabria, Italy
** E-mail: santange@ing.unirc.it tel.: +39.(0)965.875305; fax:+39.(0)965.875201*

G.Fanchini, A.Tagliaferro

INFM, Dipartimento di Fisica, Politecnico di Torino, corso Duca degli Abruzzi 24, 10129 Torino, Italy

Abstract In this work, demonstration is given that a proper analysis of the Raman spectra of amorphous carbon-nitride based alloys can give information on the carbon- and nitrogen- distributions of the material. The existence of a correlation between the frequency-position of the G-band and the relative C content in such films is evidenced. By interpolating the experimental data, an empirical relationship is derived, allowing the quantitative estimation of the C content in films deposited at room-temperature, from their Raman spectra. The reasons for the correlation to exist are discussed in detail with reference to the example of amorphous hydrogenated carbon-nitrides [a-CN(:H)] and understood in terms of the film local-structure and Raman G-mode origin. On the basis of the indications emerging from the complementary infrared (IR) a-CN(:H) film characterisation, the G-band frequency-position is then hypothesised to be further correlated with the relative N content of amorphous carbon-nitride based alloys. Such an hypothesis is actually confirmed by the experimental data and an empirically-derived relationship is proposed for the quantitative estimation of the N content in room-temperature grown amorphous carbon-nitride based alloys. The role of the deposition temperature in determining the film relative C- and N- contents and resulting bonding characteristics is finally pointed out.

1. Introduction

Raman spectroscopy is commonly used for non-destructively characterising carbon-based materials. The shape of the spectrum strongly depends on the bond symmetry. In large diamond single-crystals, a sharp line located at 1332 cm^{-1} is detected, associated to the scattering from zone-centre triply-degenerate T_{2g} optical phonon (k=0) originating from tetrahedrally coordinated (sp^3) C atoms *[1]*. In amorphous materials, two broad partly-overlapping bands, approximately peaking at 1360 and 1580 cm^{-1}, dominate the region between 1000 and 1700 cm^{-1} of the Raman spectra. The former (D-band) is associated with A_{1g} disorder-allowed zone-edge modes of micro-crystalline graphite, becoming active due to a lack of long-range order and to the subsequent effects of grain boundaries and impurities *[2]*. The latter (G-band) corresponds to the optically allowed E_{2g} zone-centre in-plane symmetric C-C stretching modes of trigonally bonded (sp^2) C atoms *[3]*. Since frequency-position, width and intensity of these bands reflect the C bonding nature in the films, studying their evolution is the most widely utilised approach for gaining insight on the physical properties related to the bonding characteristics of disordered carbon-based thin films.

The quantitative analysis of the Raman spectra allows deriving a set of descriptive parameters, suitable to monitor the diverse film characteristics achieved under different growth conditions. The interest, however, is usually mainly centred on the frequency position of the G-band, assumed as a qualitative indicator of the "graphitisation degree" of the film: the G-band upshift is interpreted as a diminishing of the sp^3:sp^2 C bonding ratio *[2,4]*. In addition, the D/G intensity ratio is also regarded as a meaningful parameter: its variations are currently utilised to gain information about the changes in the average size of the graphitic clusters *[2,5]*.

In this preliminary work, these parameters are shown to contain information also about the material chemical-composition. The existence is demonstrated of a correlation between relative C- and N- contents and both G-band frequency-position and D/G intensity ratio in the Raman spectra of amorphous carbon-nitride based alloys. The reasons for the correlation to exist are discussed in detail with reference to the example of 100°C sputter-deposited *a*-CN(:H) films and understood in the light of the results of their complementary Raman and IR film characterisation. By interpolating the experimental data, empirical relationships are derived, allowing the quantitative estimation of the C- and N- contents in films deposited at room-temperature, from their Raman spectra. The effect of the deposition temperature on the relative C- and N- concentrations achieved and the resulting film bonding characteristics is finally discussed.

2. Sample considered

The literature samples considered include both room-temperature (RT) deposited and annealed films. Their characteristics are reported on refs. *[2,6-12]*. In addition, a set of *a*-CN(:H) films (probe-samples) was grown at 100°C in a conventional 13.56 MHz r.f. diode sputtering system. The low-gap sp^2-rich films attained were compositionally characterised by Elastic Recoil Detection Analysis (ERDA) and Rutherford Back-Scattering (RBS). Their structural and vibrational properties were investigated by Raman and IR spectroscopies.

2.1 Deposition and characterisation of a-CN(:H) films

Thin *a*-CN(:H) films were sputter-deposited on (100) *c*-silicon substrates in Ar+He+N$_2$+H$_2$ atmosphere. The 99.999% purity graphite target was placed at 25 mm from the grounded electrode holding the substrate. No external bias was used. The r.f. power ranged between 180 and 400 W and the total pressure between 2.7 and 5.1 Pa. He and Ar flow-rates were 30

Tab.1 Growth conditions of the investigated a-CN(:H) samples: W_{rf} and p_{tot} respectively indicate the r.f. power and the total pressure. Φ_{N_2} and Φ_{H_2} denote the flow rates of the reactive gases (N_2 and H_2).

Tab.2 Frequency positions (ω_D and ω_G) and widths (σ_D and σ_G) of the D- and G- bands of the investigated a-CN(:H) samples.

Film	W_{rf} (W)	p_{tot} (Pa)	Φ_{N_2} (sccm)	Φ_{H_2} (sccm)
#A	300	2.7	20.0	0.0
#B	300	2.7	10.0	7.0
#C	400	2.7	10.0	7.0
#D	300	2.7	20.0	7.0
#E	300	2.8	10.0	5.0
#F	300	2.8	7.0	5.0
#G	300	3.3	10.0	7.0
#H	300	5.1	4.2	3.0
#I	300	5.1	8.4	3.0
#L	220	5.1	4.2	3.0
#M	200	4.0	4.2	2.2
#N	180	5.1	3.0	2.2

Film	ω_D (cm^{-1})	σ_D (cm^{-1})	ω_G (cm^{-1})	σ_G (cm^{-1})
#A	1363	302	1563	187
#B	1365	240	1574	143
#C	1371	258	1573	148
#D	1368	260	1575	140
#E	1372	318	1575	148
#F	1375	313	1572	147
#G	1376	293	1574	146
#H	1394	371	1563	149
#I	1395	348	1575	123
#L	1393	383	1570	128
#M	1387	365	1562	148
#N	1396	394	1564	140

and 70 sccm (standard cubic centimetres per minute), respectively; N_2 and H_2 flow-rates were varied between 3 and 20 sccm and between 2 and 7 sccm, respectively. The growth conditions are listed in detail in Tab.1.

The Raman spectra were recorded, at RT, by using a 2.41 eV incident light. Although interest was mainly focused onto the 1000-1700 cm^{-1} region, dominated by carbon D- and G- bands, in order to obtain more reliable information about the shape evolution of these spectral features, the analysis was extended to a wider range (200-3600 cm^{-1}). Gaussian bands were used to reproduce the spectra: frequency positions, widths (FWHM) and intensities were chosen by a least-square best-fit method. The results relative to D- and G- bands are reported in Tab.2. The IR absorption spectra were recorded in the 450-4000 cm^{-1} range by using a bare substrate as a reference. Further details about film deposition and characterisation are reported elsewhere *[13,14]*. The film H-content, as determined by ERDA measurements, ranges between 9% (samples #M and #N) and 20% (sample #D). The measured C- and N- contents (x_C and x_N, respectively) are reported in Tab.3.

3. Structural and vibrational properties of a-CN(:H) films

Figure 1 shows the shape evolution of the Raman spectra, reflecting the structural and bonding modifications related to the changing film stoichiometry (see *Sect.4.1*), produced by the different growth conditions. However, due to the simultaneous variation of the deposition parameters, the quantitative analysis of the spectra does not allow establishing straight away the role played by each of the deposition parameters (Tab.1) in determining the film characteristics described by the obtained frequency-positions (ω_D and ω_G) and widths (σ_D and σ_G) of the D- and G- bands (Tab.2). In such a situation, in fact, properly-generated dimensionless combinations of the growth-process variables are, rather,

Tab.3 Comparison between the measured (x_C and x_N) and the calculated (x_C^{RT} and x_N^{RT}) C- and N-contents of the 100°C deposited a-CN(:H) films. The different growth temperature is the origin of the discrepancy between measured and calculated concentrations: x_C^{RT} and x_N^{RT}, in fact, are obtained from empirical relationships derived by considering RT-grown samples.

Film	x_C (%)	x_N (%)	x_C^{RT} (%)	x_N^{RT} (%)
#A	---	---	68.2	20.8
#B	71.4	7.1	57.9	37.0
#C	71.4	7.1	58.7	36.2
#D	69.9	7.0	53.9	40.4
#E	73.5	7.4	54.7	39.9
#F	74.6	7.5	60.2	34.4
#G	74.6	7.5	56.9	38.0
#H	---	---	68.4	21.2
#I	75.8	7.6	54.5	40.0
#L	73.0	7.3	63.1	30.3
#M	79.4	7.9	69.3	19.6
#N	79.4	7.9	68.0	21.9

able to effectively account for the modifications of the film properties observed with changing the deposition conditions. A detailed discussion on this argument is reported elsewhere [15].

In the following, the attention will be mainly focused on the frequency position of the G-band, usually regarded as a qualitative indicator of the "graphitisation degree" in the disordered carbon-based materials [2,4], and, in minor extent, on the D/G intensity ratio, generally utilised to gain information about the average size of the graphitic clusters [2,5].

The IR absorption spectra of some of the investigated films are shown in Fig.2. On one hand, the absence of the Csp^2H_x and Csp^1H_y contributions in the 2750-3300 cm^{-1} region of the Csp^nH_z stretching modes ('CH') signals that hydrogen preferentially binds with Csp^3 atoms. On the other hand, the analysis of the 2000-2250 cm^{-1} region of the stretching modes of the terminal and non-terminal groups involving sp-hybridised N and C atoms ('CN') indicates that nitrogen is bonded mainly with Csp^1 atoms. These findings, together with the negligible intensity of the contribution originating from N_xH_y stretching modes ('NH') at ~3200 cm^{-1}, superimposed to the peaks related to the oxygen contamination around 3500 cm^{-1}, suggests that N and H atoms are substantially located in different regions (sp^2 and sp^3, respectively) of the C skeleton of the a-CN(:H) films. The analysis of the 'CN' region, by comparison with the case of a-CN (sample #A), further evidences the indirect role played by the H addition in promoting the formation of Csp_xNsp_y terminal groups, back-bonding the Csp^2 clusters. The long-range polarisation, originated into the Csp^2 phase by the sp^1-bonded nitrogenated groups vibrating into the 'CN' region, indirectly activates the 1300-1600 cm^{-1} vibrations modes of the Csp^2 islands ('CC') [16].

4. Correlation between G-peak position and chemical composition

4.1 Relationship with the measured C-content of the film

Aiming to clearly establishing whether any correlation does exist between descriptive parameters derived from the quantitative analysis of the Raman spectra (see *Sect.2.1*) and chemical composition of the film, several sets of literature samples, grown by different techniques, are considered (namely, amorphous carbon-nitride based alloys, deposited in C_2H_2-CH_3NH_2 [8], CH_4-NH_3 [7] and N_2-O_2 atmosphere [6], deposited in N_2 and annealed [9], and, in addition, sp^2-rich a-C films [2]). As a first result, by examining the relative fitting parameters, no immediate and evident relationship with the material chemical composition is found, excepted for the G-band frequency position, that, contrarily appears to be strongly correlated with the film stoichiometry. This is demonstrated in Fig.3, where

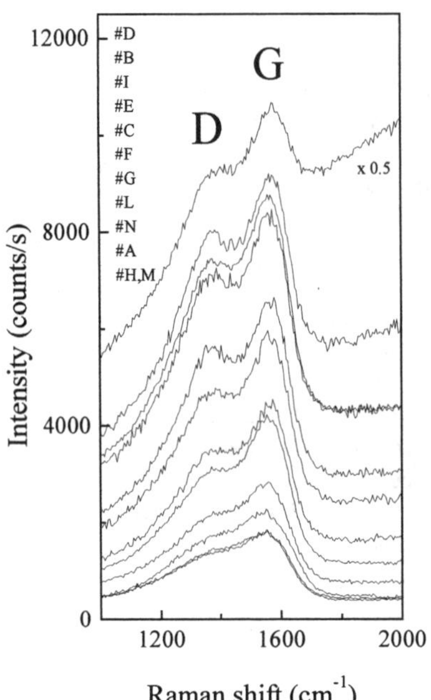

Fig.1 Evolution of D-and G- bands in Raman spectra of the *a*-CN(:H) films investigated. The samples are listed, from the top, in decreasing order of G-band maximum intensity. Spectra of samples #H and #M are nearly superimposed.

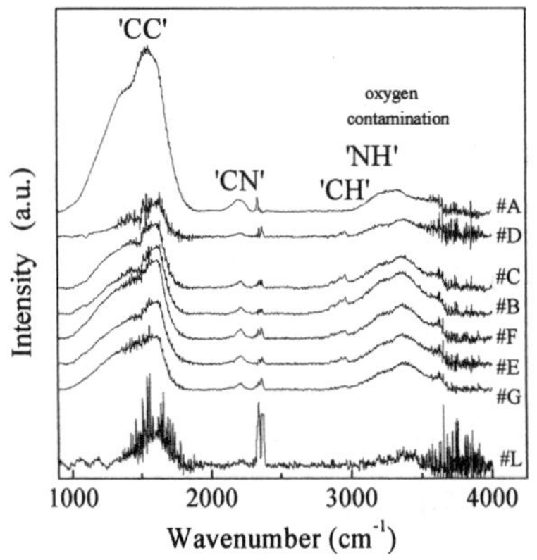

Fig.2 IR absorption spectra of some of the *a*-CN(:H) films investigated after base-line subtraction. The main spectral features are indicated. Some difference is detected in the spectrum of sample #A, the non-hydrogenated film (*a*-CN).

the ω_G values reported in literature *[2,6-9]* are plotted as a function of the measured film C-content: the ω_G representative-points are found to lie close to a single curve.

With decreasing the C-content, the G-band gradually shifts upwards, indicating the progressive sp^3:sp^2 bonding ratio diminishing *[2,4]*. In samples grown at room-temperature *[6-8]*, as the C-content (x_C) decreases from $^{max}x_C = 100\%$ down to $^{min}x_C = 45\%$, ω_G correspondingly varies from the minimum- [$\omega_G(^{max}x_C) = 1539$ cm^{-1}, peculiar of RT-grown sp^2 rich *a*-C films *[2]*] to the saturation- [$\omega_G(^{min}x_C) = 1575$ cm^{-1}] value. In annealed samples *[9]*, the G-band always remains around the saturation position, suggesting that, even when higher C-contents are recovered by N- or H-desorption during annealing, no significant change is correspondingly introduced in the sp^3:sp^2 bonding ratio, as described by the G-band position in Raman spectra.

All the elements in the films are likely to exhibit a coordination number lower than C. Nsp^1 is 1- or 2- coordinated *[17]*, Nsp^2, if present, is 2- or in very few cases 3-coordinated *[18]*, H is 1-coordinated. Thus, the ω_G decrease at a higher C-content may be interpreted in terms of a general softening of the G stretching mode when the increased average coordination-number in the film bears to more relevant internal constraints. Therefore, the G-band position is concluded to represent a meaningful parameter for the C-content estimation.

The data of Fig.3 can be interpolated by means of the expression

$$\omega_G - \omega_G(^{max}x_C) = [\omega_G(^{min}x_C) - \omega_G(^{max}x_C)] \cdot$$
$$\cdot \{1 - exp[-(100 - x_C)/A]\}, \qquad (1)$$

where x_C and ω_G represent C-content and frequency-position of the G-band in the Raman spectrum of the film; $\omega_G(^{max}x_C)$ and $\omega_G(^{min}x_C)$ denote the (minimum and saturation) ω_G values, attained in correspondence of the (maximum and minimum) C-contents, respectively; and A is a constant. Hence,

by inverting eq.(1) and utilising the experimentally-determined $A = 30\%$ value, an empirical relationship is attained allowing estimating the film C-content from the analysis

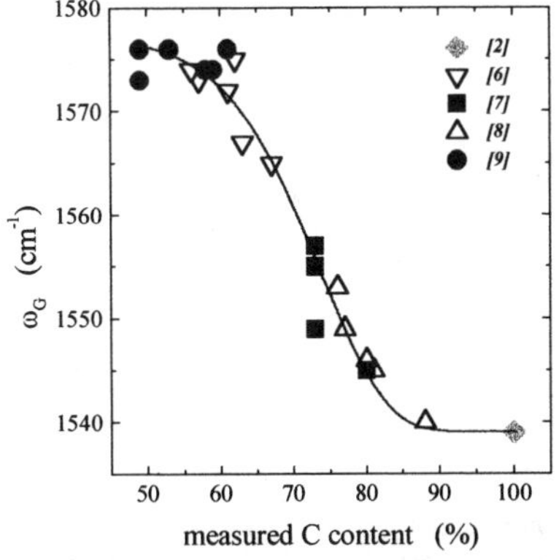
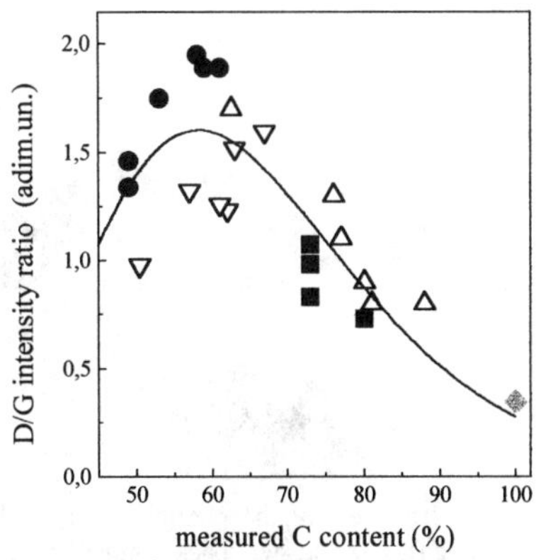

Fig.3 Relationship between frequency-position (ω_G) of the G-band in Raman spectra and measured C content in amorphous carbon-nitrides based alloys. Data relative to the considered literature samples are shown (the corresponding references are therewith indicated). The solid line represents the fitting curve derived by using eq.(1).

Fig.4 Relationship between D/G intensity ratio in Raman spectra and measured C content in amorphous carbon-nitrides based alloys. Data relative to the considered literature samples are shown (the symbols are the same as in Fig.3). The solid line is drawn for visual help.

of its Raman spectrum.

These findings further strengthen the meaningfulness of the G-band position: in disordered carbon-based materials, ω_G is a reliable indicator of the sp^3:sp^2 bonding ratio, as determined by the C-content achieved under the considered growth conditions.

Contrarily, a much more intricate role is likely to be played by the other Raman parameters. An exception is, in some extent, represented by the D/G intensity ratio. This is shown in Fig.4, where the D/G values relative to the same literature samples already considered in this Section [2,6-9] are plotted as a function of the measured film C-content. Although, indeed, the data are widespread, they roughly draw a single curve, suggesting that the C-content variation has some influence on determining the changes in the average size of the graphitic clusters formed within the film. The picture which seems to emerge from the comparative analysis of the data reported in Fig.3 and Fig.4 is that, at the highest C contents, very small graphitic clusters are formed (Fig.4) which are likely to be embedded in the network of the sp^3 bonded C atoms [19-21]. As the decrease of the film C content causes a diminishing of the sp^3:sp^2 bonding ratio (Fig.3), the mean dimension of the Csp^2 cluster islands monotonically increases until a critical C concentration is reached. Below 55%, the grown films are so C-poor as a trend inversion occurs.

4.2 Relationship between IR-active sp^1 nitrogenated groups and sp^2 C clusters

A detailed discussion of the results of the IR analysis carried out on the investigated a-CN(:H) films is reported elsewhere [16]. Briefly, the most intense bands, dominating the region between 1300 and 1600 cm^{-1} of the spectra, are necessarily related with C groups (or, at least, with mainly-containing-C groups), because they i) do not shift when isotopic substitutions of ^{14}N with ^{15}N [18,22] are made, and ii) are detected also in samples containing only C and N, with no [18,23] (or very little [9]) H. This is the reason why these bands, currently improperly-indicated as 'D$_{IR}$' and 'G$_{IR}$' even if they cannot be straightly

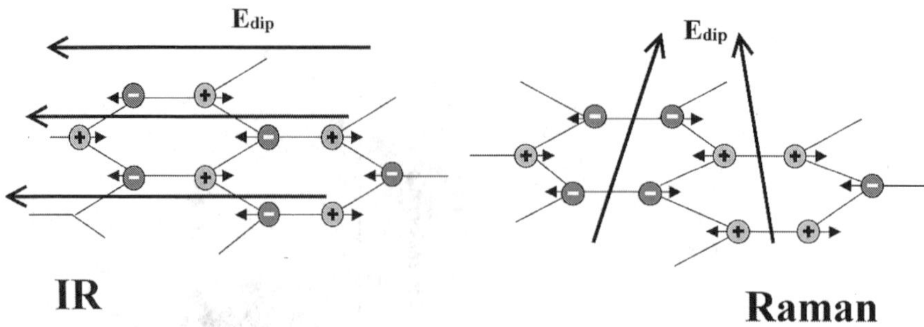

Fig.5 Schematic view of the charge transfer effects onto longitudinally-polarised (left) and transversely-polarised (right) Csp^2 clusters back-bonded to Csp_xNsp_y groups. The polarisation field (generated by external sp^1 nitrogenated groups) is sketched in terms of mean dipolar field (E_{dip}).

assimilated (as f.i. in *[18,23]*) to the Raman-active vibrations, are here simply referred as C=C stretching modes ('CC'). Their IR optical density ($^{opt}N_{\cdot CC'}$) has been recently shown to depend on the IR optical density of Csp_xNsp_y terminal groups vibrating at ~2200 cm^{-1} ($^{opt}N_{\cdot CN'}$) *[16]*. Such a dependence has been understood by assuming that the field, locally generated by nitrogenated sp^1 groups (behaving like dipoles *[17]*), polarises the π-electrons of Csp^2 clusters, which they are back-bonded to, and, as a result, 'CC' modes in the clusters become IR active. Hence, $^{opt}N_{\cdot CC'}$ is proportional to the total polarisation of the sp^2 phase and the saturating trend of the $^{opt}N_{\cdot CC'}$ vs $^{opt}N_{\cdot CN'}$ curve can be interpreted as indicative of a full polarisation of the sp^2 phase (in other words, all the clusters are back-bonded to Csp_xNsp_y groups) *[16]*. As back-bonded clusters 'feel' a different mean local field, depending on their longitudinal or transversal direction of polarisation (Fig.5) with respect to the dipolar mean-field induced by the Csp_xNsp_y groups, their polarisation involves a degeneracy-removal among the G-modes *[16]*. The resulting downshift of transversal symmetric Raman-active modes

$$\omega_{Ram} = \omega_G [1 - \beta_{ion}/(1 - \beta_{el})]^{1/2}$$

and upshift of longitudinal asymmetric IR-active modes

$$\omega_{IR} = \omega_G [1 + 2\beta_{ion}/(1 + 2\beta_{ion})]^{1/2},$$

with respect to the unperturbed frequency-position ω_G=1592 cm^{-1}, has been shown to be controlled by the electronic (β_{el}) and ionic (β_{ion}) polarisabilities *[16]*.

4.3 Relationship with the measured N-content of the film

On the basis of the above findings, it can be reasonably envisaged that the G-band frequency-position (i.e. the C=C bond strength), controlled by the material C-content, might, in some extent, be influenced also by the amount of the N incorporated within the film. In order to clearly establish whether any correlation between G-band position and N-content does actually exist, the data relative to a very large variety of literature samples, grown by quite different technique, are examined. Besides to those *[2,6-9]* utilised in *Sect.4.1*, additional samples are considered (namely, amorphous carbon-nitride based alloys, deposited in Ar-N$_2$ *[10]*, CH$_4$-N$_2$ *[11]* and C$_6$H$_{12}$-N$_2$ atmosphere *[12]*). As in the previous case, the G-band frequency position, diversely from to all the remaining fitting

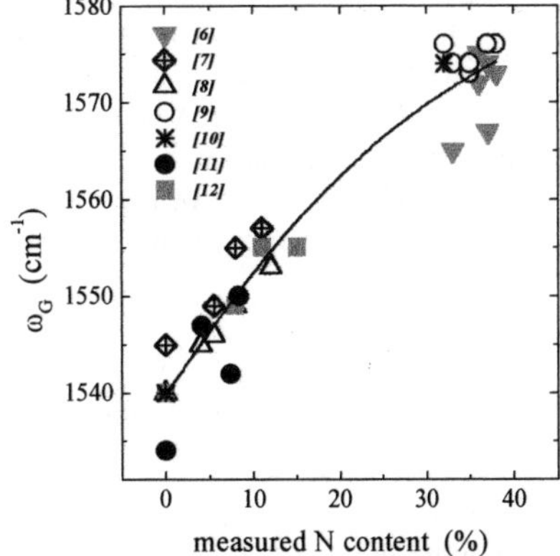

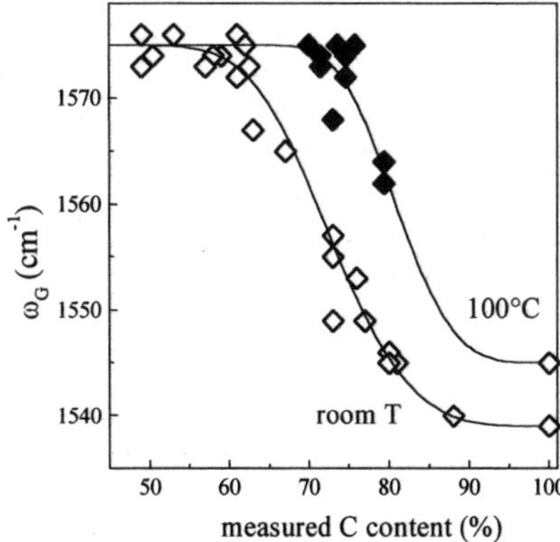

Fig.6 Relationship between frequency-position (ω_G) of the G-band in Raman spectra and measured N content in amorphous carbon-nitrides based alloys. Data relative to the considered literature samples are shown (the corresponding references are therewith indicated). The solid line represents the fitting curve derived by using eq.(2).

Fig.7 Dependence of ω_G on the measured C content at different deposition temperatures. The solid lines represent the fitting curves derived by using eq.(1). The shown data refer to literature samples ($\lozenge$) and presently-investigated *a*-CN(:H) films ($\blacklozenge$).

parameters, appears to be correlated with the measured N-content of the film. This is shown in Fig.6, where the literature ω_G values are plotted as a function of the measured film N-content: a single curve is drawn by the ω_G representative-points.

With increasing the N-content, the G-band gradually upshifts, indicating the progressive sp^3:sp^2 bonding ratio diminishing. In samples deposited at room-temperature *[6-8,10-12]*, as the N-content (x_N) increases from $^{min}x_N=0\%$ up to $^{max}x_C\approx40\%$, ω_G correspondingly varies from the minimum- [$\omega_G(^{min}x_N)$, obviously coincident with $\omega_G(^{max}x_C)$] to the saturation- [$\omega_G(^{max}x_N)\cong\omega_G(^{min}x_C)$] value; meanwhile in annealed samples *[9]*, the G-band always remains around the saturation position.

The same considerations of *Sect. 4.1* about the lowering of the film mean coordination-number, caused by the N-incorporation, apply in this case.

The data of Fig.6 can be interpolated by means of the expression

$$\omega_G - \omega_G(^{min}x_N) = [\omega_G(^{max}x_N) - \omega_G(^{min}x_N)]\cdot\{1-exp[-(x_N/B)^a]\}, \tag{2}$$

where x_N and ω_G represent N-content and frequency-position of the G-band in the Raman spectrum of the film; $\omega_G(^{min}x_N)$ and $\omega_G(^{max}x_N)$ denote the (minimum and saturation) ω_G values, attained in correspondence of the (minimum and maximum) N-contents, respectively; and B and a are constants. Therefore, by inverting eq.(2) and utilising the experimentally-determined $B=17\%$ and $a=1.25$ values, an empirical relationship is attained allowing estimating the film N-concentration from the analysis of its Raman spectrum.

For the sake of completeness, it has to be pointed out that, the plot of the D/G intensity ratio values of the above considered samples as a function of the measured film N-content suggests that, in disordered carbon-nitride based materials, the average size of the graphitic clusters is influenced by the changes in the C content of the film (*Sect.4.1*), as well as by the variation in the N-incorporation levels. The data, once again, roughly draw a single curve: the mean dimension increases up to reach a saturation value and then falls-down abruptly at the highest N-contents. Such a trend seems to indicate that the progressive increase of the concentration of nitrogenated groups promotes the formation of increasingly larger Csp^2 clusters (finding which would be consistent with the hypothesis that N atoms act as the nucleation centres for the graphitic phase *[24,25]*). However, above a critical N concentration, films become so C-poor as the size of the graphitic domains cannot increase anymore.

5. The role of the growth temperature

In this *Section*, the 100°C sputter-deposited *a*-CN(:H) films are used as probe-samples. Their C- and N- contents are evaluated from eq.s (1) and (2) and compared with the experimentally-determined concentrations (Tab.3). The discrepancy observed is utilised for both clarifying the role played by the deposition temperature and establishing the validity range of the relationships proposed for the estimation of relative C- and N- contents of disordered carbon-nitride based alloys.

Although the x_C and x_N values, obtained by substituting in eq.s (1) and (2) the G-band frequency positions reported in Tab.2, are reciprocally consistent within the experimental errors [large errors pertain to concentrations deduced from the saturation regions of the $\omega_G(x_C)$ and $\omega_G(x_N)$ curves], a non-negligible difference and, worst, a clear disagreement, with respect to the corresponding measured values, are found in case of C and N, respectively.

In order to get more insight into the problem, eventually clarifying the origin of such a discrepancy, the G-band frequency position of the 100°C sputter-deposited *a*-CN(:H) films are plotted, together with those of literature samples (already shown in Fig.3), as a function of the measured C-content (Fig.7). The ω_G representative points, relative to samples deposited at room and 100°C temperature, lie close to distinct curves, exhibiting, however, analogous trends, similarly saturating at low C contents. By the way, the latter finding suggests that annealing and higher-temperature depositions could provide similar effects, in terms of film C content ultimately achieved. The ω_G dependence of the presently-investigated *a*-CN(:H) films on x_C is well reproduced by eq.(1), provided that the proper $\omega_G(^{max}x_C)$ and A values are utilised, namely $\omega_G(100°C, {}^{max}x_C)=1545$ cm^{-1} (from literature sp^2 rich *a*-C films *[2]*) and $A_{100°C}=20\%$ (experimentally-determined). The distinct behaviours at different process-temperatures are easily understood if the role of temperature in quenching the internal constraints and distortions is considered. This reflects onto stronger C=C bonds and, thus, onto higher stretching frequencies. Hence, the G-band position is concluded to be a meaningful parameter to estimate the relative C content of the film, when the growth-process temperature is known.

Analogous results are found when the D/G intensity ratio is considered. The values obtained for the 100°C sputter-deposited *a*-CN(:H) films are plotted, together with those of literature samples (already shown in Fig.4), as a function of the C-content (Fig.8). Also in this case, two distinct curves, exhibiting similar trends, are found. The influence of temperature on the average graphitic-cluster dimension is evident. As a major result, smaller clusters are likely to be formed at higher deposition temperatures. Moreover, with increasing the growth temperature from RT to 100°C, the critical C content, above which a monotonical decrease of the mean cluster-size is achieved by enriching the film of C, raises from 55% up to 80%.

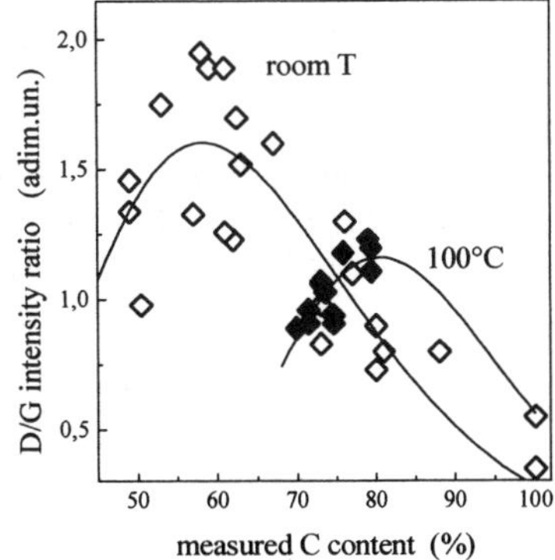

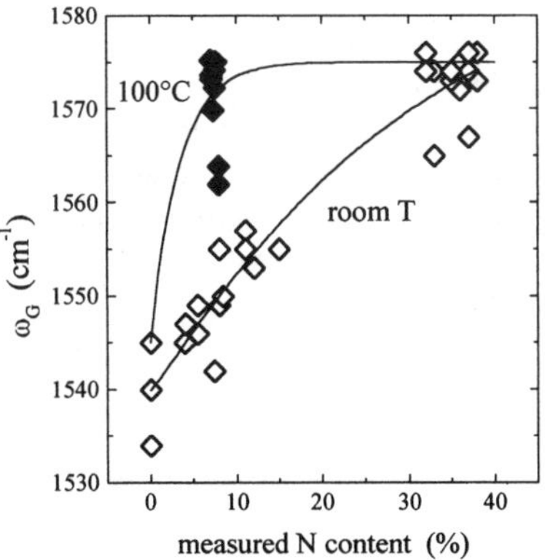

Fig.8 Dependence of the D/G intensity ratio on the measured C content at different deposition temperatures. The solid lines are drawn for visual help. The shown data refer to literature samples ($\lozenge$) and presently-investigated a-CN(:H) films ($\blacklozenge$).

Fig.9 Dependence of ω_G on the measured N content at different deposition temperatures. The solid lines represent the fitting curves derived by using eq.(2). The same symbols as in Fig.8 are used.

Finally, the validity limits of eq.(2) are explored. To this purpose, the ω_G values of the 100°C sputter-deposited a-CN(:H) films are plotted, together with those of literature samples (already shown in Fig.6), as a function of the N-content (Fig.9). The ω_G dependence of the 100°C grown a-CN(:H) films on x_N is well reproduced by eq.(2), if the proper $\omega_G(^{min}x_N)$ and B values are utilised, namely $\omega_G(100°C, {}^{min}x_N) \equiv \omega_G(100°C, {}^{max}x_C) = 1545$ cm^{-1} and $B_{100°C} = 4\%$ (experimentally-determined).

As for the influence of deposition temperature on the relationship existing between film N content and mean graphitic-cluster size, as measured by the D/G intensity ratio in Raman spectra, once again analogous trends are found at room and 100°C temperature. However, at higher temperatures, as already pointed out, the average dimensions of Csp^2 clusters are smaller and the largest cluster-size (see *Sect.4.3*) is reached at lower N-incorporation levels.

6. Conclusion

The frequency position of the G-band in Raman spectra of disordered carbon-nitride based alloys, currently assumed as a qualitative indicator of the sp^3:sp^2 C bonding ratio of the film, is here shown to contain information also about the material chemical-composition, so as, from a proper analysis of the spectra, the relative C- and N- contents can be deduced.

In this work, in fact, a wide variety of literature samples, prepared by several different deposition-techniques, under quite diverse growth-conditions, is considered and demonstration is given that the G-band position is strongly correlated with the C- and N-concentrations in such films. The reasons for the correlation to exist are discussed in detail with reference to the example of (the presently investigated) amorphous hydrogenated carbon-nitrides, deposited, at 100°C, by reactive sputtering of graphite, and understood in the light of the indications emerging from the comparative analysis of the complementary Raman and IR a-CN(:H) film characterisation results.

The variations of N-incorporation levels within the film involve changes in the mean coordination-number, ultimately resulting in diverse G-band frequency positions. The influence of the deposition temperature on the C=C bond strength in the grown films, which the G-band position is related to, is then pointed out.

By interpolating the experimental data, empirical relationships are derived, allowing the quantitative estimation of the C- and N- concentrations of the films, from their Raman spectra, if the growth-process temperature is known. In particular, called ω the frequency position of the G-band in the Raman spectrum of the film, its C- and N- contents can be respectively evaluated as

$$x_C = 100 + A_T \cdot ln\,[1-(\omega - {}^{min}\omega_T)/({}^{sat}\omega - {}^{min}\omega_T)] \tag{3}$$

and

$$x_N = -B_T \cdot ln\,\{[1-(\omega - {}^{min}\omega_T)/({}^{sat}\omega - {}^{min}\omega_T)]^{4/5}\}, \tag{4}$$

with ${}^{min}\omega_T$ denoting the lowest value of the G-band position, attained, at temperature T, in the limiting situation of $x_C = 100\%$ and $x_N = 0\%$; ${}^{sat}\omega = 1575$ cm^{-1} indicating the (independent on T) saturation value, approached at the lowest C- and highest N- contents; and A_T and B_T T-dependent parameters experimentally-determinable.

As a result of this preliminary work, by comparison with the experimental data, the values of all the T-dependent parameters entering in eq.s (3) and (4), necessary for estimating x_C and x_N in films grown at RT or 100°C, are determined. In the former case, the values of 30%, 17% and 1539 cm^{-1} are found for A_T, B_T and ${}^{min}\omega_T$, respectively; meanwhile, 20%, 4% and 1545 cm^{-1} are correspondingly found in the latter.

More general relationships, from which the stoichiometry of films grown at any temperature can be deduced, are presently under investigation and will be the argument of a future work.

Finally, it is worthwhile pointing out that analogous results are found when the D- to G-band intensity-ratio in the Raman spectra is considered. Thus, the average graphitic-cluster dimension, as measured by the D/G intensity ratio, is concluded to be mainly controlled, similarly to the sp^3:sp^2 C bonding ratio, by the film growth-temperature and chemical-composition.

References

[1] S.A.Solin, A.K.Ramdas, *Phys. Rev.* B1 (1970) 1687

[2] A.C.Ferrari, J.Robertson, *Phys. Rev.* B61 (2000) 14095

[3] R.J.Nemanich, S.A.Solin, *Phys. Rev.* B20 (1979) 392

[4] S.Prawer, K.W.Nugent, Y.Lifshitz, G.D.Lempert, E.Grossman, J.Kulik, I.Avigal, R.Kalish, *Diamond Relat. Mater.* 5 (1996) 433

[5] F.Tuinstra, J.L.Koenig, *J. Chem. Phys.* 53 (1970) 1126

[6] V.Vorlícek, P.Široký, J.Sobota,V.Perina, V.Zelezný, J.Hrdina, *Diamond Relat. Mater.* 5 (1996) 570

[7] D.F.Franceschini, F.L.Freire Jr., C.A.Achete, G.Mariotto, *Diamond Relat. Mater.* 5 (1996) 471

[8] M.M.Lacerda, D.F.Franceschini, F.L.Freire Jr., G.Mariotto, *Diamond Relat. Mater.* 6 (1997) 631

[9] M.M.Lacerda, F.L.Freire Jr., G.Mariotto, *Diamond Relat. Mater.* 7 (1998) 412

[10] F.Rossi, B.Andre, A.vanVeen, P.E.Mijnarends, H.Schut, F.Labohm, M.P.Delplancke, H.Dunlop, E.Anger, *Thin Solid Films* 253 (1994) 85

[11] G.Mariotto, F.L.Freire Jr, C.A.Achete, *Thin Solid Films* 241 (1994) 255

[12] C.Lenardi, M.A.Baker, V.Briois, L.Nobili, P.Piseri, W.Gissler, *Diamond Relat. Mater.* 8 (1999) 595

[13] G.Fusco, F.Giorgis, C.F.Pirri, A.Tagliaferro, E.Tresso, C.De Martino, P.Rava, *Defects and Diffusion Forum Vols.* 134-135 (1996) 3

[14] G.Messina, A.Paoletti, S.Santangelo, A.Tagliaferro, A.Tucciarone, *J. Appl. Phys.* 89 (2001) 1053

[15] G.Messina, S.Santangelo, G.Fanchini, A.Tagliaferro, *Vacuum* (2001) in press

[16] G.Fanchini, G.Messina, A.Paoletti, C.S.Ray, S.Santangelo, A.Tagliaferro, A.Tucciarone, *Surf. Coat. Technol.* (2001) in press

[17] S.Muhl, J.M.Mendez, *Diamond Relat. Mater.* **8** (1999) 1809

[18] N.M.Victoria, P.Hammer, M.C.DosSantos, F.Alvarez, *Phys. Rev.* B **61** (2000) 1083

[19] J.Robertson, *Adv. Phys.* **35** (1986) 317

[20] F.De Michelis, Y.C.Liu, X.F.Rong, S.Schreiter, A.Tagliaferro, *Solid State Commun.* **95** (1995) 475

[21] J.Robertson, *Phys. Rev.* B **53** (1996) 16302

[22] J.H.Kaufman, S.Metin, D.D.Saperstein, *Phys. Rev.* B **39** (1989) 13053

[23] F.Demichelis, X.F.Rong, S.Schreiter, A.Tagliaferro, C.DeMartino, *Diamond Relat. Mater.* **4** (1995) 361

[24] F.L.Freire Jr., G.Mariotto, C.A.Achete, D.F.Franceschini, *Surf. Coat. Technol.* **74-75** (1995) 382

[25] S.R.P.Silva, J.Robertson, G.A.J.Amaratunga, B.Rafferty, L.M.Brown, J.Schawn, D.F.Franceschini, G.Mariotto, *J. Appl. Phys.* **81** (1997) 2626

GNSR 2001
G. Messina and S. Santangelo (Eds.)
IOS Press, 2002

The restoration of the Ursino Castle in Catania: investigations and planning

D. Barilaro, D. Majolino, P. Migliardo

*Dipartimento di Fisica e Gruppo Operativo di Fisica Applicata,
Università di Messina C.p. 55 - S.Agata di Messina*

G. Barone, S. Ioppolo

*Istituto di Scienze della Terra, Università di Catania,
Corso Italia 55 - 95129 Catania*

A. Muscarà, N.F. Neri.

Soprintentenza Beni Culturali. Sezione Monumenti di Catania

Abstract. The object of this work is characterization of malthae and plasters drawn from Ursino Castle in Catania, to prove compositional differences between plasters installed since 1200 to today. During its story, infact, the castle underwent many rearrangements that altered original feature of the building.
FTIR analyses provide some useful information by which the Superintendence of Cultural Heritage of Catania has planned a restoring intervention of the castle. In particular malthae show differences about hardness, composition in binding and inert. Then it was hypothesized that different provision quarry were used for plasters fabrication and it denotes subsequent periods of building.

1. Historical and artistic notices about Ursino Castle in Catania

The Ursino Castle was built in 1239 under King Federico II of Svevia on a promontory inside the medieval boundary of the city. The stately monument shows a perfectly regular layout: a square perimeter, including a wide open courtyard in the middle. Four cylindrical towers are positioned at four corners of building perimeter; there were also four semicircular little towers, placed on the middle point of the four boundary walls, but only north and west towers still remain. Clearness and strength of the construction allowed permanence of much of ground floor walls and of original plan.

During the passing of time, Castle was subjected to alterations and readjustment according to subsequent new functions held from the building. Main sixteen-century changes are sited on ground floor of southern wing and on the first floor of the eastern wing. Since seventeen century the castle lost its residential function: ground floor was used as prison and then the whole building declined. Eighteen and nineteen century interventions produced overlaps upon all the areas and a flourishing of inner division walls, required by the different use of the rooms. High position preserves the castle from a huge lava flow in 1669 but it was seriously damaged during the earthquake in 1693, which provoked some collapses in the southern part. Then the castle, partially ruined and surrounded by lava, lost its prestige and its defensive function. Since 1800 the building transformed into barracks and remarkable architectural modifications were carried out: several openings were performed and all the surfaces were plastered. [1,2]

Considerable restoring works were performed in 1932; every modification as to original feature of the castle was considered completely arbitrary and therefore to be erased: many wall structure were destroyed and rough stylistic reproduction were inserted to complete the building which, actually, hadn't ever been finished. In the 1980's the monument was turned into Civic Museum and the new work of repair was devoted to improve sveve and renaissance construction and wipe stylistic additions out, then the marks of the stratifications further on during centuries were cancelled.

The purpose of the present intervention is facing a right historical interpretation of the castle and acquiring scientific data, necessary both in restoring unsafe parts and focusing on essential conservation conditions. It was realized that subsequent modifications of the structure are marks of a continuous adjustment to new historical request and the expression of several cultures proceeding in time. In particular, experimental researches, supported by historical-architectonic considerations, were performed to determine identify distinctive parameters for plasters of different historical periods in order to enhance the value of stratifications and operate on adequately. [3]

2. Macroscopic identification and classification of samples

On account of historical and artistic notices, samples were drawn from different point in Ursino Castle, in order to analyse materials belonging to different ages; they were classified and characterized first by macroscopic recognition, then by spectroscopic analysis. For many samples inert was separated from binding matrix to identify and analyse with high precision numerous components present in malhae.

Cast1, Cast3, Cast4 were drawn outside turret on the east side of the castle, dated about 1200; they show grey colour, inhomogeneous granulometry and considerable compactness. **Cast6a, Cast7** were picked up by point of different height (+1m, -3m from trample floor) on the same wall, in a room in which castle foundation was brought out by archaeological excavation. The wall structure is datable around 1200 and a clay sample (**Cast5**) was drawn from original foundation, sited about 4m under ground floor. **Cast8** belong to a wall sited crosswise original structures and then it was hypothesized that it was built later. Maltha is more friable than other samples belonging to this group and has a quite homogeneous granulometry.

Cast11, Cast14, Cast15, Cast17, Cast18, Cast19, Cast20 were drawn from several point of the parliament room, sited on second floor and dated about 1500. Maltha Cast11, coming from an inner niche, is whitish and rather friable, as the whole the malthae of this group. Cast20 was taken from the vault and contains fragment of pomice stone, used for its lightness.

Cast2 is a red plaster, picked up by an external wall on the east side of the castle. On basis of historical and artistic notices it was dated around 1800. **Cast7a** and **Cast10** show whitish and with a thin mixture, peculiar feature of plasters to make outside layer of building less porous and more rain waterproof.

Cast12, Cast13, Cast16 are rock fragment picked up from different wall structure. Cast16 belong to an original external window, after closed.

The whole set of samples were analysed by means of Fourier Transform InfraRed Spectroscopy. The measurement performed provided with useful information about analytic composition of building materials. [5]

Tab.1 Description of 1200 malthae samples

sample	sampling point	height	historical period	inert granulometry	macroscopic identification
Cast.1	outside turret	+1m	1200	1÷2cm	volcanic included, brick fragment
Cast.3	outside turret	+1m	1200	1÷2cm	schistose pebble, brick fragment
Cast.4	outside turret	+1m	1200	1÷2cm	schistose philladic rock pebble
Cast.6a	ground floor east outside wall	+1m	1200	0.1÷1cm	brick fragment
Cast.7	ground floor east outside wall	-3m	1200	0.1÷1cm	fluvial pebble, volcanic fragment
Cast.8	ground floor north outside wall	-1m		0.1÷0.3cm	earthenware, volcanic and Qz-feld included

Tab.2 Description of 1500 malthae samples

sample	sampling point	height	historical period	inert granulometry	macroscopic identificatio
Cast.11	parliament room inside niche	+2m	1500?	0.2÷0.3cm	brick and rock fragment
Cast.14	parliament room south outside wall	+1.5m	1500	1cm	volcanic included, earthenware, Qz-feld
Cast.15	parliament room south outside wall	+1.7m	1500	1cm	rock pebble, Qz-feld fragment
Cast.17	parliament room north inside wall	+1m	1500	0.5÷1.5cm	Qz-feld fragment
Cast.18	parliament room north inside wall	+0.5m	1500?	0.3÷1.5cm	Rock pebble, volcanic and Qz-feld fragment
Cast.19	parliament room east inside wall	+1.5m	1500?	0.3÷1.5cm	Qz-feld and volcanic fragment
Cast.20	parliament room vault		1500?	0.1÷0.2cm	volcanic fragment, pomice stone

Tab.3 Description of plaster samples

sample	sampling point	height	historical period	inert granulometry	macroscopic identification
Cast.2	east outside wall	+1m	1200	0.1÷0.5cm	
Cast.7a	ground floor room east outside wall	+1m	1200	0.1÷0.2cm	volcanic, metamorphic, earthenware fragment
Cast.10	parliament room inside niche	+2m	1500?	0.1÷0.2cm	

Tab.4 Description of rock samples

sample	sampling point	height	historical period
Cast.12	east wall column	+0.5m	
Cast.13	entrance column	+0.2m	
Cast.16	entrance window	+1.5m	1200?

3. FTIR spectroscopy experimental result

Cast1, **Cast3**, **Cast4** samples are mainly constituted by calcite, clay and quartz. In **Cast6a** and **Cast8** there is also gypsum, absent in **Cast7** probably for humidity under trample floor. Cast8, hypothesized subsequent 1200, show instead similar composition to Cast6a and Cast7.

Cast11, **Cast14**, **Cast15**, **Cast17**, **Cast18**, **Cast19** e **Cast20** are constituted by calcite, clay and gypsum and show differences about calcite quantity, in particular lower in Cast15 and Cast18.

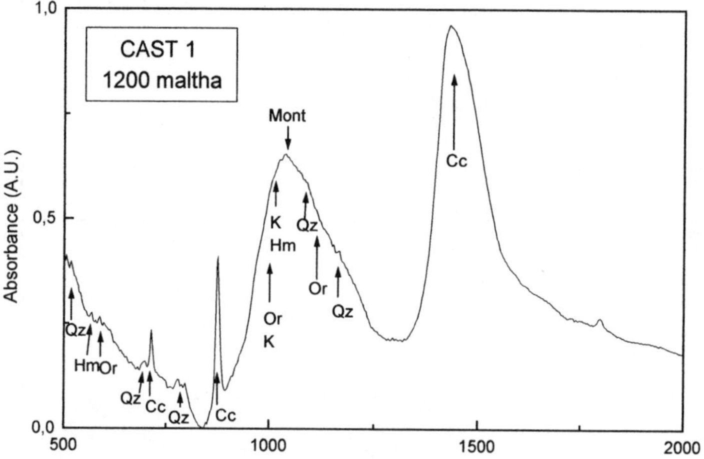

Fig.1 FTIR absorbance spectra of 1200 maltha

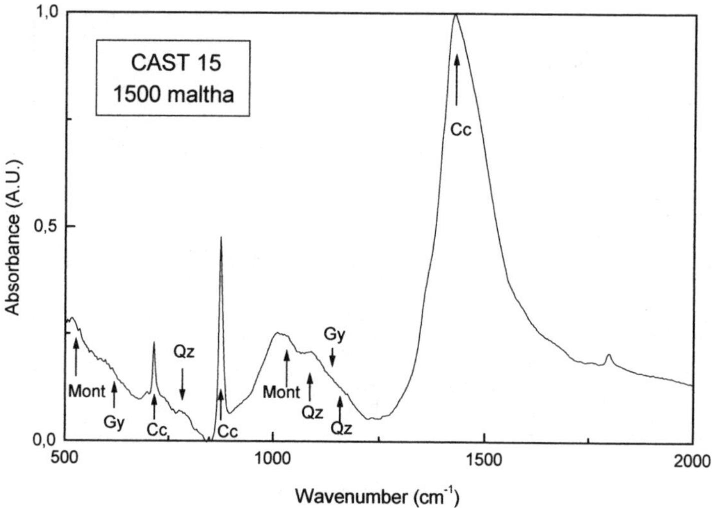

Fig.2 FTIR absorbance spectra of 1500 maltha

 D. Barilaro et al. / The Restoration of the Ursino Castle in Catania

Cast2, Cast7a and **Cast10** contain calcite, clay, quartz and gypsum. Cast2 has a red colouration due to iron oxides abundance, typical of seventeen-century constructions [6]; it also contains thenardite (Na_2SO_4), originated by reactions between calcite and sodium chloride coming from sea spray

Cast12, Cast13 and **Cast16** are constituted by calcite, clay and gypsum in different quantity. Cast12 e Cast16 contain thenardite; in Cast12 only is present also whewellite that reveal rock alteration processes

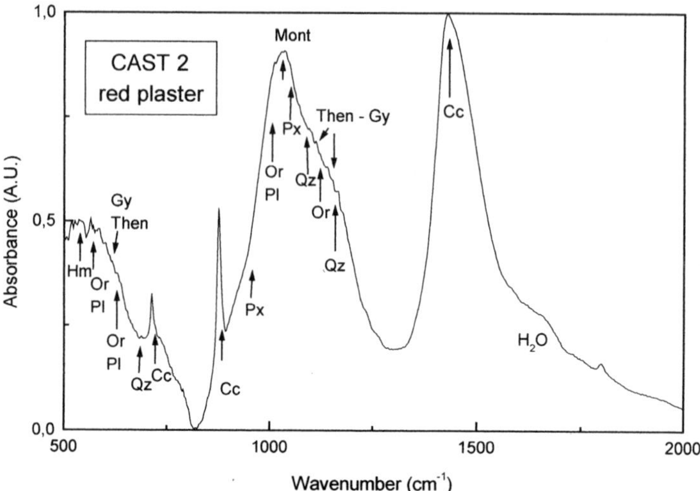

Fig.3 FTIR absorbance spectra of 1800 plaster

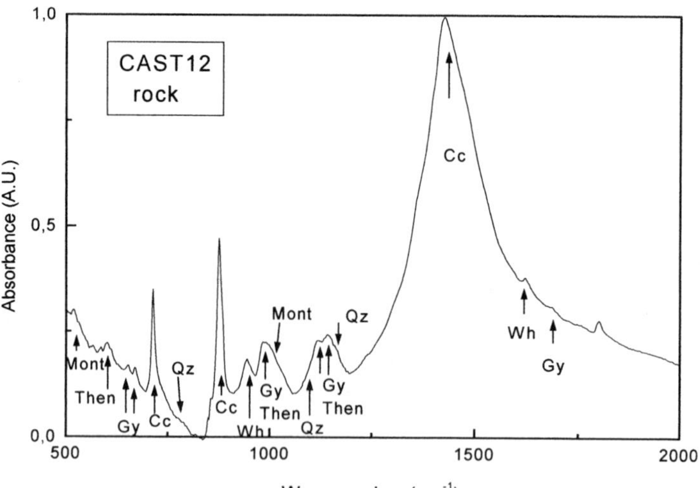

Fig 4 FTIR absorbance spectra of rock sample

4. Conclusions

The FTIR measurement performed on the samples of Ursino Castle have led to the characterization of the malthae and plasters manufactured since 1200 up to now.

The malthae Cast1, Cast3, Cast4, Cast6a, Cast7 e Cast8, picked up by castle ground floor and dated to 1200, don't show big compositional differences one from the other. The brick fragments reveal a composition similar to the clay (Cast5) found under the foundation of the castle, so it is supposed that, during the construction of 1200, in the surrounding of the castle a furnace for brick production was present, so avoiding the carriage of huge quantity of materials. Almost all 200 maltha samples are very consistent.

The malthae Cast11, Cast14, Cast15, Cast17, Cast18, Cast19 e Cast20, drawn from parliament room and dated to 1500, have a similar analytic composition but in different quantity. Most of these samples are rather friable.

Comparing FTIR absorbance spectra, malthae of different periods show a similar analytic composition but 1200 samples show a lower calcite quantity respect to 1500 samples. In 600 years the used materials and the manufacture techniques of the malthae were practically the same, then peculiarities of malthae and plasters, mainly in the past, resulted related to particular sites of provision rather to different ways of preparation [4b, 4c].

Tab.5 Compositional analysis of 1200 maltha samples

sample	sampling point	height	inert granulometry	maltha composition	inert composition
Cast.1		+1m	1÷2cm	Cc-Qz-Mont-Kfeld-Hm	Cc-Qz-Kfeld-Kaol-Pl-Px-Hm-Nep-Oliv
Cast.3	outside turret	+1m	1÷2cm	Cc-Qz-Mont	Qz-Kfeld-Kaol-Bi
Cast.4		+1m	1÷2cm	Cc-Qz-Mont-Kfeld-Hm	Qz-Feld-Mu
Cast.6a		+1m	0.1÷1cm	Cc-Qz-Mont-Ill- Gy-Kfeld	Cc-Qz-Ill-Kfeld
Cast.7	ground floor	-3m	0.1÷1cm	Cc-Qz-Mont-Kfeld	Cc-Qz-Mont-Ill-Kfeld-Bi
Cast.8		-1m	0.1÷0.3cm	Cc-Qz-Mont-Ill-Gy-Kfeld	Cc-Qz-Mont-Kfeld-Kaol-Pl-Px-Bi-Mu-oxides

Tab.6 Compositional analysis of 1500 malthae samples

sample	sampling point	height	inert granulometry	maltha composition	inert composition
Cast.11		+2m	0.2÷0.3cm	Cc-Qz-Mont-Gy	
Cast.14		+1.5m	1cm	Cc-Qz-Mont-Gy	Cc-Qz-Pl-Px-oxides (Hm-Mt)
Cast.15	parliament room	+1.7m	1cm	Cc-Qz-Mont-Gy	Cc-Qz-arg-Kfeld-Bi-Mu
Cast.17		+1m	0.5÷1.5cm	Cc-Qz-arg	Cc-Qz-Mont-Kfeld-Kaol-Bi-Mu-oxides
Cast.18		+0.5m	0.3÷1.5cm	Cc-Qz-Mont-Gy	Cc-Qz-Mont-Kfeld-Kaol-Pl-Px-Bi-Nep-Zeo-Mu-oxides
Cast.19		+1.5m	0.3÷1.5cm	Cc-Qz-Mont-Gy	Cc-Qz-Mont-Ill-Kfeld-Kaol-Pl-Px-Bi-Nep-Zeo-oxides
Cast.20	vault		0.1÷0.2cm	Cc-Qz-Mont-Gy	Cc-Qz-Kaol-Pl-Px-Hm-Nep-Zeo- Fe oxides

Tab.7 Compositional analysis of plaster samples

sample	sampling point	height	inert granulometry	composition
Cast.2	east outside wall	+1m	0.1÷0.5cm	Cc-Qz-Mont-Ill-Kfeld Pl-Px-Hm-Gy-Then
Cast.7a	ground floor	+1m	0.1÷0.2cm	Cc-Qz-Mont-Gy-Kfeld
Cast.10	parliament room	+2m	0.1÷0.2cm	Cc-Qz-Cl

Tab.8 Compositional analysis of rock samples

sample	sampling point	height	composition
Cast.12	east wall column	+0.5m	Cc-Qz-Cl-Gy- Then-Wh
Cast.13	entrance column	+0.2m	Cc-Qz-Mont-Ar
Cast.16	entrance window	+1.5m	Cc-Qz-Gy-Then

About quantitative differences the hypothesis is that most ancient malthae suffer alterations that lead to a decreasing of the calcite concentration due to the slow reaction of the silica in the sand, used as inert, together with calcium carbonate with formation of silicate that induces an hardening of the manufactures[4d]. Infact 1500 malthae are most friable than 1200 malthae. In this frame the characterization of the inert give important information to understand the origin of the materials, while the characterization of the binding furnishes useful notices regarding the historical period of the materials. Therefore the variation in the amount of calcite is a noteworthy parameter for the in direct dating of the materials.

Plasters Cast7a, Cast10, Cast2 reveal a composition similar to the malthae below, but with a thinner mixture. The presence of thenardite in the outside plaster (Cast2) is indicative of an alteration process of the carbonate.

In analysed carbonate rocks (Cast12, Cast13, Cast16) the presence of gypsum, thenardite or whewellite reveals alteration processes of carbonate.

The spectroscopic investigation has given a valid contribution to the historical and artistic knowledge of the monument and to the choice of the best method for its restoration. Besides the performed analysis have provided considerable information about degradation state of structures and better preservation conditions. Experimental results have influenced the choices of the intervention, devoted to exalt the original bond by removing the recent plaster, by cleaning and revival of the original plaster, with a differentiated treatment of each layer, to show the different periods and the different historical stages of the walling. The aim of the present work of repair is that the building itself is able to perform a critical reading of subsequent ages and cultures that live in its walls.

Tables legenda

Cc = Calcite; Qz = Quartz; Cl = Clay; Mont = Montmorillonite; Ill = Illite; Gy = Gypsum; Kfeld = Orthoclase; Kaol = Kaolinite; Pl = Plagioclase; Px = Pyroxene; Hm = Hematite; Bi

= Biotite; Nep = Nephelina; Zeo = Zeolite; Oliv = Olivine; Mu = Muscovite; Then = Thenardite; Wh = Whewellite; Ar = Aragonite.

References

[1] R.Santoro "La Sicilia dei castelli-La difesa dell'isola dal VI al XVIII secolo-Storia e Architettura" Edizioni Pegaso

[2] C.A.Willemsen "I castelli di Federico II nell'Italia meridionale" Società Editrice Napoletana

[3] Regione Siciliana- Assessorato Regionale dei Beni Culturali ed Ambientali e della P.I.-Soprintendenza per i Beni Ambientali, Architettonici, Artistici e Storici - Catania "Relazione Castello Ursino- lavori di restauro relativo alla soluzione dei problemi di funzionalità della struttura museale"

[4] Articoli tratti da : "SCIENZA E BENI CULTURALI - L'intonaco: storia, cultura e tecnologia - Atti del Convegno di studi - 1985" Libreria Progetto Editore Padova
4a) G. Alessandrini "Gli intonaci nell'edilizia storica: metodologie analitiche per la caratterizzazione chimica e fisica"
4b) M.Laurenzi Tabasso-P.Rota Rossi-Doria "Malte impiegate per un edificio storico nella zona di Tor di Nona in Roma: caratterizzazione chimico-fisica
4c) R.Franchi-F.Fratini-Manganelli "Caratterizzazione degli intonaci mediante l'impiego di tecniche mineralogico-petrografico"
4d) G.Biscontin-G.Driussi-S.Volpin "Il degrado degli intonaci: aspetti chimici e fisici"

[5] Libreria Sadtler "Mineral and Clays"

[6] Sciuto Patti "Sui materiali da costruzione più usati a Catania" – Atti del collegio di ingegneria e architettura in Catania 1896

GNSR 2001
G. Messina and S. Santangelo (Eds.)
IOS Press, 2002

Two-photon fluorescence excitation and optoacoustic spectra of polyDCHD-HS

Laura Moroni[a], Pier Remigio Salvi[a], Cristina Gellini[a], Giovanna Dellepiane[b,c], Davide Comoretto[b,c], Carla Cuniberti[c]

[a]*Laboratorio di Spettroscopia Molecolare, Dipartimento di Chimica, Universita' degli Studi di Firenze, Via Gino Capponi 9, 50121 FIRENZE*
[b]*INFM-,* [c]*Dipartimento di Chimica e Chimica Industriale, Universita' degli Studi di Genova, Via Dodecaneso 31, 16146 GENOVA*

Abstract. The two-photon absorption spectrum of red polyDCHD-HS in benzene solution has been measured by means of fluorescence excitation and optoacoustic detection in the 1600-750 nm exciting wavelength region. Two band systems have been observed and assigned to two-photon allowed $A_g \rightarrow A_g$ transitions, both at energies higher than that of the one-photon allowed $1B_u$ state. An estimate of the two-photon absorptivity is given at 406 nm using optoacoustic detection.

1. Introduction

The π-conjugated polymers form a class of one-dimensional semiconductors[1-4] whose properties can be varied by changing their chemical structure. In particular, the nonlinear optical response of these systems can be tuned to produce photonic devices such as light emitting diodes and optical waveguides.[5-7] For this reason it is of great relevance to study the electronic properties of these systems as a function of their structure.

Assuming a C_{2h} symmetry for the polymer backbone, the energy ordering of the two lowest $\pi\pi^*$ excited states has been shown to be $2A_g/1B_u$ or, alternatively, $1B_u/2A_g$ as a function of several factors, including the bond alternation parameter and on-site and nearest-neighbours Coulomb interaction terms.[2,8,9] These factors depend on the polymer chemical nature, chain length and conformation, as well as on the sample morphology.[2,3]

In polysilanes and poly(p-phenylenevinylenes) the conjugation is not extended enough to lower the $2A_g$ state below $1B_u$ and this results in an efficient fluorescence emission from $1B_u$ following optical excitation. On the contrary, in polyacetylenes (PA) and polydiacetylenes (PDA) the torsional arrangement of the polymeric chain affects the conjugation length such that $2A_g$ may shift below $1B_u$ producing a quenching of the fluorescence emission.[2,3] In fact, different PDA morphologies, the so-called blue and red form, mainly differing in the effective conjugation length, have different relative location of the two states. For instance, blue poly[1,6-bis-(N-carbazolyl)-2,4-hexadiyne] (polyDCHD) shows a two-photon resonance below the onset of the strong one-photon absorption.[10] Similarly, the lowest $\pi\pi^*$ state of blue poly[5,7-dodecadiyne-1,12-diyl-bis- ethyluretane] (PDA-4U2) as a thin film

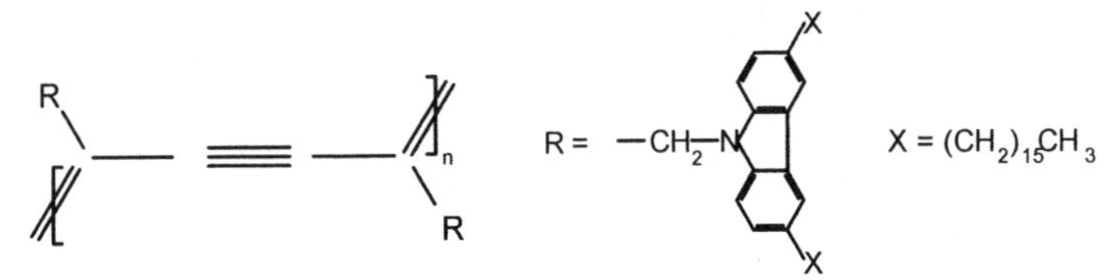

Fig.1 The molecular structure of polyDCHD-HS.

on a glass substrate has been observed at 1.9 eV and assigned to $2A_g$ state.[11] On the other hand, the situation is less definite for red polymers, probably due to the reduced conjugation length. Infact, in poly[5,7-dodecadiyn-1,12-diol-bis(n-butoxycarbonyl-methyl-urethane)] (poly(4-BCMU)) the $2A_g$ state is still found at energies lower than $1B_u$,[12,13] while in red PDA-4U2 the $1B_u$ and $2A_g$ states are almost degenerate, around 2.33 eV.[11] Many spectroscopic techniques including THG generation,[10,11] time resolved spectroscopies,[8,9,14] and two-photon absorption[15] were able to identify the lowest $\pi\pi^*$ states in a wide variety of systems. The latter technique, however, has the great advantage of a direct detection of A_g states.[16,17] For this reason we have relied on two-photon spectroscopy to investigate the electronic properties of a novel polymer, poly[1,6-bis(3,6-dihexadecyl-N-carbazolyl)-2,4-hexadyine] (polyDCHD-HS, see Fig. 1), whose peculiar property is the high solubility in common organic solvents as red form, due to the insertion of long alkyl chains on the carbazolyl substituents.[18] Previous studies on this polymer were concerned with the third order nonlinear optical susceptibility, measured by means of photoinduced absorption,[19-22] time-resolved spectroscopy[23] and guided propagation of light.[24,25] Being this system highly fluorescent at room temperature,[26] we have investigated the two-photon spectrum through excited fluorescence in the wavelength range 800-415 nm. The two-photon optoacoustic spectrum was measured up to 375 nm.

2. Experimental

PolyDCHD-HS has been synthesized according to a known procedure.[18] Solutions 10^{-3} – 10^{-5} M were prepared in benzene-h_6 and $-d_6$ following previous considerations on improved spectral quality and resolution using this solvent with respect to other organic solvents.[26] The one-photon spectrum was measured with a Cary 5 spectrometer on a 10^{-5} M solution at room temperature in benzene-h_6 in the range 280-700 nm and up to 200 nm in cyclohexane. The fluorescence spectrum of the benzene solution was taken exciting at 476 nm with a conventional spectrofluorimeter (JASCO FP-750, 10 nm resolution). The absorption and emission spectra, shown in Fig. 2, have clear mirror symmetry.

Two-photon measurements were performed by using as tunable exciting radiation the idler beam of an optical parametric oscillator pumped by the third harmonic (355 nm) of a nanosecond pulsed Nd:YAG laser with 10 Hz repetition rate. Two different probing methods were used, fluorescence excitation and optoacoustic detection. In the first case the fluorescence induced by the two-photon transition first crosses a set of appropriate optical filters stopping the incident radiation and then impinges on a photomultiplier placed 90° with respect to the direction of the exciting beam. The spectrum has been measured by exciting between 830 and 1600 nm which corresponds to an effective two-photon wavelength range 415-800 nm.

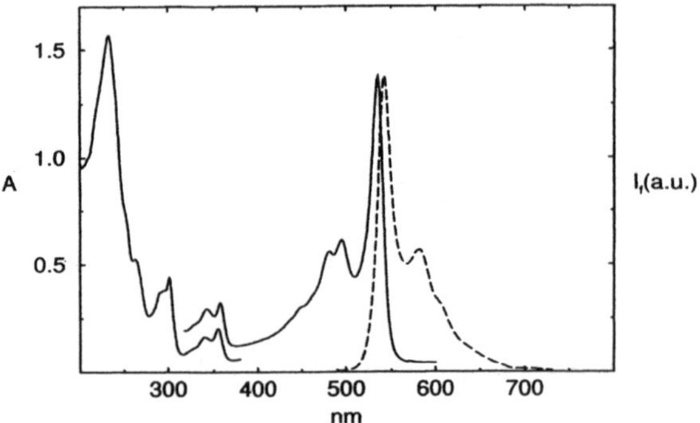

Fig.2 Absorption (solid line; absorbance units on the left vertical axis) and fluorescence (dashed line; fluorescence intensity in arbitrary units on the right vertical axis) spectra of a 10^{-5} M polyDCHD-HS solution in benzene.

The second probing method is based on the heat conversion of the non radiatively relaxed part of the absorbed energy and on the consequent expansion of the absorption volume which generates an acoustic wave with amplitude P.[27,28] An ultrasonic transducer (Panametrics V103-RM) in contact with the sample cell through a thin grease film, detects the pressure wave. The spectrum was extended up to 750 nm, i.e., to an effective two-photon wavelength 375 nm. Being this technique sensitive to the overall, i.e., one- and two-photon, absorption of the solution, interfering linear signals due to solvent C-H or C-D stretching overtones may be observed. For this reason the optoacoustic spectra of the pure solvents were measured in the same wavelength range. Finally, the contribution of the carbazolyl residue to the one- and two-photon absorption of polyDCHD-HS was estimated repeating the measurements on a solution of carbazole in benzene. Comparing one-photon results on carbazole with those of Fig. 2, the high energy part of polyDCHD-HS spectrum has been assigned to electronic excitations of the carbazolyl substituent.

3. Results and Discussion

The fluorescence excitation spectrum of polyDCHD-HS shows two absorption regions: a first weak band around 1190 nm and a structured absorption extending up to 830 nm with origin at 1030 nm (see Fig. 3). No absorption is detected above 1300 until 1600 nm. An accurate measurement of the dependence of the induced fluorescence signal on the incident intensity revealed that the order of the absorption process is different in the two regions. By exciting at 1030 and 950 nm, a quadratic dependence is found, indicating that a two-photon transition occurs at half wavelength, 515 and 475 nm, respectively. On the contrary, the 1190 nm band depends on a number of photons higher than two. The same band with equal dependence on incident intensity is observed in pure carbazole, leading to the conclusion that the 1190 nm band corresponds to multiphoton excitation localized on the carbazolyl substituents. The 535-415 nm two-photon absorption region (1070-830 nm excitation wavelength) is due to $\pi\pi^*$ transitions of the polymer chain as the lowest absorption of the carbazolyl residue is observed around 350 nm (see Fig. 2). Comparing the one-photon data with the two-photon absorption of polyDCHD-HS (see Fig. 3), the difference between the two spectra is evident, as predicted for centrosymmetric systems. The one-photon origin (0-0) $1B_u$ falls at 534 nm (see Fig. 2). A vibronic structure is observed with bands at 1512, 2072, 3595 and 4280 cm^{-1} from (0-0), assigned to double and triple C-C stretching modes and their combinations. On the other hand, the two-photon spectrum has origin at 514 nm, 730 cm^{-1} above (0-0) $1B_u$, and it shows an intensity profile as a function of decreasing wavelength that is opposite to that of the one-photon spectrum. These observations are evidence of two-photon activity of vibronic states one-photon silent.

The optoacoustic spectrum has been measured in benzene-h_6 and -d_6 between 1200 and 750 nm as exciting wavelength. It results from two overlapping contributions: one linear, due to solvent overtones and the second quadratic, related to the two-photon absorption of the polymer. The second and third C-H overtone are found at 1190 and 880 nm, while the third and fourth C-D overtone at 1150 and 930 nm. Being the overtone intensities of benzene-d_6 weaker than those of -h_6 , the spectrum measured in benzene-d_6 is less distorted, as shown in Fig. 4. Optoacoustic and fluorescence excitation data agree each with the other since the comparable yield of radiative and nonradiative relaxation processes in polyDCHD-HS makes both detection methods good probes for two-photon transitions. However, the optoacoustic spectrum shows on the whole higher resolution and we will refer to this for the assignment.

Considering an all-trans C_{2h} geometry of the polymer backbone, the lowest $\pi\pi^*$ states of polyDCHD-HS belong to A_g and B_u symmetry.[3] In the dipole moment approximation one- and two-photon selection rules are mutually exclusive: $A_g \rightarrow A_g$ transitions are two-photon, while $A_g \rightarrow B_u$ one-photon allowed. However, vibronic g states of B_u levels may be two-photon active due to coupling with B_g or A_g levels. The first coupling, i.e., $B_u \leftrightarrow B_g$ through a_u vibrations may be reasonably excluded from our discussion as B_g states have predominant $\pi\sigma^*$ (or$\sigma\pi^*$) character and therefore lie much higher in energy than $1B_u$. The second hypothesis, $1A_g \rightarrow B_u$ x $b_u = A_g$, implies that the origin of the two-photon spectrum at 514 nm is vibronic and occurs 730 cm^{-1} above (0-0) $1B_u$.

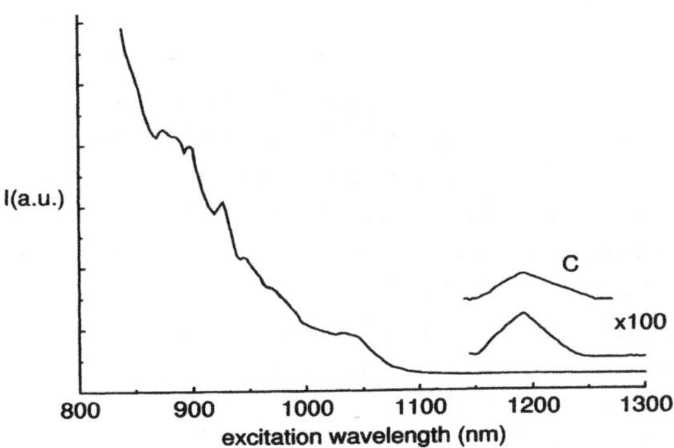

Fig.3 The two-photon fluorescence excitation spectrum of polyDCHD-HS 10^{-3} M in benzene. C, fluorescence excitation spectrum of pure carbazole.

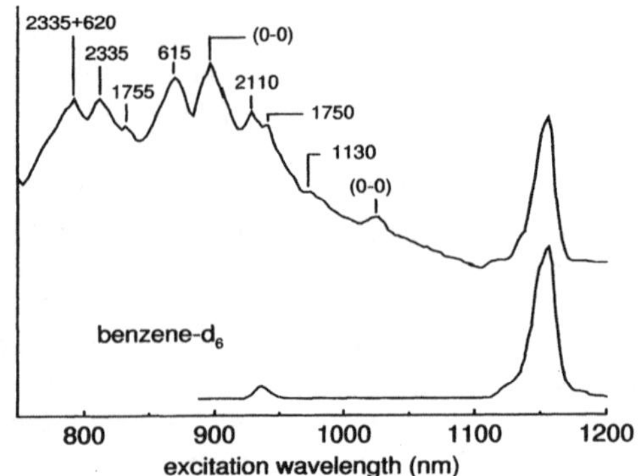

Fig.4 The optoacoustic spectrum of polyDCHD-HS 10^{-3} M in benzene-d_6. Lower: the optoacoustic spectrum of pure benzene-d_6.

Two-photon vibronic activity is well documented for the lowest $\pi\pi^*$ excited states of aromatic molecules.[16,17] There are no indications of a similar behaviour in linear polyenes.[1] This point supports the assignment of the 514 nm band to an allowed two-photon transition $A_g \rightarrow A_g$. Further, from one-photon data it is known that the fluorescence quantum yield of polyDCHD-HS is relatively high[26] and that fluorescence and one-photon absorption spectra have clear mirror symmetry (see Fig. 2). Combining these informations with our two-photon data we conclude that the $1B_u$ origin lies below (0-0) $2A_g$ state, and therefore the two-photon 514 nm band may be assigned to the (0-0) $1A_g \rightarrow 2A_g$ transition. The relative Franck-Condon structure is weak: the 1130, 1750 and 2110 cm^{-1} bands far from the 514 nm origin, related to C-C single, double and triple bond stretchings respectively,[26] are observable with difficulty.

Below 900 nm the optoacoustic spectrum shows a strong intensity increase related to the two-photon activity of a second A_g state with origin at 448.5 nm and with vibronic bands 615, 1755, 2335, 2335+620 cm^{-1} from (0-0). Our overall assignment of the two-photon spectrum as due to two allowed $\pi\pi^*$ $A_g \rightarrow A_g$ transitions strongly agrees with similar conclusions on PDA-4U2[11] and on other PDAs.[12,15,29] To confirm this assignment, the two-photon absorptivity δ,[30] proportional to the nonlinear susceptibility $\chi^{(3)}$ and to the two-photon absorption coefficient β, has been also evaluated optoacoustically by exciting around 812 nm. The value $\delta \cong 10^4$ cm^4 s molec^{-1} photon^{-1} indicates the allowedness of the transition and corresponds to a $\beta \cong 0.2$ cm/GW. β values in the range 200-700 cm/GW have been already reported for polymeric crystalline films.[15,29] Since β depends on the number density of absorbing species and in our 10^{-3} M solution the density is 3-4 orders of magnitude smaller than for crystalline systems, our results agree qualitatively with previous data and give an additional indication about high non linearities in π-conjugated systems.

4. Conclusions

In this work two-photon spectroscopy has been employed to reveal electronic states of gerade symmetry in red polyDCHD-HS. Both detection techniques, fluorescence excitation and optoacoustic probe, give comparable results. These have been discussed in terms of two allowed $\pi\pi^*$ $A_g \rightarrow A_g$ transitions. A major results of this work is the location of the lowest A_g state 730 cm^{-1} above the one-photon allowed (0-0) $1B_u$. The tuning of the $2A_g/1B_u$ location in PDAs is a good example of the opportunities offered by molecular engineering in the chemical modulation of the electronic properties of polymeric semiconductors.

References

[1] B. S. Hudson, B. E. Kohler, K. Schulten, in Excites States; Ed E. C. Lim, Academic Press, New York, **6** (1982), 1.

[2] S. Etemad, Z. G. Soos, in Spectroscopy of Advanced Materials, John Wiley and Sons, New York (1991) 87.

[3] Z. G. Soos, D. S. Galvao, S. Etemad, Adv. Mater., **6** (1994) 280.

[4] D. Beljonne, J. Cornil, Z. Shuai, J. L. Bredas, F. Rohlfing, D. D. C. Bradley, W. E. Torruellas, V. Ricci, G. I. Stegeman, Phys. Rev. B, **55** (1977) 1505.

[5] R. L. Sutherland, Handbook of Nonlinear Optics, Dekker, New York, (1996).

[6] M. Pope, C. E. Swemberg, in Electronic Processes in Organic Crystals and Polymers, 2nd. Ed. Oxford Sci. Publ., Oxford, (1999) 1182.

[7] Handbook of Organic Conductive Molecules and Polymers, **1-4**, Ed. H. S. Nalwa Wiley, New York (1997).

[8] E. Bourdin, A. Davey, W. Blau, S. Delisse, J. M. Nunzi, Chem. Phys. Lett., **103** (1997) 275.

[9] S. V. Frolov, Z. Bao, M. Wohlgenannt, Z. V. Vardeny, Phys. Rev. Lett., **85** (2000) 2196

[10] J. Le Moigne, A. Thierry, P. A. Chollet, F. Kajzar, J. Messier, J. Chem. Phys., **88** (1988) 6647.

[11] T. Manaka, T. Yamada, H. Hoshi, K. Ishikawa, H. Takezoe, Synth. Met. **95** (1998) 155.

[12] W. E. Torruellas, K. B. Rochford, R. Zanoni, S. Aramaki, G. I. Stegeman, Opt. Comm. **82** (1991) 94.

[13] D. Guo, S. Mazumdar, S. N. Dixit, Nonlinear Optics, **6** (1994) 337.

[14] S. Haacke, J. Berrehar, C. Lapersonne-Meyer, M. Schott, Chem. Phys. Lett., **308** (1999) 363.

[15] G. P. Banfi, D. Fortusini, P. Dainesi, D. Grando, S. Sottini, J. Chem. Phys., **108** (1998) 4319.

[16] L. Goodman, R. P. Rava, Acc. Chem. Res., **17** (1984) 250.

[17] P. R. Callis, Ann. Rev. Phys. Chem., **48** (1997) 271.

[18] B. Gallot, A. Cravino, I. Moggio, D. Comoretto, C. Cuniberti, C. Dell'Erba, G. Dellepiane, Liquid Crystals, **26** (1999) 1437.

[19] D. Comoretto, I. Moggio, C. Dell'Erba, C. Cuniberti, G. F. Musso, G. Dellepiane, L. Rossi, M. E. Giardini, A. Borghesi, Phys. Rev. B, **54** (1996) 16357.

[20] D. Comoretto, I. Moggio, C. Cuniberti, G. F. Musso, G. Dellepiane, A. Borghesi, F. Kajzar, A. Lorin, Phys. Rev. B, **57** (1998) 7071.

[21] D. Comoretto, I. Moggio, C. Cuniberti, G. Dellepiane, E. Giardini, A. Borghesi, Phys. Rev. B, **56**, (1997) 10264.

[22] D. Comoretto, G. Dellepiane, C. Cuniberti, L. Rossi, A. Borghesi, J. Le Moigne, Phys. Rev. B, **53** (1996) 15653.

[23] G. Lanzani, S. Stagira, G. Cerullo, S. DeSilvestri, D. Comoretto, I. Moggio, C. Cuniberti, G. F. Musso, G. Dellepiane, Chem. Phys. Lett., **313** (1999) 525.

[24] E. Giorgetti, G. Magheri, S. Sottini, X. Chen, A. Cravino, D. Comoretto, C. Cuniberti, C. Dell'Erba, G. Dellepiane, Synth. Met., **115** (2000) 257.

[25] E. Giorgetti, G. Magheri, F. Gelli, S. Sottini, D. Comoretto, A. Cravino, C. Cuniberti, C. Dell'Erba, I. Moggio, G. Dellepiane, Synth. Met., **116** (2001) 129.

[26] M. Alloisio, A. Cravino, I. Moggio, D. Comoretto, S. Bernocco, C. Cuniberti, C. Dell'Erba, G. Dellepiane, J. Chem. Soc., Perkin Trans 2, (2001) 146.

[27] C. K. N. Patel, A. C. Tam, Rev. Mod. Phys., **53**, (1981) 517.

[28]   A. C. Tam, C. K. N. Patel, Nature, **280** (1979) 304.
[29]   B. Lawrence, W. E. Torruellas, M. Cha, M. L. Sundheimer, G. I. Stegeman, J. Meth, S. Etemad, G. Baker, Phys. Rev. Lett., **73** (1994) 597.
[30]   R. J. M. Anderson, G. R. Holtom, W. M. McClain, J. Chem. Phys., **70** (1979) 4310.

GNSR 2001
G. Messina and S. Santangelo (Eds.)
IOS Press, 2002

Vibrational study by Raman and FT-IR spectroscopy of trehalose/water solutions

C. Branca[1]*, S. Magazù[1], G. Maisano[1], F. Migliardo[1], G. Romeo[1], E. Druido[2]

[1]*Dipartimento di Fisica and INFM, Università di Messina, P.O.Box 55, Papardo, 98166 S. Agata di Messina, Italy*
[2]*Dirigente Servizio Sanitario, V Rgt F. Aosta, Messina, Italy*

Abstract. We report measurements performed by Fourier Transform Infrared Spectroscopy (FT-IR) and Raman scattering on trehalose aqueous solutions. The interest on this system is mainly due to its extraordinary effectiveness as bioprotector against freezing and dehydration. The main purpose of the present work is to investigate the structural properties of trehalose and to study the effects of this disaccharide on the hydrogen bonded network of water.

1. Introduction

The structure and dynamics of hydrated disaccharides have become the subject of intense research efforts in recent years, motivated by fundamental physico-chemical problems as well as by important biomedical and biotechnological applications in which the molecular mechanisms are not well understood. Among all the sugars, trehalose has received a great attention, both because of its role in nature and its potential use as a highly efficient natural preservative [1]. Trehalose is a well-known non-reducing disaccharide, which is present in many simple organisms (among others bacteria, yeast, fungi and several species of nematodes) where the concentration can be as high as 20 % of the dry weight [2-6]. It is capable of maintaining unaltered the integrity of the membranes during exposure to severe thermal stresses, either high or very low temperature [1]. One of the most important examples of the biological role of trehalose is represented by *tardigrades* (*water bears*), that under dry conditions pass into a state of suspended life (cryptobiosis) by synthesizing trehalose, and restore full metabolic functionality when rehydrated, even after several decades [2].

The circumstance that these organisms can be reversibly dehydrated and rehydrated at room temperature without cumulating effects of functional stress suggest that trehalose, can be employed in the conservation technologies of high value added macromolecules. The presence of trehalose in several freeze-tolerant organisms has induced to concentrate on this molecule also as a cryoprotectant for freeze-drying [7]. An enormous interest is put on the possibility of using trehalose, or other semi-synthetic sugars, as a means of preserving valuable delicate biologically active proteins or living cells at ambient temperature without the need of lyophilization. However, the correct use of trehalose is still hampered since the understanding of its mechanism of action is still incomplete. Among the several hypotheses offered to explain the peculiarity of trehalose protection, the most credited ones are based on:

Glass transition hypothesis. Green and Angell [7,8] have hypothesized that the bioprotection from destruction by freezing is related to the high glass transition temperature of trehalose. On the other hand Crowe and co-workers [9] have noted that *vitrification* is not sufficient for preservation. *Vitrification* alone doesn't explain, for example, why another carbohydrate, dextran, which has a significantly higher glass transition temperature than trehalose, is a much less effective cryoprotectant than trehalose. Other researchers however argue that there must be an additional interaction between the sugars and the object of protection.

Water replacement hypothesis. The water replacement hypothesis, first proposed by Crowe [10], asserting the bonding of the hydrogen with the polar headgroups of the lipids that constitute biomembranes, accounts for how the non-reducing trehalose and sucrose preserve the integrity of biological structures. As the systems are dried or frozen, these interactions replace those of the hydration water at the membrane-fluid interface. In such a way, in the opinions of these authors, it prevents the phase transition and the accompanying leakage upon rehydration.

Trehalose has been studied from several points of view. Theoretical and experimental works [10-13] on the details of the molecular features of this anomalous sugar and of the interactions between trehalose and protein macromolecules in solution and during dehydration are ongoing in several laboratories. Our current work uses a range of techniques to study the dynamical and structural properties of disaccharide solutions. In previous and current work using viscometric [14,15], ultrasonic [16], and NMR [17] techniques, performed on aqueous solutions of trehalose, maltose and sucrose, our aim has been to quantify the basic hydration behaviour and solution structure of the main disaccharides as a function of concentration and temperature. Raman scattering measurements [18] showed that, among disaccharides, trehalose has the greatest destructuring effect on the tetrahedral hydrogen bonded network of water, the same that on lowering temperature would give rise to ice. As a result the amount of freezable water is reduced and the crystallization process is obstructed. Finally a series of Quasi Elastic Neutron Scattering (QENS) experiments [19,20] suggests that trehalose affects significantly the dynamics of water molecules in its neighbourhood consistently with the presence of a series of hydration layers or shells each of which is characterized by a relatively slow diffusion.

The main purpose of the present work is to investigate the structural differences among homologous disaccharides, which could account for their different bioprotective effectiveness. To carry out this study we have performed Fourier Transform Infrared Spectroscopy (FTIR) and Raman scattering measurements on trehalose dihydrate and on trehalose aqueous solutions at different concentration values. Trehalose spectra have been compared with the density of states obtained by applying Density Functional Theory (DFT)[21] for the gas phase and the monohydrate crystal of trehalose. The comparison with computation provides a detailed picture for the interpretation of the vibrational properties of trehalose crystals.

2. Experimental set-up

Pure sample of α,α-trehalose dihydrated was purchased from Aldrich-Chemie. The solutions at different weight fractions ϕ, $\phi=n_d/n_d+n_w$, being n_d and n_w the disaccharide and water mole numbers, respectively, were prepared by weight using double distilled deionised water. Care was taken in order to obtain stable, clear and dust-free sample; ample time was allowed for equilibration.

The samples for the Raman light scattering measurements were sealed in optical quartz cells of inner diameter of 5 mm and then mounted in an optical thermostat especially built to avoid any unwanted stray-light contribution. The measurements were performed with a temperature stability better than 0.1 °C. Raman spectra were obtained by a high resolution fully computerized Spex-Ramalog 5 triple monochromator in a 90° scattering geometry. Vertically polarized radiation of an INNOVA 70 Series Ar-Kr gas mixed laser operating in the 4579÷6764 Å was used as an excitation source. To reduce fluorescence, we chose the 6471 Å laser line. The laser power was maintained at approximately 1 W. The detection apparatus consisted of a photon counting system whose outputs were processed on line by a computer. The scattered photons were automatically normalized for the incoming beam intensity in order to ensure good data reproducibility. Isotropic scattering intensity were calculated from the parallel, I_{VV}, and perpendicular, I_{VH}, components of the scattered light by: $I_{ISO}=I_{VV}-4/3I_{VH}$. All the spectral contributions were normalized with the C-H stretching band, which does not suffer H-bond interactions.

Fourier Transform Infrared spectra were taken by a Bomem DA8 Fourier Transform Infrared (FTIR) spectrometer working with a globar lamp source, a KBr beamsplitter and DTGS/KBr detector. In this configuration it has been possible to cover the 400-4000 cm^{-1} range with a resolution of 4 cm^{-1}. A sample holder with CaF_2 windows was used and the apparatus was purged with dry nitrogen. In order to obtain a good signal to noise ratio, 32 repetitive scans were automatically added for each measurement.

For all the samples the experimental IR data were deconvoluted into the minimum number of symmetrical Voigt profiles V (ω). This distribution, the most suitable for band shape analysis, is a convolution of a Gaussian and a Lorentzian curve

$$V(\omega) = \frac{\displaystyle\int_{-\infty}^{+\infty} \frac{a_0 \exp\left(-y^2\right)dy}{a_3^2 + \left[\left(\dfrac{\omega - a_1}{a_2}\right) - y\right]^2}}{\displaystyle\int_{-\infty}^{+\infty} \frac{\exp\left(-y^2\right)dy}{a_3^2 + y^2}}$$

where a_0 and a_1 are the amplitude and the centre of the distribution, respectively, a_2 is a width related to the Gaussian half width at half maximum and a_3 is a width depending on the ratio of the Lorentzian half width at half maximum to a_2.

3. Results and discussion

3.1 Theoretical background

From a general point of view the interaction between a system under investigation, at thermal equilibrium T and composed of N particles *i* located at $\vec{r}_i$, and a probe, whose initial state is known, can be described by an Hamiltonian obtained by an expansion in the electric field [22]:

$$H_I = -\int d\vec{r}\left[\vec{E}(\vec{r})\vec{\mu}(\vec{r}) + \frac{1}{2}\vec{E}(\vec{r})\underline{\alpha}(\vec{r})\vec{E}(\vec{r}) + \text{higher order terms} \right] \tag{1}$$

The first term gives rise to dielectric and infrared absorption while the bilinear term is responsible for the Rayleigh and Raman scattering.

In particular, the infrared absorption spectrum comprises a series of absorption lines centred around the vibration frequencies each of which is proportional to the convolution of the rotational and of the vibrational spectrum:

$$C_{VV}(\omega) = \sum_{V} \left(\mu^V q^V\right)^2 G_V\left(\omega + \Omega_V\right) \otimes F_{1V}\left(\omega + \Omega_V\right) \qquad (2)$$

where q^V is the vth normal coordinate, $\mu^V = (\partial\mu/\partial q^V)_0$ the derivative dipole, corresponding to vibration ν, and the terms G_v and F_{1v} represent the convoluted vibrational and rotational contributions centred at frequencies Ω_v.

For what concerns the Raman effect, the dynamical structure factor, $S(Q,\omega)$, measured in a Raman experiment is essentially connected with the Fourier transform of the autocorrelation function of the polarisability tensor $\tilde{\alpha}$:

$$J_{VV}(\underline{Q},t) \propto \left\langle \sum_{i,j,V,V'} \left(\varepsilon_S\tilde{\alpha}_i^V(0)\varepsilon_I\right)\left(\varepsilon_S\tilde{\alpha}_j^{V'}(t)\varepsilon_I\right)q_i^V(0)q_j^{V'}(t)\exp i\underline{Q}\left[\underline{r}_i(t)-\underline{r}_j(0)\right]\right\rangle \qquad (3)$$

where ε_S and ε_I are the scattered and incoming polarization vectors respectively, and the other symbols have the usual meaning. By a matter of fact, the polarisability tensor is usually split into its isotropic part $\bar{\alpha}I$ and anisotropic part $\tilde{\beta}$, $\tilde{\alpha}^V = \bar{\alpha}^V I + \tilde{\beta}^V$, with $\bar{\alpha} = \frac{1}{3}Tr(\tilde{\alpha})$ and $\tilde{\beta} = \tilde{\alpha} - \frac{1}{3}Tr(\tilde{\alpha})*I$, I being the unit tensor.

3.2 Experimental results

FT-IR spectrum of trehalose dihydrate has been compared with that obtained by Ballone et al. [21] by applying the density functional theory for the gas phase and the monohydrate crystal of trehalose. The computation has provided a detailed picture for the interpretation of the vibrational properties of trehalose crystals and solutions despite the description of the intermolecular interactions and of the harmonic dynamics can be only moderately accurate. From the simulation of the gas-phase it has been found, in order of decreasing frequency, a narrow band (from 3502 to 3558 cm^{-1}) of O-H stretching for the groups not involved in the intramolecular hydrogen bond. At lower frequency (3388 cm^{-1}) we find the stretching mode of the OH groups involved in intramolecular hydrogen bond. The stretching modes of the C-H bonds are found at a frequency between 2874 and 3023 cm^{-1}, separated by a large gap from the band of the hybridised H-C-H, C-C-H, and C-O-H bending modes, observed between 1175 and 1433 cm^{-1}. The stretching of the C-O bonds occurs in a range between 970 and 1100 cm^{-1}, while it was difficult to assign a well-defined character to most of the modes below 950 cm^{-1}: however, the oscillations of the OH groups around the minimum of the C-C-O-H torsional potential have been localized at energies between 450 and 570 cm^{-1}, with the exception of the OH group giving rise to the intramolecular hydrogen bond, whose torsional mode has an energy of 700 cm^{-1}. The total and infrared-active vibrational densities of states are reported in Fig 1.

DFT simulation has been also performed for the monohydrate crystal structure. The main points emerged are the following:
- the OH stretching of water gives rise to two isolated peaks at 3141 cm^{-1} and 3397 cm^{-1};
- the pure OH stretching of trehalose gives rise to a band from 3155 to 3257 cm^{-1} and broader than in the gas phase molecule;

- the CH stretching modes give rise to a band extending from 2860 to 3045 cm^{-1}, which is contaminated by some OH stretching character.

Keeping in mind this peak assignment, in the FT-IR spectrum of trehalose dihydrated, see Fig. 2, one can recognize high energy bands ($\Delta\omega$>2900 cm^{-1}) corresponding to typical localized vibrations, in particular the O-H ($\approx$3350 cm^{-1}) and C-H ($\approx$2900 cm^{-1}) stretching. In particular, in the OH stretching region it is clearly visible a strong sharp band at 3500 cm^{-1} attributed to the stretching modes of the OH groups considered to be more or less free. Most bands observed in the CH stretching region correspond to the bands also observed by Raman scattering. However, a band at 2934 cm^{-1}, which is absent in the Raman spectrum, appears in the IR spectrum of trehalose dihydrate. The presence of crystal water gives rise to broad band around 2200 cm^{-1}, assigned to combination of OH librational and bending modes, as well to a band peaked around 1689 cm^{-1} attributed to the H-O-H bending of liquid water. At intermediate energies (1000-1500 cm^{-1}) we find the contributions coming from the stretching of C-C and C-O groups and from the C-H bending. In particular, the spectral contributions at 1312, 1333 and 1365 cm^{-1} have been assigned to OH deformations, while the broad band seen at 1460 cm^{-1} has been tentatively assigned to CH$_2$ scissoring vibration. At lower energies (<1000 cm^{-1}) the peaks corresponding to skeletal and collective vibrations are not very well resolved. However, the peaks observed at 998 and 956 cm^{-1} have been attributed to the antisymmetric and symmetric stretching modes of the glycosidic bond. This peak assignment attests the goodness of the performed DFT simulation.

In order to get information on the effects of disaccharides on the hydrogen bonded network of water, we also performed Raman and FTIR measurements on trehalose aqueous solutions of varying concentrations.

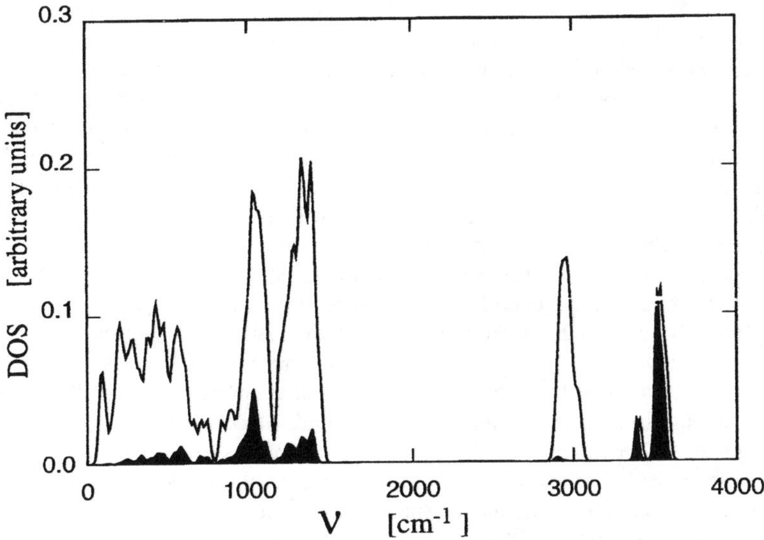

Fig.1 Vibrational spectrum of the gas phase trehalose molecule. The infrared activity is given by the black area under the lower curve (P. Ballone et al., *J. Phys. Chem. B* **104,** 2000, 6313).

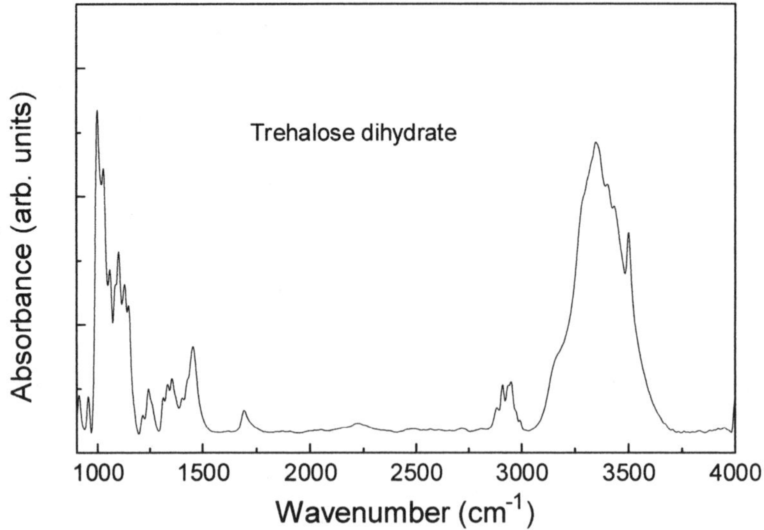

Fig.2 FTIR spectrum of trehalose dihydrate

In a previous work, our attention has been focused on the fundamental intramolecular OH stretching band. The existence of an isosbestic point in the isotropic spectrum of pure water suggested to some authors[31] the decomposition of each spectrum into an "open" and a "closed" contribution. The first one, centred near 3210 cm^{-1}, has been attributed to the O–H vibration in tetrabonded H_2O molecules that have an "intact bond" and originate low density patches in the system, while the second one, centred near 3420 cm^{-1}, would correspond to the O–H vibration of H_2O molecules that have a not fully developed hydrogen bond (distorted bond). With declining temperature, one finds that the "open" contribution increases indicating an enhanced hydrogen bonding which gives rise to patches of four bonded water molecules. What emerged from our data is that, by decreasing the water content, one observes a marked decrease of the "open" area contribution in the OH stretching band of disaccharide/water solutions. These findings have been interpreted in terms of a disaccharides destructuring effect on the H_2O tetrabonded network of molecules that originates low density conformations similar to that of supercooled water. Moreover, it has been observed that among the three disaccharides, trehalose promotes the most extensive layer of structured water around its neighbourhood, which most destroys the tetrahedral H-bond network of pure water.

In the present work our attention is devoted to analyse the intramolecular H-O-H bending mode in which contrary to the OH stretching region, there is no overlapping between the water and the sugar peaks. Fig. 3 shows, as an example, the Raman spectra relative to the spectral region within 1500 ÷1800 cm^{-1} of trehalose, aqueous solutions at different concentration values at T=293 K. The relative area of this spectral contributions, centred at 1645 cm^{-1}, scales with the water content, not evidencing any anomalous behaviour of these intramolecular modes due to the presence of trehalose.

More interesting are the FT-IR spectra obtained for the disaccharide solutions, see Fig 4. According to several reports on the crystal structure of trehalose dihydrate, one bound water molecule has four hydrogen bonds in a tetrahedral or distorted tetrahedral configuration, while the other water molecule has three hydrogen bonds in an approximately pyramidal or planar trigonal arrangement [23]. This finding suggests us the possibility to decompose the H-O-H bending mode into two distinct components. The first one, the only observable in trehalose dihydrate, is centred around 1690 cm^{-1}, while the second one, is approximately centred at 1645 cm^{-1}. On the basis of the IR data for pure water and the structural information [24,25] the bending band at 1690 cm^{-1} is assigned to that of ice-like water while the 1645 cm^{-1} band is easily assigned to that of liquid like water. As the water concentration increases, we can observe that the HOH bending band at 1690 cm^{-1} steeply decreases whereas the 1645 cm^{-1} band increases. A reasonable interpretation of this behaviour is that the environment surrounding the bound water molecules suddenly changes from the ice-like to liquid-like ones. From this point of view, this result confirms our previous hypothesis according to which the presence of trehalose obstructs the crystallization process destroying the network of water molecules compatible with that of ice.

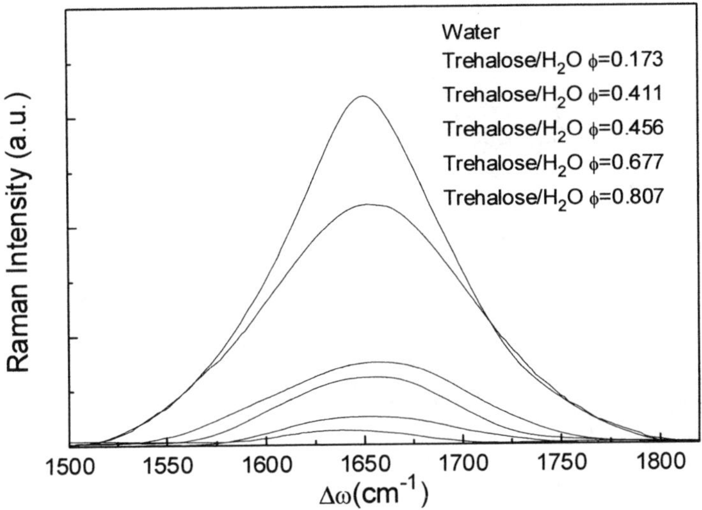

Fig. 3 Raman spectra of the intramolecular bending mode of trehalose/H$_2$O solutions at different concentration values.

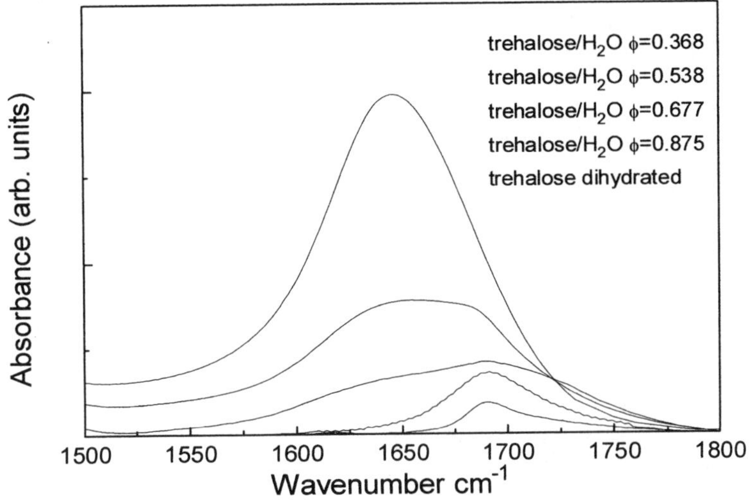

Fig. 4 FTIR spectra of the intramolecular bending mode of trehalose/H$_2$O solutions at different concentration values.

4. Concluding remarks

In this work we have reported measurements performed by Fourier Transform Infrared Spectroscopy and Raman scattering on trehalose aqueous solutions to investigate the structural properties of trehalose and to study the effects of this disaccharide on the hydrogen bonded network of water. In particular, the analysis of the intramolecular HOH bending mode confirms our previous hypothesis about a destructuring effect induced by the presence of the disaccharide on the hydrogen bonded network of pure water.

Finally, the comparison between FT-IR findings and density of states calculated starting from a DFT approach, attests the goodness of the performed simulation.

References

[1]   J.H. Crowe et al., *Arch. Biochem. Biophys.* **220** (1983) 477.
[2]   J.H. Crowe and L.M. Crowe, *Biological Membranes*, D. Chapman (Ed.), Academic Press, New York, 1984, p 57.
[3]   A.D. Elbein, *Chem. Biochem.* **30** (1974) 227.
[4]   J.S. Clegg, *Comp. Biochem. Physiol.* **20** (1967) 8.
[5]   K.C. Fox, *Science* **267** (1995) 1922.
[6]   J.H. Crowe and L.M. Crowe, *Science* **223** (1984) 701.
[7]   J. L. Green and C. A. Angell, *J. Phys. Chem.* **93** (1989) 2880.
[8]   C. A. Angell, *Science* **267** (1995) 1950.
[9]   J. H. Crowe et al., *Cryobiology,* **31** (1994) 355.
[10] J. H.Crowe et al., *Annu. Rev. Physiol.* **60** (1998) 73.
[11] P. Carnici et al., *Biochemistry* **95** (1998) 520.
[12] W. Q. Sun et al., *Biophysical Journal* **70** (1996) 1769.
[13] T. Hottiger et al., *Eur. J. Biochem.* **219** (1994) 187.
[14] S. Magazù et al., *Molecular Physics* **96** (1999) 381.
[15] C. Branca et al., *J. Phys.: Condensed Matter* **11** (1999) 3823.
[16] C. Branca et al., *J. Biol. Phys.* **26** (2000) 295.
[17] S. Magazù et al., *Molecular Physics* **96** (1999) 381.
[18] C. Branca et al., *J. Chem. Phys.* **111** (1999) 281.
[19] A. Faraone et al. *J. Chem. Phys.* **115** (2001) 3281
[20] S. Magazù et al. *Progr. of Theor. Phys.,* supplement **126** (1997) 195.
[21] P. Ballone et al., *J. Phys. Chem. B* **104** (2000) 6313.
[22] F. Volino, *Spectroscopy Methods for the Study of Local Dynamics in Polyatomic Fluids*, NATO ASI Ser. B, **33**, J. Dupuy and A. J. Dianoux (ed.), Plenum Press, New York, 1978.
[23] F. Sussich et al., *J. Am. Chem. Soc.* **120** (1998) 7893.
[24] G.E. Walrafen *J Chem Phys* **36** (1961) 1035.
[25] K. Akao et al., *Chemistry Lett.* 8, (1998), 759.

Author Index